Lecture Notes in Mathematics

Edited by A. Dold and B. Eckmann

T0171847

1389

E. Ballico C. Ciliberto (Eds.)

Algebraic Curves and Projective Geometry

Proceedings of the Conference
held in Trento, Italy, March 21–25, 1988

Springer-Verlag

Berlin Heidelberg New York London Paris Tokyo Hong Kong

Editors

Edoardo Ballico
Dipartimento di Matematica, Università di Trento
38050 Povo (Trento), Italy

Ciro Ciliberto
Dipartimento di Matematica, Università di Roma II
Via O. Raimondo, 00173 Roma, Italy

Mathematics Subject Classification (1980): 14H99, 14N05, 14N10, 14J99

ISBN 3-540-51509-7 Springer-Verlag Berlin Heidelberg New York
ISBN 0-387-51509-7 Springer-Verlag New York Berlin Heidelberg

Printed in Germany

Printing and binding: Druckhaus Beltz, Hemsbach/Bergstr.
2146/3140-543210 – Printed on acid-free paper

PREFACE

The conference on "Algebraic curves and projective geometry" was held in Villa Madruzzo, Cognola (Trento) from March 21 to 25, 1988. The meeting has been sponsored and supported by C.I.R.M. (Centro Internazionale per la Ricerca Matematica, Trento). This volume contains most of the works reported in the formal and informal lectures at the conference. The main topics of the volume are geometry of projective curves and moduli, enumerative and projective geometry, homological algebra and arithmetic surfaces.

We liked the idea of inserting at the end of this volume a list of problems and questions; most of them were asked and discussed during the conference. We collected this list mentioning the proposers of the questions and a few related references. We believe that publishing such lists may be a very useful contribution to the mathematical life, and hope this will be done more often.

We are very grateful to all participants for their enthusiasm, to the contributors of this volume for their efforts, to the referees for their precious help, and to the CIRM for the organization of the meeting and the assistance in editing this volume.

Edoardo Ballico, University of Trento

Ciro Ciliberto, University of Rome

CONTENTS

ON REGULAR AND STABLE RULED SURFACES IN $\mathbb{P}^3$

by E. Arrondo(*), M. Pedreira and I. Sols(*)

We study ruled surfaces in the geometrical sense, i.e. surfaces of the form $S = \mathbb{P}(F)$ for some rank 2 bundle F on a smooth curve C on an algebraically closed field k. The irregularity q of S is clearly the genus of C. By a ruled surface in $\mathbb{P}^3$ we mean the image of a ruled surface as a scroll of $\mathbb{P}^3$ with no multiple generators or equivalently a smooth curve C in the Grassmann variety Gr(1,3) of lines in $\mathbb{P}^3$. We denote by $Q_n \subset \mathbb{P}^{n+1}$ the smooth n-dimensional quadric, and see Gr(1,3) as the Klein's quadric Q_4.

There is a universal sequence on Q_4

$$0 \longrightarrow E \longrightarrow \mathcal{O}^4_{Q_4} \longrightarrow E'^{\vee} \longrightarrow 0$$

restricting a sequence on C

$$0 \longrightarrow E_C \longrightarrow \mathcal{O}^4_C \longrightarrow E'^{\vee}_C \longrightarrow 0$$

providing a birational embedding

$$S = \mathbb{P}(E'^{\vee}_C) \longrightarrow \mathbb{P}(\mathcal{O}^4_C) = C \times \mathbb{P}^3 \longrightarrow \mathbb{P}^3$$

with no multiple generators. Conversely, given such a birational embedding $S \longrightarrow \mathbb{P}^3$ of a ruled surface $\pi : S = \mathbb{P}(F) \longrightarrow C$ on a curve C, corresponding to a quotient $\mathcal{O}^4_C \twoheadrightarrow F$, we recover an immersion of C in Q_4 so that $E'^{\vee}_C = F$ by the universal property of the Grassmann variety. It is clear that the degree d of S in $\mathbb{P}^3$ is the degree of C in Q_4 anthat $H^1(\mathcal{O}_S(1)) = H^1(\pi_*\mathcal{O}_S(1)) = H^1(E'^{\vee}_C)$. Since $h^0(\mathcal{O}_S(1)) = h^1(\mathcal{O}_S(1))+d+2-2q$, the vanishing of $H^1(E'^{\vee}_C)$ is equivalent to the fact that the surface S is a projection of a linearly normal surface of $\mathbb{P}^{d+1-2q}$, i.e. the surface is regular according to the terminology of the classics. Let $H_{d,q}(Q_4)$ be the Hilbert scheme of smooth curves of degree d and genus q. Let $R_{d,q}(Q_4)$ and $S_{d,q}(Q_4)$ be the open subschemes of $H_{d,q}(Q_4)$ corresponding respectively to regular and stable ruled surfaces, where stable ruled surface means a smooth curve $C \subseteq Q_4$ with $E'^{\vee}_c$ stable, i.e. a ruled

(*) Partially supported by CAICYT Grant No. PB86-0036

surfaces in P^3 not having unisecants of degree less than or equal to d/2. The goal of this paper is to prove the following:

Theorem. If $d\le 2q+2$ *then* $R_{d,q}(Q_4)$ *and* $S_{d,q}(Q_4)$ *are irreducible open subschemes of dimension* $4d-q+1$ *in the same component of* $H_{d,q}(Q_4)$.

In §1 we show that $R_{d,q}(Q_4)$ is nonempty if and only if $d\ge 2q+2$ and we give a proof in nowadays terms of Severi's assertion that $R_{d,q}(Q_4)$ is then irreducible. The content of this §1 belongs to the second author's thesis 5 where he has proved in addition the result that the generic regular ruled surface has maximal rank, to appear elsewhere.

In §2 we prove the irreducibility of $S_{d,q}(Q_4)$ when $d\ge 2q+2$ by using essentially the main result in 1 , and our argument consists in bounding enough the length of the cokernel of the morphism analogous to the one discussed and shown to be epimorphic in 1 . Finally we prove that the generic ruled surface in $R_{d,q}(Q_4)$ lies in $S_{d,q}(Q_4)$.

§.1. **The Hilbert Scheme of regular ruled surfaces.**

First we give a condition for a curve C of Q_4 to be smoothable, i.e. for the existence of a scheme T and a closed subscheme χ of $Q_4\times T$ with projection $\chi\subset Q_4\times T \longrightarrow T$ flat and of smooth generic fibre, having C as one of its fibres.

Proposition 1.1. *Let* C *be a nodal reduced curve of* Q_4. *If* $H^1(E_C{'}^{\vee})=0$ *then* C *is smoothable and corresponds to a smooth point of the Hilbert Scheme.*

<u>Proof</u>: Taking cohomology in the sequence

$$0 \longrightarrow E_C{'}\otimes E_C{'}^{\vee} \longrightarrow E_C{'}^{\vee 4} \longrightarrow E_C{'}^{\vee}\otimes E_C^{\vee} \longrightarrow 0$$

we see that $H^1(E_C{'}^{\vee}) = 0$ implies $H^1(T_{Q_4}\otimes \vartheta_C) = 0$, since $T_{Q_4} = E'^{\vee}\otimes E^{\vee}$. This implies that $H^1(N'_{C,Q_4}) = 0$ in the sequence

$$\begin{array}{ccccccccc} 0 & \longrightarrow & T_C & \longrightarrow & T_{Q_4}\otimes\vartheta_C & \longrightarrow & N_{C,Q_4} & \longrightarrow & T^1_C \longrightarrow 0 \\ & & & & \searrow & & \nearrow & & \\ & & & & & N'_{C,Q_4} & & & \end{array}$$

presenting the Lichtenbaum-Schlessinger sheaf T^1_C supported at the nodes of C. Therefore $H^0(N_{C.Q_4}) \longrightarrow H^0(T^1_C)$ is surjective and $H^1(N_{C,Q_4}) = 0$, and then [3] prop. 1.1. concludes the proof.

Observe that C is deformed to a smooth irreducible curve C' with $H^1(E'^{\vee}_{C'}) = 0$ and the same degree d and arithmetical genus p_a.

Proposition 1.2. *Let* $X \subset Q_4$ *be the union of a nodal reduced curve* C *with* $H^1(E'^{\vee}_C)=0$ *and a line* L *meeting* C *transversally at a point, or a smooth conic* Q_1 *meeting* C *transversally at one or two points. Then* $H^1(E'^{\vee}_X)=0$.

<u>Proof</u>: We prove, for instance, the case of the conic Q_1. Tensoring with $E'^{\vee}$ the sequence

$$0 \longrightarrow \mathcal{O}_X \longrightarrow \mathcal{O}_C \oplus \mathcal{O}_{Q_1} \longrightarrow \mathcal{O}_{C\cap Q_1} \longrightarrow 0$$

we get

$$H^0(E'^{\vee}_C) \oplus H^0(E'^{\vee}_{Q_1}) \xrightarrow{\ f\ } H^0(E'^{\vee}_{C\cap Q_1}) \longrightarrow H^1(E'^{\vee}_X) \longrightarrow$$

$$\longrightarrow H^1(E'^{\vee}_C) \oplus H^1(E'^{\vee}_{Q_1})$$

therefore, it is enough to show that f is surjective, since $H^1(E'^{\vee}_{Q_1}) = H^1(E'^{\vee}_{Q_1}) = H^1(\mathcal{O}_{\mathbb{P}^1}(1) \oplus \mathcal{O}_{\mathbb{P}^1}(1)) = 0$ (recall $Q_1 \cong \mathbb{P}^1$ and $E'_{Q_1} \approx \mathcal{O}_{\mathbb{P}^1}(1) \oplus \mathcal{O}_{\mathbb{P}^1}(1)$). This clearly follows from

$$H^0(E'^{\vee}_{Q_1}) \longrightarrow H^0(E'^{\vee}_{C\cap Q_1}) \longrightarrow H^1(E'^{\vee}_{Q_1}(-C\cap Q_1)) = 0, \text{ q.e.d.}$$

Observe that $p_a(C\cup L) = p_a(C\cup Q_1)$ is $p_a(C)$ or $p_a(C)+1$ depending on whether $C\cap Q_1$ is one or two points. Let now $d,q\geq 0$ integers such that $d\geq 2q+2$. Choosing q+1 proper conics of Q_4 each one meeting the next transversally at two points, and then adding d-2q-2 disjoint lines of Q_4 meeting the whole of these conics transversally at one point, we get a nodal reduced curve X of degree d and arithmetical genus q, with $H^1(E'^{\vee}_X)=0$, by applying inductively prop. 1.2. Then prop. 1.1. applies, proving the existence of an irreducible and smooth curve C with the given invariants $d\geq 2q+2$ and $H^1(E'^{\vee}_C)=0$.

Theorem 1.3. *The Hilbert Scheme* $R_{d,q}$ *of regular ruled surfaces of degree* d *and irregularity* q *is non empty if and only if* $d\geq 2q+2$ *and it is irreducible.*

<u>Proof</u>: We are left with the irreducibility, for which it is enough to prove that the fibres of $R_{d,q} \longrightarrow M_q$ are irreducible of same dimension 4d-4q+4. The fibre at a smooth curve $C \in M_q$ consists of all rank 2 quotient bundles of the trivial bundle $\mathcal{O}_C^4$

$$0 \longrightarrow E_C \longrightarrow \mathcal{O}_C^4 \longrightarrow E_C'^{\vee} \longrightarrow 0$$

of degree d and $H^1(E_C'^{\vee})=0$. These are smooth points of Grothendieck's Quot scheme since this vanishing implies $H^1(E_C^{\vee} \otimes E_C'^{\vee})=0$ and the tangent space has dimension $\chi(E_C^{\vee} \otimes E_C'^{\vee}) = 4d-4q+4$.

As $E_C'^{\vee}$ is generated by global sections, a general section vanishes nowhere, thus yielding a sequence

$$0 \longrightarrow \mathcal{O}_C \longrightarrow E_C'^{\vee} \longrightarrow L \longrightarrow 0$$

for some line bundle L on C of degree d. Conversely, assume it is given an element L of the variety $\mathrm{Pic}^d(C)$, irreducible of dimension q, and an extension ξ in the space $\mathbb{P}(\mathrm{Ext}^1(L,\mathcal{O}_C)^{\vee})=\mathbb{P}(H^1(L^{\vee})^{\vee})$ of dimension d+q, and a surjection $\phi: \mathcal{O}_C^4 \longrightarrow\!\!\!\!\rightarrow E_C'^{\vee}$ in the space $\mathbb{P}(\underline{Hom}(\mathcal{O}_C^4, E_C'^{\vee})^{4\vee})$ of dimension 4(d-2q+2)-1. Then we get an element of the fibre of $R_{d,q}$ at C. The data (L,ξ,ϕ) are parametrized by an irreducible variety U dominating the fibre at C, which proves that this fibre is irreducible.

More precisely, U can be defined as follows: let P^d be a Poincaré line bundle on $X=C\times\mathrm{Pic}^d(C)$ and $pr: X \longrightarrow \mathrm{Pic}^d(C)$ the projection. For sheaves G_1, G_2 on X, we denote $\mathrm{Hom}_{pr}(G_1,G_2) = pr_*\underline{Hom}(G_1,G_2)$, and $\mathrm{Ext}^i_{pr}(G_1,-) = R^i\mathrm{Hom}_{pr}(G_1,-)$.

We claim that the functor $\mathrm{Sch}/\mathrm{Pic}^d(C) \longrightarrow \mathrm{Set}$ assigning to $S \xrightarrow{\sigma} \mathrm{Pic}^d(C)$ the set $\mathrm{Ext}^1(P^d_{X_s}, \mathcal{O}_{X_s})$ is represented by the irreducible scheme $Y = \mathbb{P}(\mathrm{Ext}^1_{pr}(P^d,\mathcal{O}_X)^{\vee}) \longrightarrow \mathrm{Pic}^d(C)$; i.e. there is a natural isomorphism: $\mathrm{Ext}^1(P^d_{X_s}, \mathcal{O}_{X_s}) \cong \mathrm{Hom}_{\mathrm{Pic}^d(C)}(S,Y)$. (We denoted $X_S = X \times \mathrm{Hom}_{\mathrm{Pic}^d(C)} S$ and by $P^d_{X_s}$ the lifting of P^d to X_S). An element of $\mathrm{Hom}_{\mathrm{Pic}^d(C)}(S,Y)$ is given by a line quotient bundle $\sigma^*\mathrm{Ext}^1_{pr}(P^d,\mathcal{O}_X)^{\vee} \longrightarrow\!\!\!\!\rightarrow L$. This is a section of $\sigma^*\mathrm{Ext}^1_{pr}(P^d,\mathcal{O}_X) \otimes L$ which is isomorphic to $\mathrm{Ext}^1_{pr_s}(P^d_{X_S},\mathcal{O}_{X_S})$ for the lifting

$$\begin{array}{ccc} X_S = C \times \mathrm{Pic}^d(C) \times S & \xrightarrow{pr_\sigma} & S \\ \downarrow & & \downarrow \sigma \\ X = C \times \mathrm{Pic}^d(C) & \xrightarrow{pr} & \mathrm{Pic}^d(C) \end{array}$$

of pr by σ: This follows by comparing the low terms sequences of the spectral sequences of composite functors $pr_*\underline{Hom} = Hom_{pr}$ and $(pr)_*\underline{Hom} = Hom_{pr_\sigma}$

$$\begin{array}{ccccccc} \sigma^* R^1(pr)_*\underline{Hom}(P^d,\vartheta_X) & \longrightarrow & \sigma^* Ext^1_{pr}(P^d,\vartheta_X) & \longrightarrow & \sigma^*(pr)_*Ext^1(P^d,\vartheta_X) & \longrightarrow & 0 \\ \| & & \downarrow \beta & & \| & & \\ 0 \longrightarrow R^1(pr_\sigma)_*\underline{Hom}(P^d_{X_S},\vartheta_{X_S}) & \longrightarrow & Ext^1_{pr_\sigma}(P^d_{X_S},\vartheta_{X_S}) & \longrightarrow & (pr_\sigma)_*Ext^1(P^d_{X_S},\vartheta_{X_S}) & \longrightarrow & 0 \end{array}$$

By the "five lemma", the morphism β is, indeed, an isomorphism.

From the spectral sequence of composite functor $\Gamma Hom_{pr_\sigma} = Hom$, we get the isomorphism

$$0 = H^1(S, Hom_{pr_\sigma}(P^d_{X_S},\vartheta_{X_S})) \longrightarrow Ext^1_{X_S}(P^d_{X_S},\vartheta_{X_S})) \longrightarrow$$
$$\longrightarrow H^0(S, Ext^1_{pr_\sigma}(P^d_{X_S},\vartheta_{X_S})) \longrightarrow H^2(S, Hom_{pr_\sigma}(P^d_{X_S},\vartheta_{X_S}))=0$$

since $Hom_{pr_\sigma}(P^d_{X_S},\vartheta_{X_S})=0$ and this ends the proof of our claim.

This provides a bundle F on $Y \xrightarrow{\rho} \mathrm{Pic}^d(C)$ as universal extension

$$0 \longrightarrow \vartheta_Y \longrightarrow F \longrightarrow P^d_Y \longrightarrow 0$$

The functor assigning to schemes $S \xrightarrow{\sigma} Y$, the set of all epimorphism $\vartheta_{CxS} \otimes H^0(\vartheta_{\mathbb{P}^3}(1))^\vee \twoheadrightarrow \sigma^* F$ is represented by an open subset U of $\mathbb{P}(Hom_\rho(\vartheta_{CxY} \otimes H^0(\vartheta_{\mathbb{P}^3}(1))^\vee, F)$. Therefore, there is a universal epimorphism

$$\vartheta_{CxU} \otimes H^0(\vartheta_{\mathbb{P}^3}(1))^\vee \twoheadrightarrow F_{CxU}$$

The wanted morphism from U to the fibre of $R_{d,q} \longrightarrow M_q$ at a smooth curve $C \in M_q$, appears now in a unique way, by just recalling that this fibre is the scheme $\mathrm{Quot}(\vartheta^4_C, 2, d)$ representing the functor assigning to a scheme S all rank 2 quotient bundles $\mathcal{O}^4_{C\times S} = H^0(\mathcal{O}_{\mathbb{P}^3}(1))^\vee \otimes \mathcal{O}_{C\times S} \twoheadrightarrow G$ such that for all geometric points $s \in S$, the bundle $G\otimes k(s)$ has degree d and $H^1(G\otimes k(S))=0$. (This is a representable subfunctor of the Quot-functor because this vanishing is an open property). We have already checked that this morphism is set-theoretically a surjection. We conclude that the fibres of $R_{d,q} \longrightarrow M_q$ are irreducible and thus the whole of $R_{d,q}$ is irreducible. (q.e.d.).

§.2. The Hilbert scheme of stable ruled surfaces of high degree.

We complete in this section the proof of our main theorem. It can be seen, as we will show at the end of the paragraph that the generic element of $R_{d,q}(Q_4)$ is stable. Therefore we assume there is a component $X \subseteq H_{d,q}(Q_4)$ whose generic C has $E'^\vee_C$ stable and $\delta = h^1(E'^\vee_C) > 0$ (thus $r+1 = h^0(E'^\vee_C) = \delta + d + 2 - 2q$) and will prove our theorem by finding some contradiction. We assume throughout the proof that $q \geq 2$ since curves $C \subseteq Q_4$ of genus q=0,1 and $d \geq 2q+2$ have $h^1(E'^\vee_C)=0$.

We fix an integer $m \geq 2$ and denote by $M_{q,m}$ the moduli of curves C of genus q endowed with a level m structure as will always be assumed. This moduli is fine, i.e. equipped with a universal curve $\mathcal{C} \longrightarrow M_{q,m}$. Let X_m the irreducible component dominating X in the Hilbert scheme $H_{d,q,m}$ of curves of Q_4 of degree d and genus q with a level m structure.

There is, after Maruyama [4], a moduli $\mathcal{W}_{d,q}$ of stable bundles of rank 2 and degree d on the curve $\mathcal{C}$ relative to the scheme $M_{q,m}$. Let $\mathcal{W}^r_{d,q} \subseteq \mathcal{W}_{d,q}$ correspond to those $(C, E'^\vee_C) \in \mathcal{W}_{d,q}$ with $h^0(E'^\vee_C) \geq r+1$. We consider the natural map

$$S_{d,q,m}(Q_4) \longrightarrow \mathcal{W}_{d,q}$$

sending a curve $C \subseteq Q_4$ into $(C, E'^\vee_C)$ and denote Y the irreducible component of $\mathcal{W}^r_{d,q}$ where the image of X lies. By Mori's theorem we can bound its dimension

$$4d-q+1 = \chi(N_{C,Q_4}) \leq \dim X = \dim X_m \leq \dim Y + 4(\delta+d+2-2q) - 1$$

since $4h^0(E'^{\vee}_C)$-dim $\mathrm{Hom}(E'^{\vee}_C,E'^{\vee}_C)=4(\delta+d+2-2q)-1$ is the dimension of the generic fibre of X over Y. We will show however, that dim $Y\leq 7q-4\delta-7$, a contradiction.

Let $C\in X_m$ be such that $h^1(E'^{\vee}_C)=\delta$. Since C is generated by global sections we have an epimorphism

$$\mathcal{O}_C^{r+1}\simeq \mathcal{O}_C\otimes H^0(E'^{\vee}_C)\longrightarrow\!\!\!\!\!\rightarrow E'^{\vee}_C$$

embedding the ruled surface $S=\mathbb{P}(E'^{\vee}_C)$ in $\mathbb{P}^r=\mathbb{P}(H^0(E'^{\vee}_C)^{\vee})$ as a surface S' without multiple lines (since it projects to the surface of $\mathbb{P}^3$ parametrized by $C\subseteq Gr(1,3)$, which has no multiple lines). This provides an embedding $C\subseteq Gr(1,r)=G$ as a smooth curve in the Grassmann variety of lines in $\mathbb{P}^r$.

Let F be the rank 5 bundle on C defined as a push-out in the diagram

$$\begin{array}{ccccccccc}
 & & 0 & & 0 & & & & \\
 & & \downarrow & & \downarrow & & & & \\
 & & N^{\vee}_{C,G}\otimes K_C & = & N^{\vee}_{C,G}\otimes K_C & & & & \\
 & & \downarrow & & \downarrow & & & & \\
0 & \longrightarrow & \Omega_G\otimes K_C & \longrightarrow & H^0(E'^{\vee}_C)\otimes E'_C\otimes K_C & \longrightarrow & E'_C\otimes E'^{\vee}_C\otimes K_C & \longrightarrow & 0\\
 & & \downarrow & & \downarrow & & \| & & \\
0 & \longrightarrow & K^2 & \longrightarrow & F & \longrightarrow & E'_C\otimes E'^{\vee}_C\otimes K_C & \longrightarrow & 0\\
 & & \downarrow & & \downarrow & & & & \\
 & & 0 & & 0 & & & &
\end{array}$$

Here, $H^0(K_C^2)=H^1(T_C)^{\vee}$ is the cotangent space to $M_{g,m}$ at C, $H^0(E'_C\otimes E'^{\vee}_C\otimes K_C)$ is the cotangent space at $E'^{\vee}_C$ to the moduli $W_d(C)$ of stable rank 2 bundles of degree d on C, and

$$0\longrightarrow H^0(K^2)\longrightarrow H^0(F)\longrightarrow H^0(E'_C\otimes E'^{\vee}_C\otimes K_C)\longrightarrow 0$$

is the sequence of cotangent spaces associated to the map $\mathcal{W}_{d,q}\longrightarrow M_{q,m}$ at the point $(C,E'^{\vee}_C)\in\mathcal{W}_{d,q}$ lying in the fibre $W_d(C)\subseteq W_{d,q}$ of $C\in M_{q,m}$. The middle term $H^0(F)$ is the cotangent space to $\mathcal{W}_{d,q}$ at $(C,E'^{\vee}_C)$. (Although no needed for the sequel, we would like to point as a conjecture that F can be viewed as follows: Each line of the ruled surfaces $\mathbb{P}(E_C'^{\vee}) \subseteq \mathbb{P}^r$ spans together with the neighboring line a $\mathbb{P}^3$, namely $\mathbb{P}(P^1E_C'^{\vee}) \longrightarrow \mathbb{P}(H^0(E_C'^{\vee})^{\vee}) = \mathbb{P}^r$, since $H^0(E_C'^{\vee})\otimes\mathcal{O}_C \twoheadrightarrow E_C'^{\vee}$ factors through the rank 4 bundle $P^1E_C'^{\vee}$ of principal parts of $E_C'^{\vee}$. This yields a section of the fibrations $Gr(2,P^1E_C'^{\vee}) \subseteq \mathbb{P}(\Lambda^2P^1E_C'^{\vee})$ in grassmannians $Gr(1,3)$ on C. We have a few reasons to conjecture that $\mathbb{P}(F^{\vee})$ is the $\mathbb{P}^4$-fibration tangent to $Gr(2,\mathbb{P}^1E_C'^{\vee})$ into $\mathbb{P}(\Lambda^2P^1E_C'^{\vee})$ along this section)

From the middle row of the above diagram, we get an exact sequence

$$H^0(E_C'^{\vee})\otimes H^0(E_C'\otimes K_C) \xrightarrow{\mu} H^0(F) \xrightarrow{\nu} H^1(N_{C,G}^{\vee}\otimes K_C) \to H^0(E_C'^{\vee})\otimes H^1(E_C'\otimes K_C) \to H^1(F) \to 0$$

The space $H^1(F) = H^1(E_C' \otimes E_C'^{\vee} \otimes K) = H^0(E_C' \otimes E_C'^{\vee})^{\vee}$ is $\mathbb{C}$ by stability of $E_C'^{\vee}$, so we get by duality

$$0 \longrightarrow H^0(E_C'^{\vee})^{\vee} \otimes H^0(E_C'^{\vee})/\mathbb{C} \longrightarrow H^0(N_{C,G}) \longrightarrow (\mathrm{Im}\ \varphi)^{\vee} \longrightarrow 0$$

The middle term is the tangent space of $H_{d,q,m}(G)$ at $C \subseteq G$ the first term is the tangent space at $C \subseteq G$ to its fibre $\mathrm{Aut}\ \mathbb{P}(H^0(E_C'^{\vee})^{\vee})$ in the map $H_{d,q,m}(G) \longrightarrow W^r_{d,q}$, and this is the sequence of tangent spaces associated to this map and to $C \in H_{d,q,m}(Q_4)$. The last term $(\mathrm{Im}\ \nu)^{\vee}$ is the tangent space of $W^r_{d,q}$ at $(C, E_C'^{\vee})$, the image of μ being the anihilator of the tangent space of $W^r_{d,q}$ at this point.

Let t be an integer, $1 \le t \le r-2$, and let $q_1,\dots,q_t$ be points of the smooth surface $S = \mathbb{P}(E_C'^{\vee})$ lying on distinct points $p_1,\dots,p_t$ of C and applying by the birational morphism

$$S = \mathbb{P}(E_C'^{\vee}) \longrightarrow\!\!\!\!\!\rightarrow S' \subseteq \mathbb{P}(H^0(E_C'^{\vee})^{\vee}) = \mathbb{P}^r$$

into smooth points of S' spanning a $\mathbb{P}^{t-1}$ which intersects S' transversaly at such points. Applying the direct image functor of $S \longrightarrow \mathbb{P}(E_C'^{\vee})$ to the diagram

$$\begin{array}{ccc} H^0(E_C'^{\vee}) \otimes \mathcal{O}_S & \longrightarrow\!\!\!\!\!\rightarrow & \mathcal{O}_S(1) \\ \downarrow & & \downarrow \\ \oplus\, H^0(\mathbb{C}(q_i))\otimes\mathcal{O}_S & \longrightarrow\!\!\!\!\!\rightarrow & \oplus\, \mathbb{C}(q_i) \end{array}$$

we get a diagram

$$
\begin{array}{ccccc}
 & 0 & & 0 & \\
 & \downarrow & & \downarrow & \\
 & V_{r+1-t} \otimes \vartheta_C & \xrightarrow{\alpha} & T & \\
 & \downarrow & & \downarrow & \\
(D_t) \quad & H^0(E_C'^{\vee}) \otimes \vartheta_C & \longrightarrow & E_C'^{\vee} & \longrightarrow 0 \\
 & \downarrow & & \downarrow & \\
 & \bigoplus_i H^0(\mathbb{C}(q_i)) \otimes \vartheta_C & \longrightarrow & \oplus\, \mathbb{C}(p_i) & \longrightarrow 0 \\
 & \downarrow & & \downarrow & \\
 & 0 & & 0 &
\end{array}
$$

Geometrically, $\mathbb{P}(V_{r+1-t}{}^{\vee}) = \mathbb{P}^{r-t}$ is the space on which S is projected from the center $\mathbb{P}^{t-1}$ spanned by $q_1,\dots,q_t$. The homomorphism α between kernels is surjective by the proof of lemma 1.(b) in [1] (This is read directly from [1] only in case t=r-2, but Ein's argument clearly extends to all values of t).

We use now the inclusions

$$x_i = (q_i,q_i) \in S\mathrm{x}S \subseteq \mathbb{P}(E_C' \otimes E_C'^{\vee} \otimes K_C) \subseteq \mathbb{P}(F) \subseteq \mathbb{P}(H^0(E_C'^{\vee}) \otimes E_C' \otimes K_C) = Z$$

in order to factor the natural morphism $\vartheta_Z(1) \twoheadrightarrow \bigoplus_{1}^{t} \mathbb{C}(x_i)$ in two different ways:

$$
\begin{array}{ccc}
\vartheta_Z(1) & \longrightarrow\!\!\!\!\rightarrow & \vartheta_{\mathbb{P}(F)}(1) \\
\downarrow & & \downarrow \\
\oplus\, \vartheta_{\{q_i\}\mathrm{x}S}(1) & \longrightarrow\!\!\!\!\rightarrow & \oplus\, \mathbb{C}(x_i)
\end{array}
$$

Applying to this diagram the direct image functor of $Z \longrightarrow C$ we get

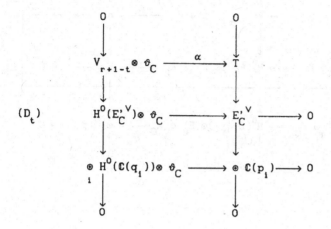

(D_i)

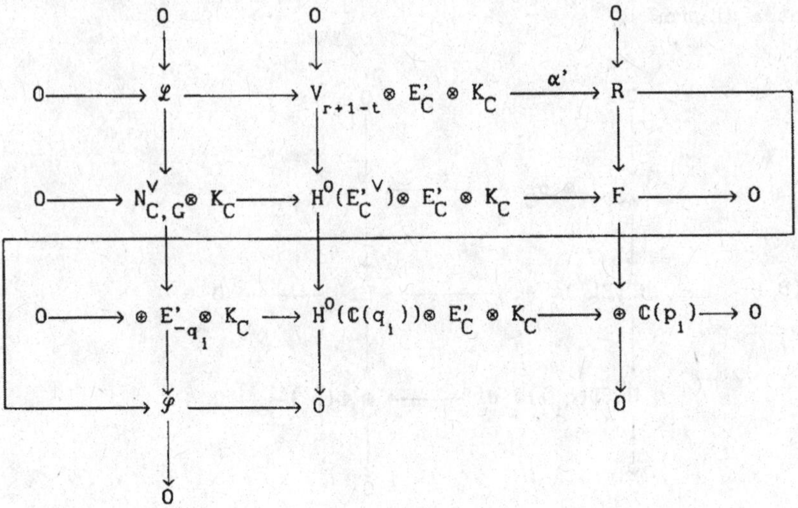

Let us show that α' is an epimorphism outside $p_1,\ldots,p_t$. Tensor diagram (D_t) with $H^0(E_C'^{\vee}) \otimes K_C \otimes \Lambda^2 E'$ and map it into (D_t') by epimorphisms which are natural in each case:

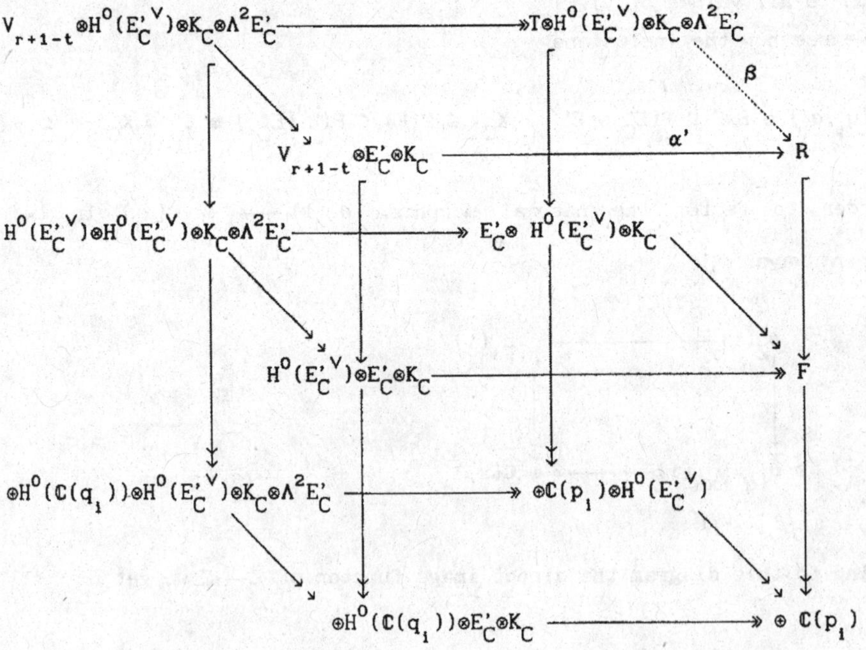

(Here the map $H^0(E_C'^{\vee}) \longrightarrow\!\!\!\!\!\rightarrow \mathbb{C}(p_i)$ corresponds to the choice of $q_i \in S$, considered in $\mathbb{P}(H^0(E_C'^{\vee}))=\mathbb{P}^r)$. Outside $p_1,\dots,p_t$, the map β agrees with the epimorphism $E_C\otimes H^0(E_C'^{\vee})\otimes K_C \longrightarrow\!\!\!\!\to F$, so α' must be there an epimorphism, as wanted.

We show next that the torsion sheaf $\mathcal{S}$ supported at the points $p_1,\dots,p_t$ has length $\leq 2t$. The elementary transform R of F by $x_1,\dots,x_t$ is clearly the elementary transform by x_i of the elementary transform R_i of F by $x_1,\dots,x_{i-1},x_{i+1},\dots,x_t$:

$$\begin{array}{ccccccccc}
 & & 0 & & 0 & & 0 & & \\
 & & \downarrow & & \downarrow & & \downarrow & & \\
0 & \longrightarrow & \mathcal{L} & \longrightarrow & V_{r+1-t}\otimes E'_C\otimes K_C & \xrightarrow{\alpha'} & R & & \\
 & & \downarrow & & \downarrow & & \downarrow & & \\
0 & \longrightarrow & \mathcal{L}_i & \longrightarrow & V_{r+2-t}\otimes E'_C\otimes K_C & \xrightarrow{\alpha'_i} & R_i & & \\
 & & \downarrow & & \downarrow & & \downarrow & & \\
0 & \longrightarrow & E'_{-q_i}\otimes K_C & \longrightarrow & H^0(\mathbb{C}(q_i))\otimes E'_C\otimes K_C & \longrightarrow & \mathbb{C}(p_i) & \longrightarrow & 0 \\
 & & \downarrow & & \downarrow & & \downarrow & & \\
 & & \mathcal{S}_i & & 0 & & 0 & & \\
 & & \downarrow & & & & & & \\
 & & 0 & & & & & &
\end{array}$$

We recall that in $U_i=C\setminus\{p_1,\dots,p_{i-1},p_{i+1},\dots,p_t\}$ the map α'_i is epimorphic, so we can apply Snake lemma to the restriction of this last diagram to the neighborhood U_i of p_i and get then that the torsion sheaf $\mathcal{S}_i|_{U_i}$ supported at p_i and of length $\leq$ rank $E'_{-q_i}\otimes K_C = 2$ is isomorphic to the cokernel $\mathcal{S}|_{U_i}$ of $\alpha'|_{U_i}$. This proves that length $\mathcal{S}\leq 2t$.

From now on, we take $t=r-2$. We observe that the line bundle $\mathcal{L}$ has negative degree, so it has no sections. Indeed,

$$\begin{aligned}
\deg \mathcal{L} &= \deg (N^{\vee}_{C,G}\otimes K_C) - \sum_{1}^{r-2} \deg (E'_{-q_i}\otimes K_C) + \text{length}\, \mathcal{S} \leq \\
&\leq (r+1)(-d-4q-4)-12q+12-(r-2)(-d+4q-3)+2(r-2) = \\
&= 3\,(r-2-d) = 3\,(\delta-2q+1)
\end{aligned}$$

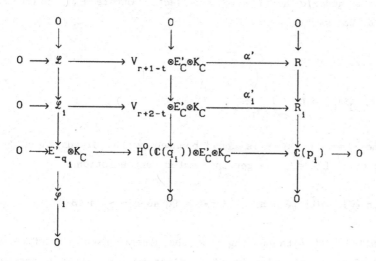

This last number is strictly negative, since $\delta\leq q$ as read from the cohomology sequence

$$H^1(\mathcal{O}_C) \longrightarrow H^1(E'^{\vee}_C) \longrightarrow H^1(\Lambda^2 E'^{\vee}_C) = 0$$

obtained from the exact sequence

$$0 \longrightarrow \mathcal{O}_C \longrightarrow E'^{\vee}_C \longrightarrow \Lambda^2 E'^{\vee}_C \longrightarrow 0$$

associated to a generic section of the vector bundle $E'^{\vee}_C$, which is generated by global sections.

This implies that

$$h^0(N^{\vee}_{C,G} \otimes K_C) \le h^0(\oplus E'_{-q_i} \otimes K_C) = (r-2)(\delta-1)$$

Since the dimension of Y is upper bounded by the dimension of its tangent space at $(C, E'^{\vee}_C) \in Y$, we get the wanted contradiction

$$\dim Y \le h^0(F) - \dim \operatorname{Im} \mu \le 7q-3\delta-r-4 = 9q-4\delta-d-5 \le 7q-4\delta-7$$

We are still left with proving that the generic regular surface of degree $d \ge 2q+2$ is stable. In the first place we prove that a regular ruled surface which as curve of Q_4 is connected union of q+1 generic conics and d-2q-2 lines as constructed inductively in prop. 1.2 has a unisecant of minimal degree [d+q/2]. Look at a generic conic of Q_4 as one of the twofamilies of lines in a smooth quadric of $\mathbb{P}^3$ and a union of conics meeting in two points corresponds to a union of smooth quadrics of $\mathbb{P}^3$ sharing two disjoint lines. It is easy to see that the rest of the intersection of the two quadrics must consist of two disjoint lines shared by the opposite reguli of both quadrics (observing, for instance, that the opposite regulus is obtained by polarity of conics of Q_4, i.e. planes of $\mathbb{P}^5$, respect the quadric $Q_4 \subseteq \mathbb{P}^5$). We look at a line of Q_4 as a pencil of lines in $\mathbb{P}^3$, i.e. all lines of a plane of $\mathbb{P}^3$ passing through one of its points. This is a degenerate ruled surface of $\mathbb{P}^3$ whose corresponding abstract ruled surface is the $F_1 = \mathbb{P}(\mathcal{O}_{\mathbb{P}^1} \oplus \mathcal{O}_{\mathbb{P}^1}(-1))$ obtained by blowing up the center of the pencil. Remark that the exceptional divisor is a unisecant of degree 0 (since it applies into a point of $\mathbb{P}^3$), while curves of type (1,1) in F_1 (i.e. equivalent to the exceptional divisor plus a generator) have degree 1 since they apply into lines of $\mathbb{P}^3$. The family of these (1,1) curves has dimension 2.

The first two quadrics share two lines of the same regulus, thus also share two lines of the opposite regulus. We take one of these two lines as unisecant of the first and of the second quadric, as abstract ruled surfaces. The union of both unisecants is a unisecant of the union ruled surface, and it is of minimal degree 1+1 since it is union of two lines applying into the same line of $\mathbb{P}^3$.

Next, we prolong this to a unisecant of minimal degree 1+1+2 of the union with the third quadric by adding a conic of it passing through the two points where the unisecant line of the second quadric meets the two lines shared with the third, and such that intersects the two disjoint lines shared with the fourth in two points lying in a line L of the fourth. We can prolong it to a unisecant of minimal degree 1+1+2+1 of the union with the fourth quadric by adding the line L of this quadric, and so forth, we pick line or conics, alternatively, in each quadric we added at the first q+1 steps of the induction, i.e. until we get a ruled surface of degree 2q+2.

Recall, that we add then d-2q-2 lines of Q_4 intersecting mutually in one point, i.e. d-2q-2 pencils of lines of $\mathbb{P}^3$ sharing mutually one line, until getting our reducible ruled surface of degree d.

1) If q is odd, the chosen unisecant of the last quadric was a line, and we choose as unisecant at this step, adding a unity to the minimal degree, the curve of type (1,1) passing through the point shared with the last unisecant and through the center of the second pencil. The unisecant we choose in the second pencil is the exceptional divisor, adding no unity to the minimal degree, and so forth we keep adding (1,1) curves alternating with exceptional divisors, adding 1+0+1+0+.. to the minimal degree of the unisecant.

2) If q is even, the chosen unisecant of the last quadric was a conic, passing through fixed points (those shared with the unisecant of the former quadric) and we can assume this conic chosen as to pass also through the center of the first pencil. Then we start in this case by choosing the exceptional divisor as unisecant of the first pencil, and then keep adding as before (1,1) curves and exceptional divisors alternatively, thus adding in this case 0+1+0+1+... to the minimal degree of the unisecant.

We end up with a unisecant of minimal degree (expressed as sum of q+1 terms plus d-2q-2 terms):

$$(1+1+2+1+2+1+\ldots+1)+(1+0+1+0+\ldots+0) = \frac{d+q-1}{2} \text{ if } q \text{ odd, } d \text{ even}$$

$$(1+1+2+1+2+1+\ldots+2)+(0+1+0+1+\ldots+1) = \frac{d+q}{2} \text{ if } q \text{ even, } d \text{ odd}$$

$$(1+1+2+1+2+1+\ldots+1)+(1+0+1+0+\ldots+1) = \frac{d+q}{2} \text{ if } q \text{ odd, } d \text{ odd}$$

$$(1+1+2+1+2+1+\ldots+2)+(0+1+0+1+\ldots+0) = \frac{d+q-1}{2} \text{ if } q \text{ even, } d \text{ odd.}$$

In the second place, we conclude that the generic regular ruled surface cannot have a unisecant of degree strictly lesser than $\left[\frac{d+q}{2}\right]$. Consider a one parameter flat deformation of the reducible regular ruled surface S_0 which we have constructed, having as generic element S_t an irreducible regular ruled surface. If S_t has a unisecant of degree strictly lesser than $\left[\frac{d+q}{2}\right]$ we can apply prop. 9.8 chap. III [2] and conclude that S_0 has also a unisecant of this same degree, in contradiction with the estimation above.

This observation, which can be restated by asserting that the minimal selfintersection of a unisecant of the generic ruled surface is -e=g, g-1, completes the proof of the theorem.

As a byproduct of the proof, we show now that $R_{d,q}(Q_4)$ dominates the moduli of abstract ruled surfaces of irregularity q. The regular reducible ruled surface $C_0 \subseteq Q_4$ we have constructed, made out of conics and lines of Q_4, satisfies not only $H^1(E'^{\vee}_{C_0})=0$, but also $H^1(E^{\vee}_{C_0})=0$ (since $E'^{\vee}_C$ can obviously be replaced by E_C in the proof of prop. 1.2.) so this also holds for the generic regular ruled surface C. For the restriction to C of the tangent bundle $T_{Q_4}=E^{\vee}\otimes E'^{\vee}$ of Q_4 we get

$$0= H^1(E^{\vee}_C)^4 \longrightarrow H^1(E^{\vee}_C\otimes E'^{\vee}_C) \longrightarrow 0$$

$$H^0(N_{C,Q_4}) \longrightarrow H^1(T_C) \longrightarrow H^1(T_{Q_4}\otimes\mathcal{O}_C)=0$$

so C has "general moduli", i.e. $R_{d,q}(Q_4)$ dominates the moduli M_q of curves of genus q. Combining with the result we have just got, we obtain the following

<u>Theorem</u>: *The Hilbert scheme* $R_{d,q}(Q_4)$ *dominates the moduli of abstract ruled surfaces of irregularity* q.

REFERENCES.

[1] Ein, L. :Hilbert scheme of smooth space curves. Ann. Sci. Ecole Norm. Sup (4) 19 (1986) 469-478.
[2] Hartshorne, R.: Algebraic Geometry. Springer-Verlag (1977).
[3] Hartshorne, R.; Hirschowitz, A.: Smoothing algebraic space curves. In Algebraic Geometry. Sitges 1983. Springer LNM 1124, 98-131.
[4] Maruyama, M. :Moduli of stable sheaves II. J. Math. Kyoto Univ. 18 (1978), 557-614.
[5] Pedreira, M. : Sobre las superficies regladas regulares". Tesis doctoral. Universidad Complutense.
[6] Severi, F.: Sulla classificazione delle rigate algebriche. Rend. Mat., 2 (1941), 1-32.

Enrique Arrondo and
Ignacio Sols

Departamento de Algebra
Facultad de Matemáticas
Universidad Complutense de Madrid
Madrid 28040. SPAIN.

Manuel Pedreira

Departamento de Algebra
Facultad de Matemáticas
Universidad de Santiago
La Coruña. SPAIN.

APPENDIX TO "ON REGULAR AND STABLE RULED SURFACES IN $\mathbb{P}^3$"

R. Hernandez

Let k be an algebraically closed field of characteristic 0 and C a complete smooth curve over k of genus $q \geq 2$.

We denote by $R(n,r,d)$ the open set in the scheme Quot parametrizing quotient bundles $\vartheta_C^n \longrightarrow E$ with E of rank r and degree d verifying $H^1(E)=0$. We have a natural morphism $R(n,r,d) \longrightarrow \text{Jac}^d C$ and we denote by $R_L(n,r,d)$ the fiber over a point representing a line bundle L.

Using a fact proved in [2] and the idea of Mattuck's proof of rationality of the field of multisymmetric functions we prove the following:

Theorem: *If* $d > \text{Max}(2q-1,4)$, *then* $R_L(n,r,d)$, *if non empty, is rational.*

This result was proved by E. Ballico for some values of n,d,q (namely when $d>n(2q-1)$ and every prime dividing (n,d) divides q, [1] Lemma 6) and, as a consequence, he obtained the stable rationality of the moduli schemes of vector bundles on curves.

In particular, the above theorem proves that, once you fix a curve C and a line bundle L on C, the variety of regular ruled surfaces in $\mathbb{P}^3$ with directrix C and $\vartheta_{\mathbb{P}^3}(1)|_C \approx L$ is rational.

Proof of the theorem. It is known that $R(n,r,d)$ is irreducible and smooth ([4],[2]). Let $H(n,r,d)$ be the open set in $\text{Hilb}^d \mathbb{P}(\vartheta_C^{n-r})$ corresponding to those 0-dimensional subschemes of d points of $C \times \mathbb{P}^{n-r-1} = \mathbb{P}(\vartheta_C^{n-r})$ lying in different fibers of the projection to C.

Although not explicitly stated, the following proposition was proved in [2] $\S.2$ steps, 2,3.

Propositon. *There is a birational map between* $R(n,r,d)$ *and a geometrical vector bundle over the scheme* $H(n,r,d)$. *This map is compatible with the natural morphisms to* $\text{Jac}^d C$.

Then, our theorem would be proved if we show that the fiber $H_L(n,r,d)$ of the natural projection $\pi':H(n,r,d) \longrightarrow \text{Jac}^d C$ is rational. Moreover, there exists a birational equivalence between $H(n,r,d)$ and the symmetric power $S^d(C \times \mathbb{P}^{n-r-1})$ compatible with the projections to $\text{Jac}^d C$, and it is enough to prove that the fibers $S_L^d(C \times \mathbb{P}^{n-r-1})$ are rational.

We consider the diagram:

$$\begin{array}{ccc} C^d & \xleftarrow{\;\pi_1\;} & (C \times \mathbb{P}^{n-r-1})^d \\ \downarrow{\scriptstyle p_1} & & \downarrow{\scriptstyle p_2} \\ S^dC & \xleftarrow{\;\pi_2\;} & S^d(C \times \mathbb{P}^{n-r-1}) \\ & {\scriptstyle \vartheta_1}\searrow \quad \swarrow{\scriptstyle \vartheta_2} & \\ & Jac^dC & \end{array}$$

where ϑ_1 is $\mathbb{P}^{d-q}$-bundle $(d>2q-2)$, p_1 and p_2 are the quotients by the action of the symmetric group, $S_d=:G$, and π_1, π_2 are the projections. Let us denote by Σ the function field of C^d and by $\Sigma' = \Sigma(X_{1,1},\dots,X_{1,n-r-1},\dots, X_{d,1},\dots,X_{d,n-r-1})$ the one of $(C\times\mathbb{P}^{n-r-1})^d$. The group G acts on Σ' by the rule $\sigma(X_{i,j}) = X_{\sigma(i),j}$ and the natural action on Σ. Now, we want to compute the fixed field Σ'^G.

Let U be the open set in S^dC of points $x=\{x_1,\dots,x_d\}$ with $x_i \neq x_j$ $(i\neq j)$ and $a\in U$ a point. Let us choose a rational function on C verifying $f(a_i)\neq f(a_j)$ $(i\neq j)$ and consider the linear system

$$\sum_{j=0}^{d-1} (f(x_i))^j\, t_{j,k} = X_{i,k}\,,\quad k=1,\dots,n-r-1,\ i=1,\dots,d$$

The open set U' in U where the van der Monde matrix $m_{i,j}=(f(x_i))^j$ has maximal rank contains the point a and, over U', we can write the functions $t_{j,k}$ as linear combinations of the $X_{i,k}$ with coefficients in Σ. Then, $\Sigma' = \Sigma(t_{1,1},\dots,t_{1,n-r-1},\dots,t_{d,1},\dots,t_{d,n-r-1})$ and the functions $t_{j,k}$ are invariant by G because the linear system itself is G-invariant. Therefore, Σ'^G is a purely trascendental extension of Σ^G.

This proves the theorem for points in the image of U' in Jac^dC, but open sets like U' cover U and the map from U to Jac^dC is surjective. This finishes the proof.

I want to thank the referee of a previous version of this paper, who suggested a drastic simplification.

REFERENCES.

[1] E. BALLICO, "Stable rationality for the variety of vector bundles over an algebraic curve". J. London Math. Soc. (2) 30 (1984) 21-26.

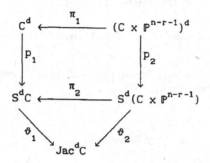

[2] R. HERNANDEZ, "On Harder-Narasimhan stratification over Quot schemes". J. reine angew. Math. Band 371 (1986) 115-124.

[3] A. MATTUCK, "The field of multisymmetric functions". Proc. Amer. Math. Soc. 19 (1968), 764-765.

[4] P.E. NEWSTEAD, "Introduction to moduli problems and orbit spaces" TIFR Bombay, Berlin-Heidelberg-New York, 1978.

R. Hernández.

Departamento de Algebra y Geometría.

Facultad de Ciencias.

47005 - VALLADOLID. SPAIN.

CONFIGURATIONS OF LINEAR PROJECTIVE SUBVARIETIES

by

GIORGIO BOLONDI and JUAN CARLOS MIGLIORE*

Dip. di Matematica e Fisica
Univ. di Camerino
I 62032 CAMERINO (MC)
Italy

Dept. of Mathematics
Drew University
MADISON, NJ 07940
U.S.A.

Introduction

A famous open problem is whether every smooth curve in $\mathbf{P}^3$ can be deformed to a nodal union of lines (a so-called "stick-figure"-- cf. for instance [HH]). The goal of this paper is to show how, even in the more general setting of codimension two subschemes of $\mathbf{P}^n$ ($n \geq 3$), this problem can often be resolved using liaison techniques.

Our main tool is to take advantage of the structure of an even liaison class (called the LR-Property). This was introduced in [LR] and studied further in [BM2] and [BM3]. In these papers this structure has been verified for many even liaison classes, and it has been conjectured that it holds in every even liaison class in codimension two ([BM3]). Briefly, the LR-Property says first that there is a well-defined notion of a "minimal" element in the liaison class; and second that every element can be obtained from a minimal one by performing a sequence of "basic double links" and then deforming. Furthermore, it was shown in [BM3] that without loss of generality these "basic double links" can be performed in a very simple way. These facts and ideas are reviewed in §1.

§2 begins with a generalization to codimension two of the notion of a "stick-figure". Then our main result (Theorem 2.8) is that if an even liaison class L in codimension two possesses the LR-Property, and if among the minimal elements there is a union of two-codimensional linear subspaces with certain "nice" properties, then every element of L specializes to a "nice" union of two-codimensional linear subspaces. In the case of curves in $\mathbf{P}^3$, these are exactly the stick-figures.

In §3 we give several applications. First we show that every arithmetically Buchsbaum curve in $\mathbf{P}^3$ specializes to a stick-figure. (In an earlier version of [BM2] we showed that every arithmetically Buchsbaum curve in $\mathbf{P}^3$ specializes to a union of lines with at worst triple points.) Then, passing to $\mathbf{P}^4$, we show that every surface S with the property that only one

---

* This material is based upon work done while this author was supported by the North Atlantic Treaty Organization under a grant awarded in 1987.

group $H^i(\mathbf{P}^4, I_S(t))$ is non-zero (i = 1,2) specializes to a "nice" union of planes. Finally, we show how this idea can be used more generally to deform curves in $\mathbf{P}^3$ to "simple" unions of curves.

Both authors would like to thank the Department of Mathematics of the University of Trento (Italy) for its hospitality during the preparation of this paper.

§1 Preliminaries.

Throughout this paper, k shall denote an algebraically closed field, $\mathbf{P}^n = \mathbf{P}^n_k$ with $n \geq 3$, and unless otherwise indicated, all schemes are assumed to be locally Cohen-Macaulay and equidimensional.

Let X be a two-codimensional subscheme of $\mathbf{P}^n$ and $(M^i)(X) = \bigoplus_{t \in \mathbf{Z}} H^i(\mathbf{P}^n, I_X(t))$, for $1 \leq i \leq n-2$. $(M^i)(X)$ is a graded $k[X_0, \ldots, X_n]$-module of finite length, and the collection $\{(M^i)(X)\}$ is an even liaison invariant up to shift (cf. [S] or [M2]). There is a well-defined notion (see [BM3], 1.5) of a leftmost shift of $\{(M^i)(X)\}$ that can actually occur for some X, and each subsequent shift of $\{(M^i)(X)\}$ actually occurs. This allows us to partition a non-arithmetically Cohen-Macaulay even liaison class L in codimension two (and in fact in any codimension other than 1 and n) into disjoint subsets as follows:

<u>Definition 1.1</u> Let L be a non-arithmetically Cohen-Macaulay even liaison class in codimension two. For a subscheme $X \in L$, we say $X \in L^h$ $(h \geq 0)$ if the collection $\{(M^i)(X)\}$ is shifted h places to the right of the leftmost possible one (cf. [BM3]).

An important tool in the study of liaison in codimension two is the notion of *Liaison Addition* introduced by Schwartau (cf. [Sw]). That says that if X and Y are two-codimensional subschemes of $\mathbf{P}^n$ and $F \in I_X$, $G \in I_Y$ with no common components, then the ideal $G \cdot I_X + F \cdot I_Y$ is the saturated ideal of a locally Cohen-Macaulay subscheme Z with

$$(M^i)(Z) \cong (M^i)(X)(-\deg G) \oplus (M^i)(Y)(-\deg F).$$

As sets (thinking of F and G as hypersurfaces) we have $Z = X \cup Y \cup (F \cap G)$.

<u>Definition 1.2</u>

Let X be a two-codimensional subscheme of $\mathbf{P}^n$. We define the function

$$F_X: \{0,1,\ldots,n-1\} \times \mathbf{Z} \to \mathbf{Z} \text{ by}$$

$$F_X(i,t) = h^i(\mathbf{P}^n, I_X(t)) = \dim H^i(\mathbf{P}^n, I_X(t))$$

We will call F_X the *cohomology function* of X, and we will say that X and Y *have the same cohomology* if $F_X = F_Y$.

Definition 1.3

Let X be a two-codimensional subscheme of $\mathbf{P}^n$, and let $F \in I_X$. Choose a general form A of degree d, and consider the subscheme Z whose homogeneous ideal is $I_Z = A \cdot I_X + (F)$. Z is a *basic double link* of X via the hypersurfaces F and A.

Remark 1.4

It is easy to see that Z is evenly linked to X, and that the cohomology function $F_Z(i,t)$ depends only on $F_X(i,t)$, on deg(F)=f and on deg(A)=d. Hence we will say that Z is obtained from X via a basic double link with hypersurfaces of degrees f and d, and we will write $X:(f,d) \to Z$. This construction can be thought of as a special case of Liaison Addition, taking Y to be the void set.

This notion of basic double link is crucial in the description of the structure of an even liaison class:

Definition 1.5

Let L be an even liaison class of two-codimensional subschemes of $\mathbf{P}^n$. We say that L has the *LR-property* if the following conditions hold:

a) If $M,N \in L^0$, then there is a deformation from one to the other through subschemes all in L^0

b) Given $V_0 \in L^0$ and $V \in L^h$, ($h \geq 1$), then there exists a sequence of subschemes $V_0, V_1, \ldots, V_r$, such that for all i, $1 \leq i \leq r$, V_i is a basic double link of V_{i-1}, and V is a deformation of V_r through subschemes all in L^h.

Let us recall from [BM3] some consequences of this property:

1) The elements of L are distributed into disjoint sets $F_{(b;g_2,\ldots,g_r)}$, where $b \geq 0$, $\alpha < g_2 < \ldots < g_{r-1} < g_r$, and α is the minimal degree of a hypersurface containing an element $N \in L^0$ (it is also possible that there are no g_i's: in this situation we assume r=1.); for every two elements of the same $F_{(b;g_2,\ldots,g_r)}$ there is an irreducible flat family of elements of $F_{(b;g_2,\ldots,g_r)}$ to which both belong (that is to say, roughly speaking, one can be deformed to the other one).

2) Two elements X and Y are in the same set $F_{(b;g_2,\ldots,g_r)}$ if and only if they have the same cohomology, that is to say $F_X = F_Y$. (In particular they have the same natural numerical invariants, such as degree, arithmetic genus and so on.)

3) In $F_{(b;g_2,\ldots,g_r)}$ lie the subschemes Y which may be obtained from a minimal element $N \in L^0$ via sequences of basic double links as follows:

$$N:(\alpha(N),b) \to Y_1:(g_2,1) \to Y_2: \ldots \to Y_{r-1}:(g_r,1) \to Y;$$

hence $L^{b+r-1} \supset F_{(b;g_2,\ldots,g_r)}$.

4) In every set $F_{(b;g_2,\ldots,g_r)}$ there exist subschemes which are the result of a liaison addition of a minimal element $N \in L^0$ and an arithmetically Cohen-Macaulay two-codimensional subscheme T, via hypersurfaces of degrees $s = \begin{cases} \alpha(N) & \text{if } b>0 \\ g_2 & \text{if } b=0 \end{cases}$ and b+r-1

containing N and T respectively. The numerical character of T can be exactly determined in terms of the integers b and g_i's.

§2 Linear Configurations

Definition 2.1

A *linear configuration* in $\mathbf{P}^n$ is a locally Cohen-Macaulay reduced union of linear two-codimensional subvarieties of $\mathbf{P}^n$. A linear configuration $X = \bigcup_{i=1}^{p} L_i$ is said to be *good* if the intersection of any three of the L_i's has dimension at most n-4.

Example 2.2

If n=3, a good linear configuration is called a *stick-figure* ; that is to say it is a reduced union of lines with only nodal singularities. Note that in this case the assumption "locally Cohen-Macaulay" is unnecessary.

If n>3, to check that a configuration of two-codimensional linear subvarieties X is locally Cohen-Macaulay, we will use the following procedure. Fix a singular point P, and cut with a general hyperplane, not passing through P, the components of X to which P belongs, and call Y this intersection. Then X is Cohen-Macaulay at P if and only if Y is arithmetically Cohen-Macaulay.

Remark 2.3

It is not hard to show, for example using induction on n, that if a reduced union X of two-codimensional linear subvarieties of $\mathbf{P}^n$ is locally Cohen-Macaulay then the singular locus Sing X is equidimensional of dimension n-3.

The next definition is slightly different from the one given, for linear configurations in [BMR], and for stick figures in [B]:

Definition 2.4

A good linear configuration X in $\mathbf{P}^n$ is said to be *hyperplanar in degree* p if it is contained in a reduced union of p hyperplanes such that the intersection of any three of them has dimension exactly n-3, and the intersection of any two of them is not a component of X.

A good linear configuration X which is hyperplanar in degree $\alpha(X)$ will be called simply *hyperplanar.*

Remark 2.5

Of course, if X is hyperplanar in degree p, then it is hyperplanar in degree t, for every t $\geq$ p, since then a general choice of other t-p hyperplanes satisfies the hypothesis for every t.

From now on, for simplicity let us denote by Γ the Grassmanian of the $\mathbf{P}^{n-3}$'s contained in $\mathbf{P}^n$.

Lemma 2.6

Let X be a good linear configuration, $\Sigma=\bigcup_{i=1}^{p} H_i$ a reduced union of hyperplanes containing X such that the intersection of any three of them has dimension n-3 and the intersection of any two of them is not a component of X, and let K be a general hyperplane. Then the result of a basic double link

$$X:(p,1) \to Y$$

performed by means of Σ and K is again a good linear configuration.

Proof.

Let us call $T(\Sigma) = \{ F\in\Gamma \mid \text{three components of } \Sigma \text{ meet in } F \}$

$S(X) = \{ F\in\Gamma \mid \text{the intersection of two components of } X \text{ is contained in } F \}$

$W(X,\Sigma) = \{ F\in\Gamma \mid F \text{ is contained in } X\cap \mathrm{Sing}(\Sigma) \}$.

Choose K not passing through any $F\in T(\Sigma)\cup S(X)\cup W(X,\Sigma)$ and not containing any component of $\mathrm{Sing}(\Sigma)$ (and not containing any isolated line of X, if n=3). A general choice suffices, since $T(\Sigma)$ and $W(X,\Sigma)$ are finite sets, S(X) is a closed subset and $\mathrm{Sing}(\Sigma)$ has finitely many components. Note that these assumptions imply that K does not contain any irreducible component of X.

Clearly Y is a linear configuration, namely $Y=(\bigcup_{i=1}^{p} L_i)\cup X$, where $L_i=H_i\cap K$, and the L_i's are distinct, since K doesn't contain any component of $\mathrm{Sing}(\Sigma)$. Hence, it is enough to study all possible intersections of three irreducible components of Y.

i) If the three components were already components of X, then their intersection has dimension at most n-4 by hypothesis.

ii) If two components were already components of X, and the third is a "new" component L_i, then L_i is contained in K, and K doesn't contain the intersection of two components of X. Hence L_i cannot contain the intersection of the other two components and therefore the intersection of the three pieces has dimension at most n-4.

iii) If we have a component of X and two new components L_i and L_k, then $L_i\cap L_k$ is contained in $\mathrm{Sing}(\Sigma)$. But K doesn't contain any (n-3)-dimensional F contained both in X and in $\mathrm{Sing}(\Sigma)$. Hence the third component cannot contain $L_i\cap L_k$, and as before the dimension of the intersection of the three pieces is at most n-4.

iv) Suppose we have three new components L_i, L_m and L_k meeting in an (n-3)-dimensional linear subvariety: this means that the three hyperplanes H_i, H_m and H_k contain this subvariety. This is a contradiction since we have chosen K not containing any element of $T(\Sigma)$.

Corollary 2.7

Let X, Σ and Y be as in the previous lemma. Then K can be chosen in such a way that Y is hyperplanar in degree p.

Proof.

It is enough to show that Σ is a good (in the sense of the definition of hyperplanar) hypersurface containing Y. But, choosing K as in the previous lemma, the new components of Y are not components of Sing(Σ), and hence any two components of Σ still do not meet in a component of Y.

Theorem 2.8

Let L be an even liaison class of two-codimensional subschemes of $\mathbf{P}^n$ satisfying the LR-property, and suppose that there exists a hyperplanar good linear configuration $M \in L^0$. Then every $X \in L$ specializes to a good linear configuration.

Remark.2.9.

Our proof actually shows more: it shows that there exists a good linear configuration Y such that $F_X = F_Y$ and that there exists an irreducible flat family $\{X_t\}$ to which both X and Y belong, on which the cohomology function F_{X_t} is constant, and every X_t is in the same liaison class as X. Moreover, Y is hyperplanar in some degree .

Proof.

We know that the elements of L are distributed into disjoint sets $F(b;g_2,...,g_r)$, with $\alpha(M)<g_2<...<g_r$. Thanks to point 1) following def.1.5, it is enough to exhibit in every family $F(b;g_2,...,g_r)$ a good linear configuration, hyperplanar in degree g_r.

Starting from M, we perform a sequence of basic double links

$$M:(\alpha(M),1) \rightarrow Y_1:(\alpha(M),1) \rightarrow ... \rightarrow Y_{b-1}:(\alpha(M),1) \rightarrow Y_b:(g_2,1) \rightarrow Y_{b+1}:...$$
$$....... \rightarrow Y_{b+r-2}:(g_r,1) \rightarrow Y, \text{ and } Y=Y_{b+r-1}$$

(with obvious meaning if b=0 or r=1).

We show by induction on i (the number of basic double links) that this sequence can be performed in such a way that Y_i is a good linear configuration, hyperplanar in degree d_i, where $d_i=\alpha(M)$ if $i\leq b$, and $d_i=g_{i-b+1}$ if $i\geq b+1$. Note that the sequence of integers d_i is increasing.

For i=0, let $Y_0 = M$. Then the result follows directly from the LR-Property.

For i=1, this follows directly from lemma 2.6 and corollary 2.7 , if we choose a hypersurface of degree $\alpha(M)$ containing M satisfying the hypothesis of the quoted lemma (or degree g_2 if b = 0).

If i >1, let us suppose that Y_{i-1} is a good linear configuration, hyperplanar in degree d_{i-1}. Thanks to Remark 2.5, Y_{i-1} is hyperplanar in degree d_i. Hence it is possible to perform the next basic double link with a hyperplane and a hypersurface of degree d_i satisfying the hypothesis of Lemma 2.6. By applying again Lemma 2.6 and Corollary 2.7 we conclude the induction.

By construction, X and Y lie in the same irreducible family $\mathcal{F}_{(b;g_2,\dots,g_r)}$.

§3 APPLICATIONS.

Now we give two applications of this result. The first one is a strengthening of the results contained in [B] , about stick figures. Trying to generalize Gaeta's result that every arithmetically Cohen-Macaulay curve in $\mathbf{P}^3$ specializes to a stick figure, the first author was forced to allow triple points in the configurations of lines obtained. Now we can avoid this assumption. The second consequence of this approach is that in the same way we can study the same degeneration problem in arbitrary dimension; we give an application of it to surfaces in $\mathbf{P}^4$.

The first application concerns arithmetically Buchsbaum curves in $\mathbf{P}^3$. We first recall some definitions and facts.

Definition 3.1 A curve C in $\mathbf{P}^3$ is *arithmetically Buchsbaum* if the Hartshorne-Rao module M(C) is annihilated by the maximal ideal m of the ring $k[X_0,\dots,X_3]$. If M(C) has components of dimension $n_1 > 0$, $n_2 \geq 0$, ... , $n_{r-1} \geq 0$, $n_r > 0$ (from the first non-zero component to the last), then $\mathbf{L}_{n_1\dots n_r}$ is the corresponding even liaison class.

Remark 3.2 Let $N = n_1 + \dots + n_r$.

(a) If $C \in \mathbf{L}_{n_1\dots n_r}$ then $\alpha(C) \geq 2N$ (cf. [A] or [GM2]).

(b) If $C \in \mathbf{L}^0_{n_1\dots n_r}$ then the left-most component of M(C) occurs in degree 2N-2 and $\alpha(C) = 2N$ (cf. [GM1]).

(c) $\mathbf{L}_{n_1\dots n_r}$ possesses the LR-Property (cf. [BM2]).

Lemma 3.3

There exists a hyperplanar good linear configuration $Y \in \mathbf{L}^0_{n_1\dots n_r}$.

Proof.

We proceed by induction on $N = n_1 + \dots + n_r$. We will construct Y using Liaison Addition.

For N = 1, Y is the disjoint union of two lines. So assume that N > 1. For convenience we will assume that $n_1 > 1$. The construction for $n_1 = 1$ is similar, and we will comment on it afterwards. Let $Y_1 \in \mathbf{L}^0_{n_1-1\dots n_r}$ be a hyperplanar good linear configuration (by induction) and let $F_1 \in I_{Y_1}$ be the suitable union of 2(N-1) planes guaranteed by the definition of "hyperplanar". Sing F_1 is a union of lines having no component in common with Y_1 . Let Y_2 be a disjoint union of two lines, also disjoint from Y_1 and from Sing F_1 .

Note that Sing Y_1 , ($Y_1 \cap$ Sing F_1), and {triple points of F_1 } are all finite sets. Let F_2 be a union of two planes which contains Y_2 but avoids these three finite sets, and such that the line Sing F_2 avoids Y_1, Sing F_1, and $Y_2 \cap F_1$. (Choosing one plane first gives an additional finite number of points for the second plane to avoid. Having Sing F_2 avoid $Y_2 \cap F_1$ gives us that F_1 avoids $Y_2 \cap$ Sing F_2, which we need by symmetry.)

Let Y be the liaison addition of Y_1 and Y_2 using F_1 and F_2. We claim that Y is hyperplanar. Since no component of F_2 contains a component of Y_1 or of Sing F_1 and vice versa, Y is a reduced union of lines, and it is locally Cohen-Macaulay by Liaison Addition. An analysis similar to that given in Lemma 2.6 shows that Y has no triple points, so it is a good linear configuration.

To see that Y is in fact hyperplanar we must construct a reduced union of 2N planes containing Y such that (a) the intersection of any three is a point and (b) the intersection of any two is not a component of Y. Since F_2 does not contain any component of Sing F_1, we have by construction (and induction) that F_1 satisfies both of these conditions for the curve $Y_1 \cup (F_1 \cap F_2)$. (Of course F_1 has degree 2n-2, not 2N.) Now let F_2' be a union of two planes containing Y_2 and satisfying

1) F_2' contains no component of Sing F_1 or of Y_1 (this is forced by choice of Y_2).
2) F_2' has no component in common with F_2.
3) The line Sing F_2' does not lie on F_1.

$F_1 \cup F_2'$ is a reduced union of 2N planes containing Y. Then induction together with 1) and 3) give (a), and 1), 2) 3), choice of Y_2, and induction give (b).

Finally, we claim that $Y \in L^0_{n_1 \ldots n_r}$. Indeed, $M(Y_1)$ begins in degree 2N-4 and $M(Y_2)$ in degree 0. So $M(Y) \cong M(Y_1)(-2) \oplus M(Y_2)(-2N+2)$ and so

$$\dim M(Y)_{2N-2} = (n_1-1)+1 = n_1.$$

The case $n_1 = 1$ is similar, except that now we choose for F_2 a union of two planes containing Y_2 as above, together with a suitable number of general planes.

Proposition 3.4

Let X be a Buchsbaum curve in $\mathbf{P}^3$. Then X specializes to a stick figure.

Proof.

Every even liaison class of Buchsbaum curves in $\mathbf{P}^3$ satisfies the LR-property. Hence this is just a result of Theorem 2.8 and Lemma 3.3.

We want to employ the same technique in higher dimension. A family of even liaison classes of surfaces in $\mathbf{P}^4$ having the LR-property was studied in [BM3] and [BMR]:

<u>Definition 3.5</u>

Let p be a positive integer and i=1 or 2. We denote by $L_{p,i}(\mathbf{P}^4)$ the even liaison class of surfaces X in $\mathbf{P}^4$ such that

$$h^i(\mathbf{P}^4, \mathcal{I}_X(t)) = p \quad \text{for some } t$$

and $h^k(\mathbf{P}^4, I_X(s)) = 0$ for every $(k,s) \neq (i,t)$, k=1 or 2.

Note that $L_{1,1}(\mathbf{P}^4)$ is the even liaison class of the Veronese surface V in $\mathbf{P}^4$ (that is, of the general projection to $\mathbf{P}^4$ of the Veronese surface in $\mathbf{P}^5$) ; in [BM3] it is shown that V is in the minimal shift, that all these classes have the LR-property, that $L_{p,2}(\mathbf{P}^4)$ is the residual even liaison class of $L_{p,1}(\mathbf{P}^4)$ and that these classes can be "built up" starting from the Veronese surface, using Liaison Addition.

For instance, every surface belonging to $L_{p,1}(\mathbf{P}^4)$ (for some p), that is to say every surface having only one first cohomology group different from zero, specializes to a reducible surface having p Veronese surfaces as irreducible components, and all the other irreducible components are arithmetically Cohen-Macaulay.

What we do here is to apply Theorem 2.8 to show that these deformation results can be refined, proving that every surface in $L_{p,i}(\mathbf{P}^4)$ can be specialized to a union of planes in $\mathbf{P}^4$. This gives a higher-dimensional analog of the so-called "Zeuthen-problem" for stick-figures. Here we have the Veronese surface playing the same role as the two skew lines in the case of Buchsbaum curves in $\mathbf{P}^3$.

Lemma 3.6

Let $V \in L_{1,1}(\mathbf{P}^4)$ be the Veronese surface, and F the family of surfaces Y in $L_{1,1}(\mathbf{P}^4)$ such that $F_V = F_Y$. Then there exists a good hyperplanar linear configuration $G \in F$.

<u>Remark 3.7</u>

This implies that G is a reduced locally Cohen-Macaulay, non-arithmetically Cohen-Macaulay union of four planes, and that V specializes to G.

Proof.

Since deg(V)=4, we must find a configuration of four planes in $\mathbf{P}^4$. Take three planes in general position in $\mathbf{P}^4$; let us denote them by H_1, H_2 and H_3 and by $P=H_1 \cap H_2$, $Q=H_1 \cap H_3$, $R=H_2 \cap H_3$ the three points where two of them meet. Since the planes are chosen generally, there is exactly one plane K containing P,Q and R.

Let us call $G = H_1 \cup H_2 \cup H_3 \cup K$. We want to show that G is locally Cohen-Macaulay, hence we examine the singularities of G. Through a point $\wp$ of G pass at most three irreducible components:

i) If $\wp$ lies in only one component, it is a smooth point, hence G is locally Cohen-Macaulay at $\wp$.

ii) If $\wp$ lies in exactly two components, then these components meet along a line, and hence we can suppose $\wp$ (the problem is local) as lying in the intersection of two planes contained in the same three-dimensional projective space. Hence G is locally Cohen-Macaulay at $\wp$.

iii) If $\wp$ lies in three irreducible components of G, let us call G' the union of these three components. If we intersect G' with a general hyperplane, we see that G' is a cone, with vertex $\wp$, over the union L of two skew lines with a third line intersecting them, contained in a three-dimensional projective space. Since L is projectively Cohen-Macaulay, G' (and hence G) is locally Cohen-Macaulay at $\wp$.

On the other hand, the intersection of G with a general hyperplane S is a configuration L of three skew lines with a fourth line intersecting them. Hence L is a divisor of type (3,1) on a quadric surface.

The exact sequence of restriction

$$0 \to \mathcal{I}_G(-1) \to \mathcal{I}_G \to \mathcal{I}_L \to 0$$

and a standard cohomology computation show that $G \in L_{1,1}(\mathbf{P}^4)$. (Use the fact that only finitely many $H^1(\mathbf{P}^4, \mathcal{I}_G(n))$ and $H^2(\mathbf{P}^4, \mathcal{I}_G(n))$ can be non-zero.) More exactly, it is in the minimal shift. Since $L_{1,1}(\mathbf{P}^4)$ has the LR-property, this implies that $G \in \mathcal{F}$. From the construction it is clear that G is a good linear configuration.

Moreover, it is hyperplanar. In fact, note that $\alpha(G)=3$. Hence it is enough to find a union of three hyperplanes containing G and not passing through the same plane, such that the intersection of any two of them is not a component of G. Let us call S_1 the hyperplane containing H_1 and K, S_2 a hyperplane containing H_2 and not containing K, S_3 a hyperplane containing H_3 and not containing K. Note that S_1 doesn't contain H_2 or H_3, S_2 doesn't contain H_1 or H_3, and that S_3 doesn't contain H_1 or H_2 (remember that the intersections are points and not lines). Moreover:

i) $S_1 \cap S_2$ is a plane different from K, H_1, H_2 or H_3 containing the line $H_2 \cap K$,

ii) $S_1 \cap S_3$ is a plane different from K, H_1, H_2 or H_3 containing the line $H_3 \cap K$,

iii) $S_2 \cap S_3$ is a plane different from K, H_1, H_2 or H_3.

Hence if $S_1 \cap S_2 \cap S_3$ is a plane it must contain $H_2 \cap K$ and $H_3 \cap K$; that is, it must be the plane K. This is a contradiction. Hence we have shown that G is hyperplanar.

Corollary 3.8

There exists a good hyperplanar linear configuration in the minimal shift of $L_{p,i}(\mathbf{P}^4)$ for every $p>0$, $i=1,2$.

Proof. (sketch)

i=1) Let us work by induction on p. For p=1, this is Lemma 3.6. If p>1, take a good hyperplanar linear configuration X in the minimal shift of $L_{p-1,1}(\mathbf{P}^4)$ and a good hyperplanar linear configuration M in the minimal shift of $L_{1,1}(\mathbf{P}^4)$. Then perform a liaison addition of X and M via general hypersurfaces of degrees $\alpha(X)$ and $\alpha(M)$ respectively, split into unions of hyperplanes as in Definition 2.4. Choosing these hypersurfaces carefully, first moving X and M in the space if necessary, the result of such a liaison addition is a good hyperplanar linear configuration in the minimal shift of $L_{p,1}(\mathbf{P}^4)$, exactly as in [BM3], Example 4.11.

i=2) Let M be the good hyperplanar linear configuration in the minimal shift of $L_{1,1}(\mathbf{P}^4)$,constructed in Lemma 3.6. Then consider the hypersurface Q (containing M)

$S_1 \cup S_2 \cup S_3$ (we use the notation of the proof of 3.6), and another hypersurface of degree 3 containing M: for instance the union Z of a general plane P_1 containing H_1 and not containing K, the plane P_2 containing H_2 and K, and a general plane P_3 containing H_3 and not containing K. The residual subschemes of M in the liaison via Q and A is a locally Cohen-Macaulay linear configuration in the minimal shift of $L_{1,2}(\mathbf{P}^4)$. It is good and hyperplanar if P_1 and P_3 are chosen in a careful way.

For p>1, the induction works exactly as in the case i=1.

Theorem 3.9.

Let S be a surface in $\mathbf{P}^4$ with only one cohomology group $h^i(\mathbf{P}^4, \mathcal{I}_X(t)) \neq 0$, i=1 or 2. Then S specializes to a good hyperplanar linear configuration.

Proof.

This is just a corollary of 3.8, 2.8 and the fact that $L_{p,i}(\mathbf{P}^4)$ has the LR-property ([BM3]).

There are many other deformation results which can be obtained with these liaison techniques; the proofs are much simpler if we don't mind the kinds of singularities that arise in the configurations. We sketch one example.

Proposition 3.10.

Let X be a curve in $\mathbf{P}^3$ linked to a curve contained in a quadric surface. Then X specializes to a reducible curve whose components are lines and conics.

Proof.

Let us call $\mathcal{L}$ the even liaison class of X. $\mathcal{L}$ contains a curve Q on a quadric surface, hence a divisor of type, say, (a,b). Without loss of generality we may take a < b-1: if a=b (resp. b-1), then Q is a complete intersection (resp. an almost-complete intersection), and hence $\mathcal{L}$ is the liaison class of arithmetically Cohen-Macaulay curves, where every element specializes to a stick figure. But then Q is evenly linked to a divisor M of type (0,b-a), that is to say a union of b-a skew lines. This implies that $\mathcal{L}$ satisfies the LR-property ([LR]), since -2 = e(M) = s(M)-4.

But then X specializes to the liaison addition Z of M with an arithmetically Cohen-Macaulay curve C (cf. [BM3]). We may choose C to be a configuration of lines (see the proof of 5.10 in [BM3]), the surface containing M as the union of the quadric surface with a suitable number of planes and the surface containing C as a union of planes. With these assumptions, the irreducible components of Z are lines or conics.

Remark 3.11.

This argument shows in fact that if, in an even liaison class satisfying the LR-property, there is a minimal curve M that is a union of lines, then anything in that class specializes to a curve that is the union of lines and plane curves of degree $\alpha(M)$.

Remark 3.12.

We have seen that an even liaison class L of curves (or more generally in codimension two) can be expressed as a disjoint union of sets $F_{(b;g_2,\ldots,g_r)}$, where every two elements can be deformed one to the othe one. It is natural to ask which of these families possesses

Property P1: it contains an irreducible curve, or

Property P2: it contains a stick figure.

Also, what are the relations, if any, between these properties? All of these questions depend very much on the class L. In particular, if P1 implies P2, this would answer the Zeuthen problem (and would be a stronger statement). Unfortunately, this is not the case.

First, note that there may exist families possessing neither P1 nor P2-- for example, in the even liaison class of a double line of arithmetic genus ≤ -2 consider the family of curves in the leftmost shift (cf. [M1]). Also, P2 does not imply P1-- consider the leftmost shift of the class of an arithmetically Buchsbaum curve whose associated Hartshorne-Rao module has "diameter" $r > 1$-- we have just seen that it possesses P2 (Lemma 3.3) but it does not possess P1 (cf. [BM1]). On the other hand, a general smooth curve of large degree is the unique curve in the leftmost shift of its even liaison class (cf. [LR]), so we see that P1 does not imply P2.

Hence, while the "liaison" approach to the Zeuthen problem described in this paper is useful in many cases, it is not enough by itself to solve the problem.

References

[A] M. Amasaki, *On the structure of Arithmetically Buchsbaum curves in* $\mathbf{P}^3$, Publ. RIMS, Kyoto Univ. **20** (1984), 793-837.

[B] G. Bolondi, *Irreducible families of curves with fixed cohomology,* To appear in Ark. der Math.

[BM1] G. Bolondi and J. Migliore, *Buchsbaum Liaison Classes*, to appear in J. Alg.

[BM2] G. Bolondi and J. Migliore, *The Lazarsfeld-Rao Problem for Buchsbaum Curves* To appear in Rend. Sem. Mat. Univ. Padova.

[BM3] G. Bolondi and J. Migliore, *The Structure of an Even Liaison Class*, Preprints Matematici Univ. Trento UTM **239** (1988).

[BMR] G. Bolondi and R. Miro-Roig, *Two codimensional Buchsbaum Subschemes of* $\mathbf{P}^n$ *via their Hyperplane Sections*, To appear in Comm. in Alg.

[GM1] A.V. Geramita and J. Migliore, *On the Ideal of an Arithmetically Buchsbaum Curve*, to appear in J. Pure and App. Alg.

[GM2] A.V. Geramita and J. Migliore, *Generators for the Ideal of a Buchsbaum Curve*, to appear in J. Pure and App. Alg.

[LR] R. Lazarsfeld and P. Rao, *Linkage of general curves of large degree* . In: Algebraic Geometry- open problems (Ravello 1982). Lect. Notes Math. **997,** 267-289. Berlin, Heidelberg, New York: Springer 1983.

[HH] R. Hartshorne and A. Hirschowitz, *Smoothing algebraic space curves.* In: Algebraic Geometry, Sitges (Barcelona 1983). Lect. Notes Math. **1124**, 98-131. Berlin, Heidelberg, New York: Springer 1983.

[M1] J. Migliore, *On Linking double Lines*, Trans. AMS **294** (1986), 177-185.

[M2] J. Migliore, *Liaison of a Union of Skew Lines in* $\mathbf{P}^4$, Pac. J. Math. **130** (1987), 153-170.

[S] P. Schenzel, *Notes on liaison and duality* , J. Math. Kyoto Univ., **22-** 3 (1982) 485-498.

[Sw] P. Schwartau, *Liaison addition and monomial ideals* , Ph.D. thesis, Brandeis University (1982).

PLANE SECTIONS OF ARTHMETICALLY NORMAL CURVES IN $\mathbb{P}^3$

Luca Chiantini
Ferruccio Orecchia

Dipartimento di Matematica e Applicazioni
Università di Napoli. Via Mezzocannone 8
80129 NAPOLI (ITALY)

Let C be any smooth irreducible curve in the projective 3-space $\mathbb{P}^3$ over an algebraically closed field K; if $\pi \subseteq \mathbb{P}^3$ is a general plane, many properties of the set of points $Z = \pi \cap C$ are known: e.g. Z is in "uniform position" (see [H]), its numerical character is connected (see [GP]) and so on. On the other hand, if we have more precise informations on the geometric properties of the curve C, we would like to describe more carefully which "kind" of set of points Z may arise cutting C with any (even not general) plane; a deeper knowledge of those sets of points should be useful in the classification of space curves.

Inverting the point of view, what we are asking is the following:

(Q) Let Z be any set of d distinct points in the plane $\pi \subseteq \mathbb{P}^3$; which curves C may have Z as plane section?

Let $\mathcal{I}_Z$ be the ideal sheaf of Z in the plane π; $\mathcal{I}_Z$ has a minimal resolution of the form:

$$0 \longrightarrow \oplus^{\nu-1}\mathcal{O}_\pi(-s_i) \longrightarrow \oplus^{\nu}\mathcal{O}_\pi(-r_j) \longrightarrow \mathcal{I}_Z \longrightarrow 0 \qquad (1)$$

and the principal numerical features of Z are determined by the numbers $s_1,\dots,s_{\nu-1},r_1,\dots,r_\nu$, which depend only on the set Z and not on the resolution we choose; however if C is a general curve of $\mathbb{P}^3$ such that $C \cap \pi = Z$, then its "characteristic numbers" (even the genus) and, in particular, a resolution for its ideal sheaf $\mathcal{I}_C$, are not strictly determined by resolution (1). It seems quite natural to attach the problem restricting ourselves to those curves C having the property that a resolution for $\mathcal{I}_C$ is obtained immediately from a resolution of the ideal sheaf of any plane section; thus we are led to consider arithmetically normal curves.

A curve C is called "arithmetically normal" if for any n the obvious restriction maps $H^0(\mathcal{O}_{\mathbb{P}^3}(n)) \longrightarrow H^0(\mathcal{O}_C(n))$ surject; equivalently C is arithmetically normal (a.n. for short) if and only if its Rao module $\oplus H^1(\mathcal{I}_C(n))$ is trivial; if Z is any plane section of a (possibly singular) a.n.curve C, then by the exact sequences:

$$0 \longrightarrow H^0(\mathcal{I}_C(n-1)) \longrightarrow H^0(\mathcal{I}_C(n)) \longrightarrow H^0(\mathcal{I}_Z(n)) \longrightarrow 0$$

we see that all the generators and all the syzygies for the homogeneous ideal of Z can be lifted to generators and syzygyes for the homogeneous ideal of C, so if (1) is a resolution for $\mathcal{I}_Z$, then $\mathcal{I}_C$ has a resolution of the form:

$$0 \longrightarrow \oplus^{\nu-1}\mathcal{O}_{\mathbb{P}^3}(-s_i) \longrightarrow \oplus^{\nu}\mathcal{O}_{\mathbb{P}^3}(-r_j) \longrightarrow \mathcal{I}_C \longrightarrow 0 \qquad (2)$$

the sequence of numbers $s_1,\dots,s_{\nu-1},r_1,\dots,r_\nu$ describes many features of such curves (e.g. genus, Hilbert function); examples of a.n.curves are easily found: first any complete intersection curve is arithmetically normal, then any curve linked to an a.n. curve is still arithmetically normal (see [PS]).

<u>Example</u> Not every set of point Z is the plane section of an a.n.curve; indeed, let us look what happens when d = deg(Z) = 4: if Z is general, then it is complete intersection of two conics so a general pair of quadrics in $\mathbb{P}^3$ passing through Z intersects in a smooth a.n.curve C of degree 4 which cuts Z on π; similarly if Z is contained on a line, we can take C as a general quartic on some plane $\neq \pi$, passing through Z. On the other hand, if Z has 3 points on a line and the remaining point outside, any smooth quartic C of $\mathbb{P}^3$ passing through Z has a trisecant line, thus it is rational and it cannot be arithmetically normal for $H^1(\mathcal{I}_C(1))\neq 0$.

Hence the problem of finding the plane sections of a.n.curves is not trivial.

For a.n.curves we have a complete answer to question (Q), in terms of the numbers $s_1,\dots,s_{\nu-1},r_1,\dots,r_\nu$ arising in the resolution (1), at least when the base field has characteristic 0.

<u>Theorem</u> Let Z be a set of d distinct points in the plane $\pi\subseteq\mathbb{P}^3$ over a field K with char(K)=0. Let (1) be a resolution for the ideal sheaf $\mathcal{I}_Z$ of Z in π and order the numbers in such a way that $s_1\leq\dots\leq s_{\nu-1}$ and $r_1\leq\dots\leq r_\nu$. Then there exists a smooth arithmetically normal curve $C\subseteq\mathbb{P}^3$ such that $Z = C\cap\pi$ if and only if for $n = 1,\dots,\nu-2$ $\quad s_n > r_{n-2}$.

Note that since the curve C above is smooth and arithmetically normal, it is also irreducible.

Observe that when Z is complete intersection in π, then $\nu=2$ so the condition of the theorem is empty and we get back the well known fact that a complete intersection set of points in π is the plane section of a smooth arithmetically normal (in fact, complete intersection) curve.

Given a set of points $Z\subseteq\pi$ it is easy to find a reducible a.n.curve having Z as plane section: just take the cone C(Z) over Z with vertex in any point $P\in\mathbb{P}^3-\pi$; since two a.n.curves having the same sequence can be flatly deformed one to the other inside $\mathbb{P}^3$ (see [E] or [S]), then the question on the existence of a smooth a.n.curve cutting Z on π is a question on the smoothability of C(Z): we want a smoothing of C(Z) by a flat family which fixes the points of Z. The smoothing of a.n.curves C was studied by Peskine and Szpiro (who proved in [PS] that C is smoothable when all the numbers $s_i - r_j$ are positive) and by Sauer ([S]) who solved the problem showing that C is smoothable if and only if, in the sequence (2) above, $s_n \geq r_{n-2}$ for all n. Returning to the example of a set Z of 4 points, 3 on a line and one outside, we observe that the numbers of a resolution of $\mathcal{I}_Z$ are $\nu=3$, $r_1=r_2=2$, $r_3=3$, $s_1=3$, $s_2=4$ hence by Sauer's theorem the cone C(Z) over Z can be deformed to a smooth a.n.curve, however any such smoothing will move the set Z to a set of points in generic position.

In the proof of our theorem, we make use the technique of "liaisons" in a way similar to the one followed by Peskine, Szpiro and Sauer; however the main idea is to produce a "Bertini-type" argument on the set of all a.n.curves having Z as plane section. Indeed the injection of sequence (1) is given by a $(\nu-1)\times\nu$ persymmetric matrix A of homogeneous forms in π whose degree matrix ∂A has i,j entry $s_i - r_j$; the maximal minors of A define a minimal set of generators for the homogeneous ideal of Z; if t=0 is an equation for π in $\mathbb{P}^3$, and C is a (possibly singular) a.n.curve such that $Z = C \cap \pi$, then the matrix which defines the injection in sequence (2) is of the form A+tB where B is any persymmetric matrix of homogeneous forms in $\mathbb{P}^3$ of degrees $s_i - r_j - 1$; all the matices like A+tB are thus parametrized by an affine space $\mathbb{A}^N$ and we contruct in the product $\mathbb{A}^N \times \mathbb{P}^3 - \pi$ the "singular incidence correspondence" X given by all the pairs (A+tB,P) such that the scheme defined by the maximal minors of A+tB is not smooth of codimension 2 at P; showing that the fibers of X over any $P \in \mathbb{P}^3 - \pi$ have dimension $\leq N-4$ we get the result. Since we are dealing with 2-codimensional objects, the computations of the dimensions of the fibers is not immediate, but makes use of a careful extimation of the dimension of some auxiliary subsets of $\mathbb{A}^N$.

At the end of the paper we list briefly some consequences of the above theorem. Essentially the theorem allows us to give a description of a stratification of the Hilbert scheme of arthmetically normal smooth curves in $\mathbb{P}^3$ given by the Hilbert-Burch matrix of a curve.

Further J.Harris proved that the general plane section of a smooth curve in $\mathbb{P}^3$ is formed by points in uniform position ([H]); our result, together with a theorem of A.Geramita and J.Migliore, shows that any set of points in π in uniform position are the plane section of a smooth arithmetically normal curves.

We wish finally to thank C.Ciliberto and A.Geramita, with whom we had many fruitful discussions on this subject.

<u>Notation</u>

$\mathbb{P}^3$ = projective 3-space over an algebraically closed field K of characterististic 0;

π = plane of $\mathbb{P}^3$ defined by t=0 in some fixed coordinate system;

Z = set of d distinct points in π;
$\mathcal{I}_Z$ = ideal sheaf of Z in π.

We fix a minimal resolution for $\mathcal{I}_Z$ in π given by sequence (1) above and we reorder the numbers s_i's and r_j's so that we have: $s_1 \le \ldots \le s_{\nu-1}$, $r_1 \le \ldots \le r_\nu$.

If M is any matrix of homogeneous forms in $\mathbb{P}^m$, we set ∂M = degree matrix of M; M is called "persymmetric" if all its minors are homogeneous.

Let $A=(a_{ij})$ be the $(\nu-1)\times\nu$ matrix of forms of π which defines the injection in sequence (1); then A is persymmetric and its degree matrix is $\partial A = (s_i - r_j)$. By the Hilbert-Burch theorem (see [CGO]) the maximal minors of A define a set of minimal generators for the homogeneous ideal $\oplus H^0(\mathcal{I}_Z(n))$ of Z.

If M is any persymmetric matrix of forms in $\mathbb{P}^n$ we set Y(M) = the scheme defined by the maximal minors of M.

If C is any arithmetically normal curve in $\mathbb{P}^3$ such that $Z = C\cap\pi$ and $\mathcal{I}_C$ is the ideal sheaf of C in $\mathbb{P}^3$, as we pointed out in the introduction, all the restriction maps $H^0(\mathcal{I}_C(n)) \longrightarrow H^0(\mathcal{I}_Z(n))$ surject and this implies that $\mathcal{I}_C$ has a resolution given by sequence (2), where the injection is given by a persymmetric $(\nu-1)\times\nu$ matrix A' of forms of $\mathbb{P}^3$ whose $(\nu-1)\times(\nu-1)$ minors define a set of minimal generators for the homogeneous ideal of C; since A' modulo t, defines a minimal resolution of $\mathcal{I}_Z$, we may assume A' of the form A' = A+tB where $B=(b_{ij})$ is still a persymmetric matrix of forms of $\mathbb{P}^3$ with degree matrix $\partial B = (s_i - r_j - 1)$.

<u>Proposition</u> 1 Assume $s_n \le r_{n+2}$ for some n. Then any arithmetically normal curve C such that $Z = C\cap\pi$ is singular.

<u>proof</u> Let A'=A+tB be a $(\nu-1)\times\nu$ matrix which defines a resolution for $\mathcal{I}_C$; if $s_n \le r_{n+2}$ then $\deg(a_{n,n+2}) \le 0$ and the minimality of (1) implies $a_{n,n+2}=0$; further $b_{n,n+2}$ has negative degree, thus it also vanishes. Since the degree matrix of A' is non-increasing going upward and rightward, it follows that A' is of the form:

$$\left(\begin{array}{c|c} Q & 0 \\ \hline & Q' \end{array}\right)$$

and the argument of [S], prop.1 proves that the scheme defined by Q intersects the surface det(Q')=0 in singular points of C (see also [GM]).

<u>Remark</u> 2 Our theorem is just an easy application of standard Bertini theorems when Z is complete intersection, i.e. when $\nu=2$. Hence we <u>assume through the paper</u> $\nu>2$.

<u>Definition</u> 3 Let $\mathbb{A}^N$ be the affine space which parametrizes all the persymmetric matrices of the form A+tB where A is defined

by sequence (1) and B varies among the $(\nu-1)\times\nu$ persymmetric matices of homogeneous forms of $\mathbb{P}^3$ with fixed degree matrix $\partial B = (s_i-r_j-1)$.

Let $X \subseteq \mathbb{A}^N \times(\mathbb{P}^3-\pi)$ be the "singular incidence correspondence", $X = \{(M,P)$: the maximal minors of M define a scheme which contains P and it is not smooth of codimension 2 at P$\}$.

It is clear by the definition that X is an algebraic subset.

$\forall P\in\mathbb{P}^3-\pi$ put X_P= fiber of X over P in the obvious projection. We shall consider (by abuse) X_P as a subvariety of $\mathbb{A}^N$.

We also indicate with $PGL(\pi)$ the subgroup of the projective linear group PGL(3) which fixes the plane π. This group clearly acts on $\mathbb{A}^N$.

Remark 4 Assume we have proved the following claim:

(*) there is a fixed subvariety $\Delta\subseteq\mathbb{A}^N$, different from $\mathbb{A}^N$ and invariant under $PGL(\pi)$ such that, given one fiber X_P of X, the components of X_P of dimension $\geq N-3$ are contained in Δ.

Then the theorem is proved. In fact $PGL(\pi)$ interchanges the fibers of X over $\mathbb{P}^3-\pi$ hence the claim implies that all the components of dimension $\geq N$ of X are projected inside Δ, so p is not surjective.

Hence our task is just to prove claim (*).

Definition 5 For any $(\nu-1)\times\nu$ matrix M we define:
$f(M) = (\nu-2)\times(\nu-1)$ matrix obtained erasing the two leftmost columns in M and transposing.

If $M'= A+tB \in\mathbb{A}^N$ then $f(M)=f(A)+tf(B)$ is a persymmetric matrix of homogeneaous forms of $\mathbb{P}^3$. Restricted to $\mathbb{A}^N$ f defines a surjective map between spaces of matrices whose fibers are linear subspaces, all of the same dimension.

If L is a fiber of f, let us indicate, by abuse, $Y(L)$= the scheme $Y(f(M))$, defined by the maximal minors of $f(M)$, for any $M\in L$.

Definition 6 Let $P\in\mathbb{P}^3-\pi$; define the following subsets of $\mathbb{A}^N$:
$V(P)=\{M:\ \mathrm{rank}_P M < \nu-1$ (i.e. $P\in Y(M))\}$.
$W(P)=\{M:\ \mathrm{rank}_P f(M) < \nu-2$ (i.e. $P\in Y(f(M)))\}$.
$W'(P)= W(P)\cap V(P)$.
$\Delta=\{M:\ \dim(Y(f(M)))\geq 2\}$.

All these subset are closed and algebraic in $\mathbb{A}^N$.

Δ is invariant under $PGL(\pi)$. Observe that X_P is contained in $V(P)$.

Proposition 7 a) dim $V(P)\leq N-2$ and $V(P)$ is irreducible.
b) dim $W(P)\leq N-2$. The irreducible components of $W(P)$ correspond to the mumbers $m=1,\dots,\nu-2$ such that $m=\nu-2$ or $(\partial f(A))_{m,m+2}\leq 0$.
c) Any component of $W(P)$ intersects $V(P)$ in an irreducible variety of dimension $\leq N-3$.

proof Conditions a),b),c) only depend on the residual of M or

$f(M)$ at P, so they follows from properties of spaces of numerical matrices with some fixed 0's.

Indeed, taking residuals at $P\in\mathbb{P}^3-\pi$, it is clear that $V(P)$ maps to the set of numerical $(\nu-1)\times\nu$ matrices (u_{ij}) with u_{ij} free for $j\le i+2$, $W(P)$ maps to the set of numerical $(\nu-2)\times(\nu-1)$ with u_{ij} free for $j\le i+1$, so a),b) follows at once by well known facts about numerical matrices (see e.g. [ACGH] 2.2 or also [S]).

To see c), observe that if $f(M)$ evaluated at P has one row which depends on the others, then forgetting the corresponding column in M, we see that we need only one more irreducible condition on the two leftmost columns of M to get $\mathrm{rank}_P M<\nu-1$.

<u>Remark</u> 8 Let us try to explain b) of prop.7 with a picture. Take for example the degree matrix $\partial(f(M))$ of this form:

$$\begin{pmatrix} 2 & 1 & -1 & -1 & -3 \\ 4 & 3 & 1 & 1 & -1 \\ 4 & 3 & 1 & 1 & -1 \\ 6 & 5 & 3 & 3 & 1 \end{pmatrix}$$

so that we are looking to numerical matrices of type

$$\begin{pmatrix} a_{11} & a_{12} & 0 & 0 & 0 \\ a_{21} & a_{22} & a_{23} & a_{24} & 0 \\ a_{31} & a_{32} & a_{33} & a_{34} & 0 \\ a_{41} & a_{42} & a_{43} & a_{44} & a_{45} \end{pmatrix}$$

and the components of the locus where they drop rank are given by:

$a_{11}=a_{12}=0$ (corresponding to m=1 for $\partial(f(M))_{13}\le 0$).

$$\det\begin{pmatrix} a_{23} & a_{24} \\ a_{33} & a_{34} \end{pmatrix} = \det\begin{pmatrix} a_{11} & a_{12} & 0 \\ a_{21} & a_{22} & a_{23} \\ a_{31} & a_{32} & a_{33} \end{pmatrix} = 0 \quad (m=3 \text{ for } \partial(f(M))_{35}\le 0).$$

$$a_{45} = \det\begin{pmatrix} a_{11} & a_{12} & 0 & 0 \\ a_{21} & a_{22} & a_{23} & a_{24} \\ a_{31} & a_{32} & a_{33} & a_{34} \\ a_{41} & a_{42} & a_{43} & a_{44} \end{pmatrix} = 0 \quad (\text{corresponding to } m=\nu-2=4).$$

<u>Proposition</u> 9 a) $\Delta\neq\mathbb{A}^N$.

b) For M general in $\mathbb{A}^N$, $Y(f(M))$ is an arithmetically Cohen-Macaulay curve , whose ideal is generated by elements of degree $\le r_1-1$ (r_1 minimal degree in sequence (1)), moreover $f(M)$ gives a matrix for a minimal resolution of its ideal. Further Y is linked to $Y(M)$ by the forms given by the first and the second minor (from the left) of M, i.e. by two surfaces of degree r_1 and r_2 respectively.

c) $h^0(\mathcal{I}_Y(r_1-1))$ and $h^0(\mathcal{I}_Y(r_2-1))$ are constant in $\mathbb{A}^N$.

<u>proof</u> a) We proceed by induction on ν. If $\nu=3$ the matrix $f(M)$ is 1×2 with all entries of degree >0 by assumption, so there is nothing to prove. In the general case let $g(M)$ by the matrix obtained erasing the first and the last column in $f(M)$; then the inductive hypothesis implies that for $f(M)$ general, $Y(g(M))$ is a curve, further by [PS] §4, $Y(f(M))$ is linked to $Y(g(M))$ by two surfaces, one of which has degree greater than the maximum degree of a meximal generator for the ideal of $Y(g(M))$, so we get that for $f(M)$ general, $Y(f(M))$ is a curve.

b) Follows as above from the description of the liaison procedure given in [PS] §4; the minimality of the resolution induced by $f(M)$ is a consequence of the fact that, by hypothesis,

$(\partial f(M))_{m,m+1}$ is always >0 (see [S] §1).

c) Obvious from a) and b).

Definition 10 With the previous notation, put:
$U(P)=\{M\in V(P)-(W(P)\cup\Delta)$: surfaces of degree r_2-1 through $Y(f(M))$ do not separate tangent vectors at $P\}$.
$U'(P)=\{M\in V(P)-(W(P)\cup\Delta)$: surfaces of degree r_1-1 through $Y(f(M))$ do not separate tangent planes at $P\}$.
$W''(P)=\{M\in W'(P)-\Delta$: $Y(f(M))$ is not smooth of codimension 2 at $P\}$.

$U(P), U'(P)$ are closed and algebraic in $V(P)-(W(P)\cup\Delta)$ while $W''(P)$ is closed algebraic in $\mathbb{A}^N-\Delta$.

Observe that if a fiber L of f intersects any of the sets $W(P), W'(P), U(P), U'(P), W''(P)$, then it is contained in it.

Proposition 11 Put $Y=Y(f(M))$. Then:

a) If $h^0(\mathcal{I}_Y(r_2-1))\geq 4$ then $\dim U(P)\leq N-3$.

b) If $h^0(\mathcal{I}_Y(r_1-1))\geq 3$ then $\dim U'(P)\leq N-4$.

proof a) Consider the map given by the linear system $H^0(\mathcal{I}_Y(r_2-1))$; the base locus is Y, so since $P\notin Y$, by the theorem of generic smoothness (we use the hypothesis char $K=0$), we see that surfaces of degree r_2-1 through Y separates tangent vectors at a general point of $\mathbb{P}^3-Y$. Applying $PGL(\pi)$ if necessary, we can find a matrix in $V(P)$ which is not in $U(P)$, so the first claim follows.

b) Also follows from the theorem of generic smoothness; alternatively, we know by [PS] 4.1 that a general linkage of Y with surfaces of degree r_1-1 gives a curve which is smooth outside Y. Put $U=\{(M,P)$: $M\in V(P)$ and surfaces of degree r_1-1 through Y do not separate tangent planes at $P\}$; fixing $M\in V(P)$, the fiber of U over M has dimension at most 1: were it a surface, since Y is not complete intersection, it would intersect a general linkage of Y with surfaces of degree r_1-1 in a point outside Y, necessarily singular for that linkage. So $\dim U\leq N-1$ hence over a general $P\in\mathbb{P}^3-\pi$ the fiber over P, which is exactly $U'(P)$, has dimension $\leq N-4$; acting with $PGL(\pi)$ we get the claim.

Proposition 12 $\dim W''(P)\leq N-4$

proof Enough to prove that in any component of $W(P)$ a general element gives $Y(f(M))$ smooth at P (since both the components of $W(P)$ and $W'(P)$ surject onto the corresponding components in the space of matrices of type $f(M)$). This is obvious for $\nu=3$, since in this case $Y(f(M))$ is complete intersection, so we proceed by induction on ν, proving the following claim:
- In the space of matrices of type $f(M)$, any component W of the locus of those matrices which drop rank at P has a general element such that $Y(f(M))$ and the minimal minor of $f(M)$ both are smooth at P.

We know that this is true for $\nu=3$. For general ν, consider the matrix $g(M)$ obtained by $f(M)$ erasing the first and last column. It is clear that $(\partial g(M))_{m,m+1}>0$ for any m so we may apply the inductive hypothesis on $g(M)$, further by the description of the components in $W(P)$ we know that for $f(M)$ general in W $P \notin Y(g(M))$; taking $g(M)$ such that $P \in Y(g(M))$, it follows that for $f(M)$ general in W, its minimal minor is smooth at P, further the maximal minors separate tangent planes at P; taking $g(M)$ such that $P \notin Y(g(M))$, we get $Y(f(M))$ complete intersection, at P, of a smooth surface and a surface not tangent to the first one; so we are done.

<u>Definition</u> 13 Put Φ_1 (respectively Φ_2)= the curve of π obtained from the first (resp. the second) minor in A. Fix a fiber L of f and let $q_L(M)$ be the element of $H = H^o(\mathcal{I}_{Y(L)\cup\Phi}(r_2)) \times H^o(\mathcal{I}_{Y(L)\cup\Phi'}(r_1))$ obtained considering the first and the second minor of M.

Let Ξ be the subset of H of all pairs of elements both containing t as a factor. Then Ξ is isomorfic to $H^o(\mathcal{I}_{Y(L)}(r_\nu-1)) \times H^o(\mathcal{I}_{Y(L)}(r_1-1))$.

Put finally $Y_L(P) = (q(X_P \cap L)^- \cap \Xi) \subseteq \Xi$.

<u>Lemma</u> 14 For all fibers L not in Ξ q_L surjects to H-Y and all its fibers have the same dimension.

<u>proof</u> Given any pair (α,β) in $H-\Xi$, α is a combination of the maximal minors of $f(M)$, further the trace of α on π is Φ by construction; the just rearranging the first column of $M \in L$ we may assume that the first minor of M is exactly α; the same is true for β rearranging the last column of M. The fact that the fibers have the same dimension follows at once from the fact that $f(M)$ gives a resolution for the ideal sheaf of $Y(f(M))$.

<u>Remark</u> 15 By the lemma, it follows that $Y_L(P)$ is just the subset of Ξ formed by those pairs of surfaces linking $Y(f(M))$ to a scheme passing through P and not smooth of dimension 1 at P.

Now we are ready to prove the theorem for any Z. When the following hypothesis (of proposition 11) is not satisfied:

(**) $h^o(\mathcal{I}_{Y(f(M))}(r_2-1)) \geq 4$ and $h^o(\mathcal{I}_{Y(f(M))}(r_1-1)) \geq 3$;

the general proof will not work, so we need in these exceptional cases an "ad hoc" proof.

<u>proof of claim</u> (*) Assume the hypothesis (**) are satisfied. Let L be a fiber of f.

(i) If L is not contained in $U(P) \cup W(P) \cup U'(P)$ then $\dim(X_P \cap L) \leq \dim(L)-4$; indeed in L we have fixed $Y=Y(f(M))$ not containing P and $X_P \cap L$ is given by curves linked to Y by surfaces of degree r_2 and r_1 with traces Φ_2 and Φ_1 on π; since a general surface of degree r_1-1 through Y is smooth at P and surfaces of degree r_2-1 separate tangent vectors at P, using lemma 14 it follows that we need 4 conditions to get such a linkage singular at P. Thus any component of X_P not contained in $U(P) \cup U'(P) \cup W(P)$ has

dimension $\le N-4$.
(ii) If L is contained in $U(P)-(W(P)\cup U'(P))$ then $\dim(L\cap X_P)\le \dim(L)-3$, indeed once again, since a general surface of degree r_1-1 through Y is smooth at P and surfaces of degree r_2-1 separate tangent planes at P, by lemma 14 once again we see that we need 3 conditions to get surfaces of degree r_1, r_2 through Y and Φ_1, Φ_2 respectively which link Y to a curve singular at P. Hence any component of X_P contained in $U(P)-(W(P)\cup U'(P))$ has dimension $\le \dim U(P)-3 \le N-4$.
(iii) It follows that the unique chance for X_P to have a component of dimension $>N-4$ in $\mathbb{A}^N-\Delta$ is that this component lies in $W'(P)-W''(P)$. By prop.12 in any component of $W(P)$ the general element gives a curve $Y(f(M))$ smooth at P, so by [PS] a general linkage of $Y(f(M))=Y$ by surfaces of degree r_1-1 and r_2-1 gives a curve which is smooth in a neighbourhood of P. Using $PGL(\pi)$ moving P along this smooth part of $Y(f(P))$ it is clear that for a general matrix in the component of $W'(P)$ we may assume that a general curve linked to $Y(f(M))$ by surfaces of degree r_1, r_2 passing through Φ_1, Φ_2 is smooth at P, hence no component of $W'(P)$ outside $W''(P)$ is contained in X_P. This proves that $\dim(X_P)\le N-4$.

<u>The exceptional cases</u>

Let us return to the hypothesis (**). When $h^0(\mathcal{I}_{Y(f(M))}(r_1-1))$ is less than 3, the degree matrix ∂A is of the form $\begin{pmatrix} a & 1 & 1 \\ a & 1 & 1 \end{pmatrix}$ so that $Y(f(M))$ is a line. If $a=1$ then we get $Z=$ 3 points not on a line, and the theorem is obvious. So assume $a>1$, thus in particular $h^0(\mathcal{I}_{Y(f(M))}(r_2-1))>4$. In this case we need only to prove that for $L\subseteq U(P)-(U'(P)\cup W(P))$, $\dim(X_P\cap L)\le \dim(L)-2$ for in the other cases the general proof works; If $Y=Y(f(M))$, then $Y\cup\Phi_1$ is the union of a line with a plane conic, and it is well known that quadrics through such an union separate tangent planes everywhere outside $Y\cup\pi$, so it follows that $\dim(X_P\cap L)\le \dim L-2$ and we are done.

If $h^0(\mathcal{I}_{Y(f(M))}(r_1-1))\ge 3$ and $h^0(\mathcal{I}_{Y(f(M))}(r_2-1))<4$ then the matrix ∂A is of the form $\begin{pmatrix} 1 & 1 & 1 & 1 \\ 1 & 1 & 1 & 1 \\ 1 & 1 & 1 & 1 \end{pmatrix}$, i.e Z is formed by 6 points in general position in π; we prove the theorem directly in this case; indeed it is clear that $Y=Y(f(M))$ is a (smooth) cubic rational curve in $\mathbb{P}^3$; take a smooth cubic surface S containing $Y\cup Z$; in the usual base $\ell, e_1, \ldots, e_6$ for $Pic(S)$ we may assume that Y is of the type $5\ell-\sum 2e_i$ so that we are looking for a smooth divisor of type $4\ell-\sum e_i$ on S passing through Z; if $P_1, \ldots, P_6$ are the base points in the map $\mathbb{P}^2 \longrightarrow \mathbb{P}^3$ which gives S, then by construction $Z\cup\{P_1, \ldots, P_6\}$ (read in $\mathbb{P}^2$) is the complete intersection of a cubic with a quartic curve, so that we have a smooth quartic passing through all these points.

Let us end the paper showing some consequences of the previous result.

1) Let C be any smooth curve in $\mathbb{P}^3$; it follows from a theorem of J.Harris ([H]) that a general plane section of C is formed by points in "uniform position" (see [H] for the definition). On the other hand A.Geramita and J.Migliore proved ([GM]) that if Z is any set of points in uniform position in $\mathbb{P}^2$, the degree matrix ∂A of a resolution of $\mathcal{I}_Z$ has the property $(\partial A)_{m,m+2}>0$ for $m=1,\dots,\nu$. So we may "invert" Harris' theorem, indeed we get:

Corollary 16 If Z is any set of d distinct points in uniform position in a plane $\pi\subseteq\mathbb{P}^3$, then there is a smooth arithmetically normal curve $C\subseteq\mathbb{P}^3$ such that $Z=C\cap\pi$.

2) The theorem tells us exactly which numeric $(\nu-1)\times\nu$ matrix is the degree matrix ∂A of the resolution of the ideal (i.e. the Hilbert-Burch matrix)of a smooth arithmetically normal curve in the projective space.

Corollary 17 A set of numbers $r_1,\dots,r_\nu,s_1,\dots,s_{\nu-1}$ is the set of degrees of a minimal system of generators and syzygies for an arithmetically normal smooth curve in $\mathbb{P}^3$ if and only if $\sum r_j=\sum s_i$ (persymmetric condition) and $\forall n=1,\dots,\nu-2$ we have $u_{n,n+2}>0$.

Indeed it is not hard to see that under the above condition there is a smooth set of points Z in a plane π whose ideal sheaf has a resolution given by a persymmetric matrix A with $\partial A=(s_i-r_j)$, so by the main theorem we see that we are done.

By the description of the deformation theory of arithmetically normal curves contained in [S] and [E] it is known that the Hilbert scheme $\mathcal{H}$ of smooth arithmetically normal curves in $\mathbb{P}^3$ with fixed degree and genus is stratified by fixing the degrees of a minimal set of generators $r_1,\dots,r_\nu$ and syzygies $s_1,\dots,s_{\nu-1}$ (this is clearly the same as fixing the Hilbert-Burch matrix) then the following facts are known:

a) For a general element in any component of $\mathcal{H}$ we have $r_i\neq s_j$ $\forall i,j$.

b) Any curve of $\mathcal{H}$ for which $r_j=s_i$ can be deformed in $\mathcal{H}$ to a curve with one less generator in degree r_j and one less syzygy in degree s_i.

c) Two curves with the same Hilbert-Burch matrix belong to the same component of $\mathcal{H}$.

The above corollary tells which strata are non-empty and how they specialize.

Corollary 18 Let S be the stratum of $\mathcal{H}$ corresponding to the set of degrees $r_1\leq\dots\leq r_\nu$, $s_1\leq\dots\leq s_{\nu-1}$. Then S specializes to the non-empty stratum of $\mathcal{H}$ obtained inserting the number σ in the sequence of r's and s's if and only if, setting

$n=\min(i: \sigma\le r_i)$ $m=\min(i:\sigma<s_i)$ we get $n\ge m+3$.

3) With the same procedure of the proof of the main theorem, essentially using prop.12, one can prove:

Theorem 19 Let Z be any 0-dimensional subscheme of length d in the plane $\pi\subseteq\mathbb{P}^3$; assume the ideal sheaf of Z has a resolution as sequence (1). Then there is a reduced arithmetically normal curve C in $\mathbb{P}^3$ such that $C\cap\pi=Z$.

Indeed the hypothesis imply that $(\partial A)_{m,m+1}>0$ $\forall m=1,\dots,\nu-1$.

REFERENCES

[GAC] Arbarello A., Cornalba M., Griffith P.A., Harris J. GEOMETRY OF ALGEBRAIC CURVES Springer 1985.

[CGO] Ciliberto C., Geramita A., Orecchia F. Some remarks on a theorem of Hilbert-Burch (to appear in Boll.UMI 1988).

[E] Ellingsrud G. Sur le schéma de Hilbert des varietés algébriques de $\mathbb{P}^e$ a cône de Cohen-Macaulay. Ann.Sci.Ec. Norm.Sup. 4(8) (1975) 423-432.

[GM] Geramita A., Migliore J. Hyperplane sections of a smooth curve in $\mathbb{P}^3$. (preprint).

[GP] Gruson L., Peskine C. Genre des courbes dans l'espace projectif. Lect.Notes.Math.687 (1978) 31-59.

[H] Harris J. The genus of space curves. Math.Ann.249 (1980) 191-204.

[AG] Hartshorne R. ALGEBRAIC GEOMETRY Springer 1977.

[O] Orecchia F. Points in generic position and conductor of curves with ordinary singularities. J.London Math.Soc. (2) 24 (1981) 85-86.

[PS] Peskine C., Szpiro L. Liaison des variétés algébriques I. Inv.Math.26 (1974) 271-302.

[S] Sauer T. Smoothing projectively Cohen-Macaulay space curves. Math.Ann.272 (1985) 83-90.

SUR LES LACUNES D'HALPHEN.

Ph. Ellia
C.N.R.S. U.A.168
Département de Mathématiques
Université de Nice
Parc Valrose-06034 Nice Cedex
France.

INTRODUCTION: On travaille dans $\mathbb{P}^3$, espace projectif de dimension trois sur un corps, K, algébriquement clos, de caractéristique nulle. La motivation initiale de cet article est le problème suivant (cf [GP4] introduction): "déterminer s(d,g), le plus petit entier n tel que toute courbe lisse, connexe, de degré d, de genre g, est contenue dans une surface de degré n". Il s'agit en quelque sorte d'une quintessence des problèmes de postulation, rang maximum et des problèmes d'Halphen sur le genre maximum (cf 1.Pb2) et les lacunes. C'est surtout ce dernier aspect que nous traitons. En effet après une rapide discussion générale au paragraphe un, nous consacrons la suite de cet article à démontrer:

Théorème: Soient d, s des entiers, $d > s(s-1)$, $s > 5$.
Posons $d = ks-r$, $0 \leq r \leq s-1$. Soit g un entier tel que: $G(d,s) > g > \max\{G^{(1)}(d,s), G(d,s+1)\}$ (cf 4.1, 6.1 pour les définitions).
(1) Si $r \neq 2$ et $r \neq s-2$ alors $s(d,g) \leq s-1$. Autrement dit (s,d,g) est une lacune d'Halphen (cf 1.4).
(2) Si $r = 2$ on a $s(d,g) = s$. Plus précisement il existe une et une seule composante irréductible, *H*, de H(d,g) (l'ouvert de Hilb($\mathbb{P}^3$) formé des courbes lisses, connexes, de degré d, genre g) contenant une courbe non sur une surface de degré s-1. La courbe générale de *H* est tracée sur une surface lisse de degré s. La composante *H* est génériquement lisse.
(3) (a) Si $r = s-2$ et $k \leq s+G(d,s)-g-3$ alors $s(d,g) \leq s-1$; ceci bien qu'il existe des courbes localement Cohen- Macaulay, de degré d, genre arithmétique g, tracées sur des surfaces lisses de degré s.
(b) Si $r = s-2$ et $k \geq s+G(d,s)-g-2$ alors $s(d,g) = s$ et il existe une et une seule composante irréductible, *H*, de H(d,g) contenant des

courbes non sur une surface de degré s-1. La courbe générale de *H* est tracée sur une surface lisse de degré s. Si $k \geq s+G(d,s)-g+1$, *H* est génériquement lisse.

Ce théorème complète un résultat de Dolcetti [Do] qui traitait le cas $g = G(d,s)-1$.

La valeur de $G^{(1)}(d,s)$ est donnée en VI.1. Disons simplement que $G^{(1)}(d,s)$ est le genre sous-maximal des caractères (connexes) de degré d, longueur s.

Exposons brièvement la démonstration. Soit C une courbe intègre de degré d, genre g, non sur une surface de degré s-1. La condition $g > G(d,s+1)$ implique que C est tracée sur une surface de degré s. Le lemme de Laudal [L], [GP3,4] et l'hypothèse $g > G^{(1)}(d,s)$ implique que le caractère numérique de C (cf [GP1]) est le caractère maximum de degré d, longueur s (cf VI.2, VI.3). D'après [GP1] c'est le caractère d'une courbe (de genre G(d,s)) liée à une courbe plane de degré r par une intersection complète (k,s). On étudie donc les courbes de degré d, $d > s(s-1)$, tracées sur une surface de degré s, de caractére maximum (cf §4). La section plane générale de C a la même postulation qu'un groupe de points lié par une intersection complète (k,s) à r points alignés. Au paragraphe deux on détermine la résolution libre minimale de l'idéal homogène d'un tel groupe de points (sous l'hypothèse que la courbe de degré s est intègre). Ceci nous permet, si $r < s-2$, d'utiliser un résultat de Strano [S1,2] (cf III.1 pour un énoncé un peu plus général) et de relever cette liaison à C. Ainsi C est liée à une courbe, Y, de degré r, genre arithmétique $p(r)-G(d,s)+g$, dont la section plane générale est alignée. Ici p(r) désigne le genre d'une courbe plane de degré r. La courbe Y est localement Cohen-Macaulay mais elle n'est pas forcément réduite. Si Y est plane alors C est arithmétiquement Cohen-Macaulay, de genre G(d,s). D'autre part si Y n'est pas plane, grâce encore à III.1 (mais cf aussi III.2.1), on montre que Y est soit une droite double soit la réunion de deux droites disjointes. En particulier $r = 2$. La même approche s'applique si $s-2 \leq r \leq s-1$ mais en faisant deux liaisons: C est biliée par des intersections complètes (s,k+s-2), (s,s-1) à une courbe de degré s-r. Ceci démontre (1). Pour (2) et (3) il reste à étudier les courbes liées (resp. biliées) à des droites doubles de genre g-G(d,s). Nous avons rassemblé dans le paragraphe cinq les résultats nécessaires sur les droites doubles. Notamment nous montrons que la composante irréductible de $Hilb(P^3)$ qui paramètre les droites doubles de genre $-\alpha$, $\alpha \geq 0$, est génériquement lisse (cf V.2). Ce résultat, via un théorème de J. O. Kleppe [K] permet d'obtenir la lissité de H dans (2) et (3b). Cependant si $s+G(d,s)-g-2 \leq k \leq s+G(d,s)-g$, la lissité de H dans (3b) reste un problème ouvert. Finalement (3a) présente un phénomène particulier (cf 6.6): à partir des droites doubles de genre $g-G(d,s)$ on obtient, par des liaisons $(s,s-1)$, $(s,k+s-2)$, des courbes (localement C.M.) de degré d, genre arithmétique g, tracées sur des surfaces lisses de degré s; mais chacune de ces courbes contient la droite support de la droite double correspondante. Ceci est dû au fait que si Y est une droite double, toute section de $\mathcal{O}_Y(n)$, $n < 0$, s'annule sur $Y_{réd}$. On en déduit que ces courbes ne sont pas limites plates de courbes intègres. A fortiori elles ne sont pas lissifiables.

Le théorème suggére le problème suivant: "Lorsque (s,d,g) est une lacune d'Halphen a-t-on $s(d,g) = s-1$?".

Pour conclure signalons que le point de vue adopté ici permet de déterminer $G(d,s)$ pour $s^2-2s+1 \geq d \geq s^2-3s+5$ (cf [E]).

C'est avec plaisir que je remercie les organisateurs pour l'agréable conférence qu'ils nous ont offerte.

Mes remerciements vont également à E. Mezzetti pour une lecture critique de ce manuscrit.

I) GENERALITES.

Problème 1: "Déterminer $s(d,g)$, le plus petit entier n tel que toute courbe lisse, connexe, de degré d, genre g, est contenue dans une surface de degré n".

Ce problème est étroitement lié au suivant:

Problème 2 (Halphen [Ha]):
"Déterminer $G(d,s)$ où $G(d,s):= \max\{g(C)$, pour $C \subseteq \mathbb{P}^3$ lisse, connexe, de degré d, vérifiant $h^0(\mathcal{I}_C(s-1)) = 0\}$".

Rappelons d'abord ce que nous appelerons le principe d'Halphen:
"Pour d fixé, $G(d,s)$ est une fonction strictement decroissante de s".
Comme il est clair que $G(d,s+1) \leq G(d,s)$, le principe est équivalent à l'énoncé suivant ([Ha] p.402, cf aussi [GP2] p.221):
"Toute courbe de degré d, genre $G(d,s)$, non sur une surface de degré s-1 est tracée sur une surface de degré s".
Admettant ce principe on peut, pour (presque, cf (ii) ci-dessous) tout g, trouver s tel que: $G(d,s) \geq g > G(d,s+1)$ (*). L'inégalité $g > G(d,s+1)$ implique que toute courbe de degré d, genre g est tracée sur une surface de degré s et il semble naturel de penser que, dans cette situation, il existe une telle courbe non sur une surface de degré s-1. Si cela est bien le cas, $s(d,g) = s$. Cette analyse rapide se heurte à quelques difficultés:
(i) Le principe d'Halphen n'est pas démontré
(ii) $G(d,s)$ n'est pas défini pour tout s (sauf pour ceux qui connaissent $\max\emptyset$); l'encadrement (*) n'est pas toujours possible
(iii) Même si l'encadrement (*) a lieu, on n'a pas toujours $s(d,g) = s$.

<u>(i) Le problème 2 et le principe d'Halphen:</u> Il semble difficile de démontrer le principe d'Halphen sans résoudre le problème 2. A l'heure actuelle le problème 2 est résolu si:
(i) $d \geq s^2-3s+5$ ([GP1,3] jusqu'à s^2-2s+2, [E] pour le reste)
(ii) $\lceil (s^2+6s+20)/3 \rceil \geq d \geq (s^2+4s+6)/3$ ([rH2], [HH]), $\lceil x \rceil$ désigne le plus petit entier supérieur à x.
(iii) $\lceil (s^2+4s+3)/3 \rceil \geq d \geq \lceil (s^2+s)/3 \rceil$ ([rH2], [HH]).
(iv) les valeurs de (s,d) pour lesquelles $d < (s^2+4s+6)/3$ et $d \geq (3G_A(d,s)+12)/4$ où $G_A(d,s) = d(s-1)+1-h^0(\mathcal{O}_{\mathbb{P}}(s-1))$ ([BE1,2])
(v) $d < (s^2+4s+6)/3$ et $s >> 0$ ([BEx])
(vi) $s \leq 9$ ([BE1,2], [rH], [HH])
Pour tous les cas non traités l'on dispose de conjectures ([rH1,2],

[HH]). Ceci dit le principe d'Halphen est suffisamment étayé pour que l'on ose s'en servir.

(ii) Courbes de petit genre:

I.1:Définition: Soit $d \geq 3$ un entier. On pose $s^-(d) := \max\{s \in \mathbb{N} / (s^2+4s+6)/6 \leq d\}$.

I.2:LEMME: Toute courbe lisse, connexe, de degré d est contenue dans une surface de degré $s^-(d)$.

Dém: cf [rH] Thm.3.3

Ce lemme montre que le problème d'Halphen, pour d fixé, n'a de sens que si: $2 \leq s \leq s^-(d)$. En particulier si $g \leq G(d,s^-(d))$, l'encadrement (*) n'a pas lieu. Pour pallier cet inconvénient on observe d'abord que: $G(d,s^-(d)) \leq 1 + d(s^-(d)-1) - h^o(\mathcal{O}_{\mathbb{P}^3}(s^-(d)-1))$, (cf par ex. [rH], en fait on conjecture l'égalité). De là un simple calcul montre que, pour $s \geq 3$, $G(d,s^-(d)) \leq 2d-9$. Ceci dit, l'on sait que si $g \leq 2d-9$, il existe une courbe lisse, connexe, de degré d, genre g, C, avec $h^1(\mathcal{O}_C(2)) = 0$ (cf par ex. [BE2] II.1, II.2). Finalement le théorème 2 de [BE2] montre que l'on peut raisonnablement conjecturer l'existence d'une courbe remplissant les conditions ci-dessus et de rang maximum. Ceci justifie:

Conjecture 1: Soient $d \geq 3$, $g \geq 0$ des entiers. On pose $v(d,g) = \min \{ k \in \mathbb{N} / k \geq 2, h^o(\mathcal{O}_{P^3}(k)) > kd-g+1 \}$. Si $g \leq G(d,s^-(d))$ alors $s(d,g) = v(d,g)$.

Le théorème du rang maximum permet de donner une réponse partielle:

I.3:PROPOSITION: Soit $\rho(d,g,3) := 4d-3g-12$ le nombre de Brill-Noether. Si $\rho(d,g,3) \geq 0$ alors $s(d,g) = v(d,g)$.

Dém: Soit P(d,g) la composante à modules généraux. D'après [BE1,2] si C est suffisamment générale dans P(d,g) alors C est de rang maximum. L'on peut aussi supposer $h^1(\mathcal{O}_C(2)) = 0$ (cf par ex. [BE2] II.1, II.2). Bien entendu $v(d,g) = s(C)$. Soit Y une courbe de degré d, genre g. Comme $\rho(d,g,3) \geq 0$, on a certainement $h^1(\mathcal{O}_Y(3)) = 0$. Comme $s(C) \geq 2$, le théorème du rang maximum permet de conclure: $s(Y) \leq s(C)$

(iii) Lacunes d'Halphen.
Halphen semble avoir été le premier à observer que pour certains d, g, s vérifiant l'encadrement (*) il n'existait aucune courbe de degré d, genre g non sur une surface de degré s-1 ([Ha]). Pour cela nous proposons:

I.4:Définition: Un triplet d'entiers (d,g,s) est une lacune d'Halphen si: $G(d,s) \geq g > G(d,s+1)$ et $s(d,g) \leq s-1$.

I.4.1:Remarque: Vu le point (ii) cette définition n'a de sens que si $g > G(d,s^-(d))$ (cf cependant la conjecture 1). L'alternative consiste à définir comme lacunes pour (d,s) les nombres g, $g \leq G(d,s)$, tels que $s(d,g) \leq s-1$ (cf [GP2] p.222).

L'existence de lacunes complique le problème initial. Il semble heureusement que, pour des raisons liées au caractère numérique, ces lacunes n'apparaissent que lorsque d est suffisamment grand devant s ("grands genres") et pour des genres proches de $G(d,s)$. Par exemple ce que l'on appelle couramment le "domaine A" (cf [rH]) ne devrait pas présenter de lacunes. Plus précisement:

I.5:Définition: Pour tout entier d on pose $s_A(d) := \min\{ s \in \mathbb{N} \,/\, d < (s^2+4s+6)/3 \}$ et $G^+(d,A) := 1+d(s_A(d)-1)-h^0(\mathcal{O}_{\mathbb{P}^3}(s_A(d)-1))$ (ce qui est la valeur conjecturée de $G(d,s_A(d))$).

Alors:

Conjecture 2: Pour $G(d,s^-(d)) < g \leq G^+(d,A)$ il n'existe pas de lacunes d'Halphen: si $G(d,s) \geq g > G(d,s+1)$ alors $s(d,g) = s$.

A l'heure actuelle l'on dispose d'une démonstration d'une version asymptotique du problème d'Halphen dans le domaine A et de la conjecture 2 [BEx]. Les conjectures 1 et 2 couvrent le problème initial pour $g \le G^+(d,A)$ (nb: $G^+(d,A)$ est de l'ordre de $cste.d^{3/2}$). Pour les genres plus grands (et donc lorsque d est grand devant s) une résolution complète du problème 1 nécessite la détermination des lacunes d'Halphen. Dans la suite de cet article l'on se propose de déterminer ces lacunes lorsque $d > s(s-1)$ et $g > G^{(1)}(d,s)$ (cf 6.1). Sauf quelques exceptions $G^{(1)}(d,s)$ est de l'ordre de $G(d,s)$-cste.s. Pour situer ce résultat et terminer ce panorama signalons l'énoncé suivant (cf [GP2] p.222):

Énoncé (Halphen): "Pour tous $s \ge 3$ et $g \le G(d,s)-(s^2-3s+2)/2$ il existe une courbe lisse, connexe, de degré d, genre g, non sur une surface de degré s-1."

II) GROUPES DE POINTS DE $\mathbf{P}^2$ DE CARACTERE MAXIMUM ($d > s(s-1)$).

II.0: Notation: Dans tout ce paragraphe on pose $\mathbb{P}^2 := \mathbf{P}$.

II.1: Résolutions libres minimales: Soit E un groupe de points de $\mathbf{P}$. La résolution libre minimale de l'idéal homogène gradué $\oplus H^o(\mathcal{J}_E(k))$ induit une suite exacte:

$0 \to \oplus \mathcal{O}_{\mathbf{P}}(-n_{2i}) \to \oplus \mathcal{O}_{\mathbf{P}}(-n_{1i}) \to \mathcal{J}_E \to 0$ que nous appelerons la résolution libre minimale de $\mathcal{J}_E$. Nous utiliserons le fait suivant bien connu:

II.1.1:LEMME: Avec les notations ci-dessus posons: $n^+(j) := \max \{n_{ji}\}$, $n^-(j) := \min \{n_{ji}\}$. Soit $\tau(E) := \max \{k/\ h^1(\mathcal{J}_E(k)) \neq 0\ \}$.

(i) $n^-(2) > n^-(1)$, $n^+(2) > n^+(1)$

(ii) $n^+(2) = \tau+3$ et $\mathrm{card}\{\ n_{2i} = n^+(2)\ \} = h^1(\mathcal{J}_E(\tau))$.

II.2: Caractère: Soit $\mathsf{F} = (m_o,\dots,m_{s-1})$ la suite maximale pour l'ordre lexicographique parmi les suites $(n_o,\dots,n_{s-1})$ vérifiant:

(1) $n_o \geq n_1 \geq \ldots \geq n_{s-1} \geq s$

(2) $n_i \leq n_{i+1}+1$, $0 \leq i \leq s-2$

(3) $\Sigma(n_i - i) = d$

Si $d > s(s-1)$ posons $d = ks-r$, $0 \leq r \leq s-1$.

Si $r = 0$, $\mathsf{F} = (s+k-1, s+k-2, \ldots, k)$; si $r \neq 0$, $\mathsf{F} = (s+k-2, s+k-3, \ldots, s+k-r-1, s+k-r-1, s+k-r-2, \ldots, k)$ (cf [GP1] Thm.2.7).

Finalement si $\chi = (n_o, \ldots, n_{s-1})$ est un caractère de degré d, longueur s (i.e satisfait (1), (3) ci-dessus) le genre de χ est $g(\chi) = \Sigma_{n\geq 1} h_\chi(n)$ où $h_\chi(n) := \Sigma[(n_i-n-1)_+ - (i-n-1)_+]$. Insistons sur le fait que, dans ce travail, par définition, tout caractère est connexe. Ceci est justifié par [GP1] 3.2.

II.3:LEMME: Soit $E \subsetneq \mathbf{P}$ un groupe de d points vérifiant les conditions suivantes:

(i) $d > s(s-1)$, $s \geq 2$

(ii) le caractère de E est le caractère maximum, F, de degré d, longueur s.

(iii) E est contenu dans une courbe intègre de degré s

Posons $d = ks-r$, $0 \leq r \leq s-1$. Alors E est lié à un groupe de r points alignés par une intersection complète (k,s). En particulier $\mathcal{I}_E$ admet la résolution minimale suivante:

$r = 0$: $0 \to \mathcal{O}_{\mathbf{P}}(-s-k) \to \mathcal{O}_{\mathbf{P}}(-s) \oplus \mathcal{O}_{\mathbf{P}}(-k) \to \mathcal{I}_E \to 0$

$r \neq 0$: $0 \to \mathcal{O}_{\mathbf{P}}(-k-s+r) \oplus \mathcal{O}_{\mathbf{P}}(-k-s+1) \to \mathcal{O}_{\mathbf{P}}(-k-s+r+1) \oplus \mathcal{O}_{\mathbf{P}}(-s) \oplus \mathcal{O}_{\mathbf{P}}(-k) \to \mathcal{I}_E \to 0$.

Dém: Comme E est contenu dans une courbe intègre de degré s on a: $h^o(\mathcal{I}_E(m)) = h^o(\mathcal{O}_{\mathbf{P}}(m-s))$ si $m < k$. D'autre part: $h^o(\mathcal{I}_E(k)) = (k+2)(k+1)/2 - d + h^1(\mathcal{I}_E(k))$. Or $h^1(\mathcal{I}_E(k)) = h_{\mathsf{F}}(k)$ et on obtient: $h^o(\mathcal{I}_E(k)) - h^o(\mathcal{O}_{\mathbf{P}}(k-s)) \geq 1$. Il s'ensuit que E est contenu dans une intersection complète (k,s). En particulier si r = 0, E est intersection complète. Supposons r > 0 et désignons par D le lié à E par l'intersection complète (k,s). Montrons que D est aligné. La résolution minimale de $\mathcal{J}_E$ a la forme suivante:

$0 \to \oplus \mathcal{O}_{\mathbf{P}}(-n_{2i}) \oplus x.\mathcal{O}_{\mathbf{P}}(-k-s+1) \to \oplus \mathcal{O}_{\mathbf{P}}(-n_{1i}) \oplus y.\mathcal{O}_{\mathbf{P}}(-k) \oplus z.\mathcal{O}_{\mathbf{P}}(-s) \to$

$\mathcal{J}_E \to 0$

où $n_{2i} < k+s-1$, $x = h^1(\mathcal{J}_E(k+s-4)) > 0$, $k+1 \leq n_{1i} \leq k+s-2$ (cf II.1.1).

Par mapping cone on obtient une résolution de $\mathcal{J}_D$:

$0 \to \oplus \mathcal{O}_{\mathbf{P}}(n_{1i}-k-s) \oplus y.\mathcal{O}_{\mathbf{P}}(-s) \oplus z.\mathcal{O}_{\mathbf{P}}(-k) \to \oplus \mathcal{O}_{\mathbf{P}}(n_{2i}-k-s) \oplus \mathcal{O}_{\mathbf{P}}(-s) \oplus$

$\mathcal{O}_{\mathbf{P}}(-k) \oplus x.\mathcal{O}_{\mathbf{P}}(-1) \to \mathcal{J}_D \to 0$. Comme $s \geq 2$ et $n_{1i} \leq k+s-2$ on a $h^0(\mathcal{J}_D(1)) \geq x > 0$. Ainsi D est intersection complète (1,r). A partir de la résolution minimale d'une intersection complète (1,r) on obtient, par mapping cone, la résolution de $\mathcal{J}_E$ annoncée

II.3.1:Remarque: (1) Il n'est pas vrai en général que la postulation détermine la résolution minimale.
(2) Pour les groupes de points ayant la postulation d'une intersection complète, voir par exemple [DGM], [D].

III) SECTIONS PLANES GENERALES DES COURBES GAUCHES.

Dans ce paragraphe on énonce un résultat (Thm.III.1) essentiellement démontré par Strano [S1,2], sur la section plane générale d'une courbe gauche. L'on donne ensuite une première application concernant les courbes (non forcément intègres) dont la section plane générale est alignée (cf aussi III.2.2).

III.1:Lemme: Soit $C \subsetneq \mathbb{P}^3$ une courbe (i.e un schéma de dimension un équidimensionnel et localement Cohen-Macaulay). Si H est un plan général et si $\mathrm{Tor}^1(J, K)_{t+2} = 0$ (où J est l'idéal $\oplus H^0(\mathcal{J}_{C\cap H}(k))$), la multiplication par l'équation de H induit une application injective: $m_H(t-1): H^1(\mathcal{J}_C(t-1)) \to H^1(\mathcal{J}_C(t))$.

Dém: Dans [S1,2] ce résultat est énoncé et démontré sous l'hypothèse C intègre. On vérifie que cette hypothèse est superflue

III.1.1:Remarque: Avec les notations du paragraphe 2 la condition $\mathrm{Tor}^1(J,\ K)_{t+2} = 0$ est équivalente à: $n^-(2) \geq t+3$.

III.1.2:Remarque: Comme me l'a fait observer Ch. Peskine le lemme 1 peut aussi s'obtenir à partir de la démonstration de [GP3,4] du lemme de Laudal. Si $m_H(t-1)$ a un noyau non nul on a une suite exacte:

$0 \rightarrow E \rightarrow \Omega_H(1) \rightarrow \mathcal{I}_{X,Z}(t) \rightarrow 0$ où $X = C \cap H$ et où $Z \subsetneq H$ est une courbe de degré t.

Si $h^0(E(1)) = 0$ on a un diagramme:

$$3.\mathcal{O}_H(-1) \longrightarrow \mathcal{I}_{X,Z}(t), \quad \mathcal{O}_H(-1) \searrow \Omega_H(1) \nearrow \mathcal{I}_{X,Z}(t)$$

qui montre l'existence d'une relation de degré t+2.

Si $h^0(E(1)) = 1$ on a:

$$2.\mathcal{O}_H(-1) \rightarrow \Omega_H(1) \longrightarrow \mathcal{I}_p, \quad \Omega_H(1) \searrow \mathcal{I}_{X,Z}(t) \swarrow \mathcal{I}_p$$

où p est un point de H.

En ecrivant la résolution de $\mathcal{I}_p$, on récolte une relation de degré t+2.

Si $h^0(E(1)) = 2$:

$$\mathcal{O}_H(-1) \rightarrow \Omega_H(1) \longrightarrow \mathcal{O}_L(-1), \quad \Omega_H(1) \searrow \mathcal{I}_{X,Z}(t) \swarrow \mathcal{O}_L(-1)$$

En multipliant l'image de $H^0(\mathcal{O}_L) \rightarrow H^0(\mathcal{I}_{X,Z}(t+1))$ par l'équation de la droite L, on obtient une relation de degré t+2.

Finalement $h^0(E(1)) < 3$ car $\oplus(\Omega_H(k))$ est engendré par ses éléments de degré deux.

III.2:Corollaire: Soit $Y \subsetneq \mathbb{P}^3$ une courbe localement Cohen-Macaulay dont la section plane générale est alignée. L'un des cas suivants a lieu:

(a) Y est plane

(b) Y est la réunion de deux droites disjointes

(c) Y est une structure double sur une droite.

Dém: Il suffit de montrer que si Y n'est pas plane alors $d \leq 2$ ($d := \deg(Y)$). Considérons pour H général la suite exacte de restriction: $0 \to \mathcal{I}_Y \to \mathcal{I}_Y(1) \to \mathcal{I}_{Y\cap H}(1) \to 0$. Comme Y n'est pas plane, m_H: $H^1(\mathcal{I}_Y) \to H^1(\mathcal{I}_Y(1))$ n'est pas injective. D'après III.1, ceci implique $n^-(2) \leq 3$. Or, par hypothèse, $n^-(2) = d+1$

III.2.1:Remarque: Le corollaire III.2 peut se démontrer directement de la façon suivante: on montre d'abord que $Y_{réd}$ est soit une courbe plane soit la réunion de deux droites disjointes. Dans le second cas, en raisonnant comme dans [Mi] Prop.2.4, on montre $Y = Y_{réd}$. Ensuite en utilisant la description locale d'une structure triple sur une droite ([BF], [BM] §6) l'on montre que si une telle structure triple a sa section plane alignée alors elle est plane. Nous nous sômmes ramenés à considérer une structure de multiplicité m, Y, sur une courbe plane, Y', que l'on peut toujours supposer réduite. Si Y' est une droite l'on suppose $m \geq 3$. Cette dernière condition détermine un plan (le plan contenant la structure triple $Y\cap Y^{(2)}$, cf ci-dessus). Dans tous les cas nous avons donc un plan, H, uniquement déterminé. Soit p un point de H-Y' et L une droite rencontrant H transversalement en p. Soit H' un plan par L et désignons par D la droite $H\cap H'$. Si $Y'\cap D = \{p_1,\dots,p_d\}$, par hypothèse: $Y\cap H' = \{p_1^*,\dots,p_d^*\}$ où p_i^* désigne l'unique m-uplet de D de support p_i. Ceci montre $Y^m \subseteq Y$ où Y^m est la courbe Y' "m-uplée" dans H. Pour des raisons de degré et comme Y est localement Cohen-Macaulay: $Y = Y^m$.

III.2.2:Remarque: Comme en III.2 l'on démontre qu'une courbe, non sur une quadrique, dont la section plane générale est sur une (et une seule) conique est de degré au plus cinq, etc...

IV) COURBES DE CARACTERE MAXIMUM ($d > s(s-1)$).

IV.1: Notation: Dans tout ce paragraphe, C désigne une courbe intégre de $\mathbb{P}^3$, de degré d avec $s(C) = s$, $d > s(s-1)$ et $\chi(C) = F$ (cf 2.2). On pose $d = ks-r$, $0 \leq r \leq s-1$.

Rappelons que si $d > s(s-1)$ alors $G(d,s) = g(F)$; c'est le genre d'une courbe liée à une courbe plane de degré r par une intersection complète (k,s).

IV.2: PROPOSITION: Soit C comme en IV.1. Si $0 \leq r \leq s-3$, $s \geq 3$, alors ou bien C est arithmétiquement Cohen-Macaulay, de genre $G(d,s)$ ou bien $r = 2$ et C est liée à une courbe non plane de degré deux par une intersection complète (k,s).

Dém: Pour $t \leq k-1$, $h^o(\mathcal{I}_C(t)) = h^o(\mathcal{O}_{\mathbb{P}}(t-s))$ pour des raisons de degré. Considérons la suite exacte:

$$0 \to H^o(\mathcal{I}_C(k-1)) \to H^o(\mathcal{I}_C(k)) \to H^o(\mathcal{I}_{C\cap H}(k)) \to H^1(\mathcal{I}_C(k-1)) \to H^1(\mathcal{I}_C(k)).$$

On a $h^o(\mathcal{I}_C(k)) \geq h^o(\mathcal{O}_{\mathbb{P}}(k-s))$ et (cf 2.3) $h^o(\mathcal{I}_{C\cap H}(k)) \geq h^o(\mathcal{O}_H(k-s))+1$. Si $h^o(\mathcal{I}_C(k)) = h^o(\mathcal{O}_{\mathbb{P}}(k-s))$ alors $m_H(k-1)$ n'est pas injective. D'après II.3, III.1 ceci est impossible si $s \geq 3$, $0 \leq r \leq s-3$. Donc C est contenue dans une intersection complète (k,s). En particulier si $r = 0$, C est intersection complète. Supposons $r \neq 0$ et soit Y la liée (de degré r) à C par cette intersection complète (k,s). Comme C est localement C.M., il en est de même de Y. La section plane générale de Y est liée à celle de C par une intersection complète (k,s). En utilisant II.3 on obtient par mapping cone la résolution suivante:

$$0 \to \mathcal{O}_H(-k) \oplus \mathcal{O}_H(-s) \oplus \mathcal{O}_H(-r-1) \to \mathcal{O}_H(-r) \oplus \mathcal{O}_H(-k) \oplus \mathcal{O}_H(-s) \oplus \mathcal{O}_H(-1) \to$$

$\mathcal{I}_{Y\cap H} \to 0$ qui montre que $Y\cap H$ est aligné. On conclut avec III.2

IV.3: PROPOSITION: Soit C comme en IV.1. On suppose $s \geq 4$ et $r = s-1$. Si $g(C) > G(d,s)-(s^3+3s^2+4s)/6$ alors C est arithmétiquement Cohen-Macaulay et $g(C) = G(d,s)$.

Dém: Par hypothèse C est contenue dans une surface intègre, S, de degré s.

(a) C est contenue dans une intersection complète (s,k+s-2).

Il suffit de voir que les surfaces de degré k+s-2 contenant C ne sont pas toutes multiples de S. Il suffit pour cela de vérifier: $h^o(\mathcal{O}_{\mathbb{P}}(k+s-2)) - h^o(\mathcal{O}_C(k+s-2)) > h^o(\mathcal{O}_{\mathbb{P}}(k-2))$ (*). Comme $\chi(C) = F$, on a $e(C) \leq k+s-5$, par suite $h^o(\mathcal{O}_C(k+s-2)) = d(k+s-2)-g(C)+1$. Ceci dit, un petit calcul montre que (*) est équivalent à: $g(C) > G(d,s)-(s^3+3s^2+4s)/6$.

(b) Soit Y la courbe liée à C par l'intersection complète (s,k+s-2). On a $\deg(Y) = s^2-s-1$. La section plane générale de Y est liée à celle de C, par II.3 et par mapping cone on obtient la résolution:

$0 \to \mathcal{O}_H(-k-s+2) \oplus 2.\mathcal{O}_H(-2s+2) \to \mathcal{O}_H(-2s+1) \oplus \mathcal{O}_H(-s+1) \oplus \mathcal{O}_H(-s) \oplus$

$\oplus \mathcal{O}_H(-k-s+2) \to \mathcal{J}_{Y \cap H} \to 0$. On a $n^-(2) = 2s-2$. Si $h^o(\mathcal{J}_Y(s-1)) = 0$ alors $m_H(s-2): H^1(\mathcal{J}_Y(s-2)) \to H^1(\mathcal{J}_Y(s-1)$ n'est pas injective. D'après III.1 il vient $n^-(2) < s+2$. Donc si $s \geq 4$ on peut supposer $h^o(\mathcal{J}_Y(s-1)) \neq 0$. Comme S est intègre, Y est contenue dans une intersection complète (s,s-1). Cette intersection complète lie Y à une droite. Donc Y est arithmétiquement C.M. Par suite C l'est aussi et $g(C) = g(\chi(C))$

IV.4:PROPOSITION: Soit C comme en IV.1. On suppose $s \geq 5$ et $r = s-2$. Si $g(C) > G(d,s)-(s^3+3s^2-10s+6)/6$ alors C est biliée à une courbe, Y, de degré deux par des intersections complètes (s,k+s-2), (s,s-1). En particulier: $g(C) = p_a(Y)+G(d,s)$.

Dém: Elle est semblable à la précédente. On montre d'abord que, sous l'hypothèse $g(C) > G(d,s)-(s^3+3s^2-10s+6)/6$, C est contenue dans une intersection complète (s,k+s-2). Soit Z la courbe liée. On a $\deg(Z) = s^2-s-2$. En utilisant II.3 et III.1 l'on voit que pour $s \geq 5$, Z est contenue dans une intersection complète (s,s-1) et la courbe liée, Y, a degré deux. On obtient la relation entre g(C) et $p_a(Y)$ par la formule de liaison

IV.5:Remarque: Les propositions IV.2, IV.3, IV.4 sont des extensions naturelles du résultat de Strano [S1,2] selon lequel une courbe intègre dont la section plane générale est intersection complète (a,b), $a \geq b \geq 3$, est intersection complète (Prop.IV.2 avec $r = 0$).

V) DROITES DOUBLES.

V.0:Notation: On pose $\mathbb{P}^3 := \mathrm{Proj}(K[x,y,z,t])$.

V.1:LEMME: Soit $Y \subseteq \mathbb{P}^3$ une droite double de genre arithmétique $-\alpha$, $(\alpha \geq 1)$ de support la droite L d'équations $x = 0$, $y = 0$. Posons $\mathbf{I}_Y := \oplus H^0(\mathcal{I}_Y(k))$.

(1) $\mathbf{I}_Y = (x^2, y^2, xy, xF(z,t)-yG(z,t))$ où F, G sont des polynômes homogènes de degré α, sans zéros communs.

(2) La résolution libre minimale de $\mathbf{I}_Y$ induit une suite exacte:

$$0 \to \mathcal{O}_{\mathbb{P}}(-3-\alpha) \xrightarrow{g} 2\mathcal{O}_{\mathbb{P}}(-2-\alpha) \oplus 2\mathcal{O}_{\mathbb{P}}(-3) \xrightarrow{n} \mathcal{O}_{\mathbb{P}}(-1-\alpha) \oplus 3\mathcal{O}_{\mathbb{P}}(-2) \xrightarrow{f} \mathcal{I}_Y \to 0$$

où: $g = \begin{pmatrix} y \\ -x \\ F \\ G \end{pmatrix}$, $n = \begin{pmatrix} x & y & 0 & 0 \\ -F & 0 & y & 0 \\ G & -F & -x & -y \\ 0 & -G & 0 & -x \end{pmatrix}$, $f = (xF-yG, x^2, xy, -y^2)$

Dém: (1) est traité dans [Mi2] modulo quelques détails. La donnée de Y correspond à la donnée d'un quotient du fibré conormal $N^* = 2.\mathcal{O}_L(-1)$, $N^* \xrightarrow{p} \mathcal{O}_L(\alpha-1) \to 0$. On a $p = (F, G)$; $F(z,t)$, $G(z,t)$ homogènes de degré α, sans zéros communs. On a une suite exacte: $0 \to \mathcal{I}_{L'} \to \mathcal{I}_Y \to \mathcal{O}_L(-\alpha-1) \to 0$, qui montre $H^0(\mathcal{I}_Y(k)) = H^0(\mathcal{I}_{L'}(k))$ pour $k \leq \alpha$ (L' désigne le premier voisinage infinitésimal de L dans $\mathbb{P}^3$). La même suite exacte montre que l'image de $H^0(\mathcal{I}_Y(\alpha)) \otimes H^0(\mathcal{O}_{\mathbb{P}}(1)) \to H^0(\mathcal{I}_Y(\alpha+1))$ est de codimension un. Il apparait donc un nouveau générateur en degré $\alpha+1$. La suite exacte $0 \to \mathcal{I}_Y(\alpha+1) \to \mathcal{I}_L(\alpha+1) \to \mathcal{O}_L(2\alpha) \to 0$ montre que l'on peut prendre

$xF(z,t)-yG(z,t)$. Finalement comme I_Y est $(\alpha+1)$-régulier on a bien $I_Y = (x^2, xy, y^2, xF(z,t)-yG(z,t))$.

(2) Il suffit de vérifier que la suite est exacte

V.2:PROPOSITION: Les droites doubles de genre arithmétique $-\alpha$ ($\alpha \geq 2$) sont paramètrées par un ouvert irréductible, génériquement lisse, de dimension $2\alpha+5$ de $Hilb(\mathbb{P}^3)$.

Dém: Comme déjà dit plus haut la donnée d'une droite double de genre arithmétique $-\alpha$ et de support L correspond à la donnée d'un quotient du fibré conormal. La famille maximale est donc irréductible, de dimension $2\alpha+5$ (4 paramètres pour choisir L, $2\alpha+1$ paramètres pour le quotient de N^*). Pour terminer la démonstration il suffit d'exhiber une droite double, Y, de genre $-\alpha$, vérifiant $h^0(N_Y) = 2\alpha+5$. Pour ce faire nous utilisons V.1: en appliquant $Hom(-,\mathcal{O}_Y)$ à la résolution de V.1(2) on obtient:

$$0 \to H^0(N_Y) \to H^0(\mathcal{O}_Y(\alpha+1)\oplus 3\mathcal{O}_Y(2)) \xrightarrow{{}^t n} H^0(2\mathcal{O}_Y(\alpha+2)\oplus 2\mathcal{O}_Y(3)).$$

Il suffit donc de calculer $\dim(\ker({}^t n))$. Pour simplifier les calculs prenons, dans les notations de V.1, $F(z,t) = z^\alpha$, $G(z,t) = t^\alpha$. Ainsi un élément de $\ker({}^t n)$ est un quartuplet $(\xi, \delta, \beta, \gamma)$, $\xi \in H^0(\mathcal{O}_Y(\alpha+1))$; $\delta,\beta,\gamma \in H^0(\mathcal{O}_Y(2))$ vérifiant:

$x.\xi - z^\alpha.\delta + t^\alpha.\beta = 0$ (dans $H^0(\mathcal{O}_Y(\alpha+2))$) **(1)**

$y.\xi - z^\alpha.\beta - t^\alpha.\gamma = 0$ (") **(2)**

$y.\delta - x.\beta = 0$ (dans $H^0(\mathcal{O}_Y(3))$) **(3)**

$-y.\beta - x.\gamma = 0$ (dans $H^0(\mathcal{O}_Y(3))$) **(4)**

Rappelons que $H^0(\mathcal{O}_Y(k)) = \{ A(z,t) + B(z,t).x/G(z,t)$; A et B homogènes de degrés k et $k+\alpha-1$ $\}$ (cf [Mi2]). Autrement dit une base de $H^0(\mathcal{O}_Y(2))$ est fournie par: $t^2, zt, z^2, xt, xz, yt, yz, \Theta_1, \ldots, \Theta_{\alpha-2}$ où $\Theta_i := x.z^{i+1}/t^i = y.t^{\alpha-i}/z^{\alpha-i-1}$; $\Theta_i = 0$, $i \geq 1$ si $\alpha = 2$. De même une base de $H^o(\mathcal{O}_Y(\alpha+1)$ est donnée par: $z^{\alpha+1}, z^{\alpha}.t, \ldots, z.t^{\alpha}, t^{\alpha+1}$ et $Y_i = x.t^{\alpha-i}.z^i$, $0 \le i \le 2\alpha$. Si $0 \le i \le \alpha$: $Y_i = x.t^{\alpha-i}.z^i$; si $\alpha+1 \le i \le 2\alpha$: $Y_i = y.t^{2\alpha-i}.z^{i-\alpha}$. Posons: $\delta = \delta_1.t^2 + \delta_2.z.t + \delta_3.z^2 + \delta_4.x.t + \delta_5.x.z + \delta_6.y.t + \delta_7.y.z + \Sigma_{1\le j\le\alpha-2}(\delta_{j+7}.\Theta_j)$ et de même pour β, γ. Les relations (3), (4) montrent: $\delta_i = \beta_i = \gamma_i = 0$, $1 \le i \le 3$ (nb: $x.\Theta_j = y.\Theta_j = 0$, $1 \le j \le \alpha-2$). Vu que: $z^{\alpha}.\Theta_j = y.z^{j+1}.t^{\alpha-j}$, $t^{\alpha}\,\Theta_j = x.z^{j+1}.t^{\alpha-j}$ et vu l'identité $x.z^{\alpha} = y.t^{\alpha}$, la relation (1) peut s'ecrire:

$$x.[\xi - \delta_4.z^{\alpha}.t - \delta_5.z^{\alpha+1} + \beta_4.t^{\alpha+1} + \beta_5.z.t^{\alpha} + \beta_6.t.z^{\alpha} + \beta_7.z^{\alpha+1}$$
$$+ \Sigma_{1\le j\le\alpha-2}(\beta_{j+7}.z^{j+1}.t^{\alpha-j})] - \delta_6.z^{\alpha}.y.t - \delta_7.y.z^{\alpha+1} +$$
$$+\Sigma_{1\le j\le\alpha-2}(\delta_{j+7}.y.z^{j+1}.t^{\alpha-j}) = 0.$$

Posons $\xi = \Sigma_{0\le m\le\alpha+1}(\xi_m.z^{\alpha+1-m}.t^m) + \Sigma_{0\le n\le 2\alpha}(\xi_n.Y_n)$. On a $x.Y_n = 0$ (car x^2 et xy s'annulent sur Y). Ceci dit la relation (1) s'ecrit:

$$x.z^{\alpha+1}(\xi_0 - \delta_5 + \beta_7) + x.z^{\alpha}.t(\xi_1 - \delta_4 + \beta_6) + \Sigma_{2\le m\le\alpha-1}(\xi_m +$$
$$\beta_{\alpha+7-m}).x.z^{\alpha+1-m}.t^m + x.z.t^{\alpha}(\xi_{\alpha} + \beta_5) + x.t^{\alpha+1}(\xi_{\alpha+1} + \beta_4) +$$
$$\Sigma_{1\le j\le\alpha-2}(\delta_{j+7}.y.z^{j+1}.t^{\alpha-j}) - \delta_6.z^{\alpha}.y.t - \delta_7.y.z^{\alpha+1} = 0.$$

On en déduit: $\delta_k = 0$, $k \ge 6$ et:

$$\xi_0 = \delta_5 - \beta_7 \tag{1.1}$$

$$\xi_1 = \delta_4 - \beta_6 \tag{1.2}$$

$$\xi_m = -\beta_{\alpha+7-m},\ 2 \le m \le -1 \tag{1.3}$$

$$\xi_{\alpha} = -\beta_5 \tag{1.4}$$

$$\xi_{\alpha+1} = -\beta_4 \tag{1.5}$$

De la même façon (2) fournit: $\gamma_4 = \gamma_5 = 0$; $\gamma_k = 0$, $k \ge 8$ et:

$$\xi_0 = \beta_7 \tag{2.1}$$

$$\xi_1 = \beta_6 \qquad (2.2)$$

$$\xi_m = \beta_{\alpha-m+7},\ 2 \leq m \leq \alpha-1 \qquad (2.3)$$

$$\xi_\alpha = \beta_5 + \gamma_7 \qquad (2.4)$$

$$\xi_{\alpha+1} = \beta_4 + \gamma_6 \qquad (2.5)$$

Les relations (1.i), (2.i) permettent de déduire:

$$\xi = \beta_7.z^{\alpha+1} + \beta_6.z^{\alpha}.t - \beta_5.z.t^{\alpha} - \beta_4.t^{\alpha+1} + \Sigma_{0\leq n\leq 2\alpha} (\xi_n.Y_n)$$

$$\delta = (0,\ 0,\ 0,\ 2\beta_6,\ 2\beta_7,\ 0,\ldots,0)$$

$$\beta = (0,\ 0,\ 0,\ \beta_4,\ \beta_5,\ \beta_6,\ \beta_7,\ 0,\ldots,0)$$

$$\gamma = (0,\ 0,\ 0,\ 0,\ 0,-2\beta_4,\ -2\beta_5,\ 0,\ldots,0)$$

On a donc bien $\dim(\ker({}^t n)) = 2\alpha+5$

V.3:LEMME: Avec les notations de 1, la surface générale de degré $\alpha+1$ contenant Y est lisse.

Dém: D'après la description de $\mathbf{I}_Y$ on voit que $\mathcal{I}_Y(k)$ est engendré par ses sections globales si $k \geq \alpha+1$. D'après Bertini la surface générale de degré $\alpha+1$ contenant Y est lisse hors de Y. Si S est la surface d'équation $x.F(z,t)-y.G(z,t) = 0$ (cf V.1), on vérifie facilement que S est lisse dans un voisinage de Y

V.4:LEMME: On garde les notations de 1. Soient a, b des entiers et F_a, F_b des éléments suffisamment généraux de $H^0(\mathcal{I}_Y(a))$, $H^0(\mathcal{I}_Y(b))$. On suppose $a \geq b \geq \alpha+1$ ou $a = \alpha+1$, $b \geq 2$. Les surfaces F_a, F_b s'intersectent proprement et la courbe liée à Y dans l'intersection complète $F_a \cap F_b$ est lisse. Si de plus $a + b - 4 \geq \alpha$, la liée est connexe.

Dém: D'après V.4 on peut supposer F_a lisse. Grâce à Bertini il suffit de montrer que $\mathcal{I}_{Y,Fa}(b)$ est engendré par ses sections globales. Si $b \geq \alpha+1$ ceci est clair car $\mathcal{I}_{Y,F_a}(b)$ est un quotient de $\mathcal{I}_Y(b)$. Si $a = \alpha+1$ on peut supposer que F_a est un générateur minimal de I_Y. Dès lors $\mathcal{I}_{Y,F_a}$ est le faisceau associé à l'idéal saturé $(x^2, y^2, xy)/(F_a)$ et il est clair que $\mathcal{I}_{Y,F_a}(b)$ est engendré par ses sections si $b \geq 2$. Pour conclure on observe que $h^1(\mathcal{I}_X) = h^1(\mathcal{I}_Y(a+b-4))$ (X désigne la liée à Y) et que $h^1(\mathcal{I}_Y(n)) = 0$ si $n \leq -\alpha$ ou si $n \geq \alpha$

VI) LACUNES D'HALPHEN.

VI.1:Définition: Soient d, s des entiers $d > s(s-1)$. On pose $d = ks - r$, $0 \leq r \leq s-1$ et on définit:

$$G^{(1)}(d,s) = \begin{cases} G(d,s) - (s-2) & \text{si } r = 0,\ s \geq 3 \\ G(d,s) - (s-r-2) & \text{si } r = 1,\ s \geq 4 \text{ ou } r = 2,\ s \geq 5 \\ G(d,s) - (r-2) & \text{si } 3 \leq r < s/2,\ s \geq 6 \\ G(d,s) - (s-r-2) & \text{si } s/2 \leq r \leq s-3,\ s \geq 6 \\ G(d,s) - (r-2) & \text{si } r \geq s-2,\ s \geq 5. \end{cases}$$

VI.2:LEMME: Soit χ un caractère connexe de degré d, longueur s, avec $d > s(s-1)$. Si $g(\chi) > G^{(1)}(d,s)$ alors $\chi = F$ (F le caractère maximum, cf II.2).

Dém: Cela suit de [BE3] 3.11 si $r = 0$ et de [D] 1.8, 1.9 si $r > 0$

VI.3:Corollaire: Soit $C \subsetneq \mathbb{P}^3$ une courbe intègre de degré d, genre g. Posons $s(C) = s$. Si $d > s(s-1)$ et $g > G^{(1)}(d,s)$ alors $\chi(C) = F$.

Dém: D'après le lemme de Laudal [L], [GP3,4], $\sigma(C) = s$. D'autre part $\chi(C)$ est connexe [GP1] et $g(\chi(C)) > g(C)$, on conclut avec VI.2

VI.4:PROPOSITION: Soient d, g, s des entiers vérifiant $d > s(s-1)$, $G(d,s) > g > \max\{G^{(1)}(d,s), G(d,s+1)\}$, $s \geq 6$. Posons $d = ks-r$, $0 \leq r \leq s-1$. Si $r \neq 2$ et $r \neq s-2$, (d,g,s) est une lacune d'Halphen.

Dém: Soit C une courbe (intègre) de degré d, genre (arithmétique) g, avec s(C) = s. D'après VI.3, $\chi(C) = F$. D'après IV.2, IV.3, C est arithmétiquement C. M. avec g(C) = G(d,s)

VI.5:PROPOSITION: Soient k, s, g des entiers, $k \geq s \geq 6$ et $G(d,s) > g > \max\{G(d,s)-(s-4), G(d,s+1)\}$. On pose $d = ks-2$.
Il existe une et une seule composante irréductible, *H*, de H(d,g) contenant une courbe non sur une surface de degré s-1. La courbe générale de *H* est tracée sur une surface lisse de degré s. En particulier s(d,g) = s.
La composante *H* est génériquement lisse.

Dém: Soit C une courbe (intègre) de degré d, genre (arithmétique) g, avec s(C) = s. D'après VI.3, $\chi(C) = F$. D'après IV.2 C est liée à une courbe de degré deux par une intersection complète (k,s). D'autre part si l'on pose $g = G(d,s) - a$, $1 \leq a \leq s-5$, d'après V.4 une intersection complète (k,s) générale contenant une droite double de genre -a lie cette droite double à une courbe lisse, connexe, de degré d, genre g. Ceci établit l'existence de *H* . L'unicité suit de ce qui précéde et de l'irréductibilité du schéma de Hilbert des droites doubles de genre -a. Finalement la lissité se déduit de V.2 et de [K] 3.9

VI.5.1:Remarque: La dimension de *H* peut facilement se calculer.

VI.6:PROPOSITION: Soient k, s, g des entiers, $k \geq s \geq 6$ et $G(d,s) > g > \max\{G(d,s)-(s-4), G(d,s+1)\}$. On pose $d = ks-(s-2)$ et $\alpha := G(d,s) - g$.

(a) On suppose $k \leq s + \alpha - 3$. Il existe des courbes localement Cohen-Macaulay, de degré d, genre arithmétique g, tracées sur des surfaces lisses de degré s. Ces courbes ne sont pas lissifiables et (s,d,g) est une lacune d'Halphen.

(b) Si $k \geq s + \alpha - 2$ il existe une et une seule composante irréductible, H, de H(d,g) contenant une courbe non sur une surface de degré s-1. Par suite s(d,g) = s. La courbe générale de H est tracée sur une surface lisse de degré s.

Si $k \geq s + \alpha + 1$, H est génériquement lisse.

Dém: Soit C une courbe (intègre) de degré d, genre (arithmétique) g, avec s(C) = s. D'après VI.3, $\chi(C) = F$. D'après IV.4 C est biliée par des intersections complètes (s,k+s-2), (s,s-1) à une courbe localement Cohen-Macaulay, Y, de degré deux avec $p_a(Y) = -\alpha$, $1 \leq \alpha \leq s-5$. La situation peut se décrire de la façon suivante:

(i) On lie une droite double de genre $-\alpha$, Y, à une courbe, Z, par une intersection complète (s,s-1). En général on peut supposer les surfaces liantes lisses (V.3) et la courbe Z lisse, connexe (V.4). D'après V.2 et [K] 3.9 on obtient une (et une seule) composante irréductible, génériquement lisse, H^*, de H(d,g) paramètrant les courbes Z ainsi obtenues. Les courbes que nous cherchons s'obtiennent à partir des courbes de H^* par liaisons (s,k+s-2).

(ii) Si Z est suffisamment générale dans H^*, Z est contenue dans une intersection complète (k+s-2,s).

En effet désignons par U l'intersection complète (s,s-1) contenant Z (cf i). La suite exacte de liaison s'ecrit:

$0 \to \mathcal{I}_U \to \mathcal{I}_Z \to \omega_Y(5-2s) \to 0$. Or $\omega_Y = \mathcal{O}_Y(-\alpha-1)$. On en déduit la suite exacte:

(+) $0 \to H^0(\mathcal{I}_U(k+s-2)) \to H^0(\mathcal{I}_Z(k+s-2)) \xrightarrow{g} H^0(\mathcal{O}_Y(2+k-s-\alpha)) \to 0$.

Comme $2+k-s-\alpha \geq -\alpha+1$, $h^0(\mathcal{O}_Y(2+k-s-\alpha)) \neq 0$ et Z est bien contenue dans une intersection complète (s,k+s-2).

(iii) On obtient donc une famille <u>irréductible</u> (cf [K] 3.4), K, "liée" à H^* par des intersections complètes (s,k+s-2). La courbe générale de K est localement C.M. De plus d'après ce qui précéde, si C est une courbe <u>intègre</u>, de degré d, genre arithmétique g, avec s(C) = s alors $[C] \in K$ (ceci n'est pas vrai à priori si C est seulement réduite: $\chi(C)$ pourrait ne pas être connexe).

(a) Si $k \leq s+\alpha-3$ K ne contient pas de courbes intègres.

La courbe générale, C, de K est liée à la courbe générale, Z, de H^* par une intersection complète (s,k+s-2). En reprenant les notations de (ii), si $U = F_s \cap F_{s-1}$, la suite exacte de liaison (+) montre que tout élément de $H^o(\mathcal{I}_Z(s))$ est de la forme $H.F_{s-1}+c.F_s$, donc contient la droite double Y. Soit $G \in H^o(\mathcal{I}_Z(k+s-2))$. Considérons la suite (+).La section g(G) s'annule sur $L = Y_{réd}$ (car toute section de $\mathcal{O}_Y(n)$, $n < 0$, s'annule sur L). Il s'ensuit que si C est la résiduelle à Z dans une intersection complète (s,k+s-2) alors C contient L et donc n'est pas intègre. Ceci prouve, vu (iii), que C n'est pas limite plate de courbes intègres donc, à fortiori, n'est pas lissifiable.

(b) Avec les notations précédentes, si $k \geq s+\alpha-2$, $\mathcal{O}_Y(k-s-\alpha+2)$ est engendré par ses sections. Il en est donc de même de $\mathcal{I}_Z(k+s-2)$ (cf +). On a: $Z \cup C = F_s \cap G$ où $\deg(G) = k+s-2$ et où F_s est lisse (cf i), de degré s. Comme $\mathcal{I}_Z(k+s-2)$ est engendré par ses sections, $\omega_C(4-s)$ l'est aussi (suite exacte de liaison). Par conséquent le systéme linéaire $|C|$ sur F_s est sans points base. Par Bertini l'élément général de ce systéme est lisse. Finalement on vérifie facilement $h^1(\mathcal{I}_C) = 0$. Donc la courbe générale de K est lisse, connexe, tracée sur une surface lisse de degré s et donc d'indice de postulation s. D'aprés (iii), K est contenue dans une unique composante irréductible, H, de H(d,g) et est dense dans H. Pour terminer observons que H est génériquement lisse si $k \geq s+\alpha+1$ (cf V.2, [K] 3.9 et (i))

VI.6.1:Remarque: On peut facilement calculer la dimension de H dans (b). Cependant si $s+\alpha-2 \leq k \leq s+\alpha$ on ne peut pas utiliser les résultats de [K] pour calculer $h^1(N_C)$ par liaison. Dans ce cas la lissité de H reste donc un problème ouvert.

BIBLIOGRAPHIE.

[BE1] E.Ballico-Ph.Ellia: "The maximal rank conjecture for non special curves in $\mathbb{P}^3$", Invent. Math.79, 541-555 (1985).

[BE2] E.Ballico-Ph.Ellia: "Beyond the maximal rank conjecture for curves in $\mathbb{P}^3$", in "Space curves, Proceedings Rocca di Papa, 1985" Lect. Notes 1266, 1-23 (1987), Springer-Verlag.

[BE3] E.Ballico-Ph.Ellia: "A program for space curves", in "Algebraic varieties of small dimension" Rendiconti del Seminario Mat.-Univ. e Politec. di Torino, vol. spe. 1986, 25-42 (1986)

[BEx] E.Ballico-Ph.Ellia: ? en préparation.

[BF] C.Banica-O.Forster: "Multiplicity structures on space curves" in "Lefschetz centennial conference" A.M.S. Contemporary Math., vol.58 (1986)

[BM] C.Banica-N.Manolache:"Moduli space $M_{\mathbb{P}^3}(-1,4)$: minimal spectrum" Preprint (1983)

[Do] A.Dolcetti: "Halphen's gaps for curves of submaximum genus", à paraitre Bull. Soc. Math. France.

[D] E.Davis:"0-dimensional subschemes of $\mathbb{P}^2$: inverting the Bezout-Jacobi theorems" Preprint (1985).

[DGM] E. Davis-A.G. Geramita-P. Maroscia:"Perfect homogeneous ideals:Dubreil's theorems revisited" Bull. Sc. Math. Fr. 2eme série 108, 143-185 (1984).

[E] Ph.Ellia:"Sur le genre maximal des courbes gauches de degré d non sur une surface de degré s-1" Preprint (1988).

[GP1] L.Gruson-Ch.Peskine: "Genre des courbes de l'espace projectif" in "Algebraic geometry, TromsØ, 1977", Lect. Notes 687, 31-60 (1978) Springer-Verlag

[GP2] L.Gruson-Ch.Peskine: "Théorème de spécialité" in "Les équations de Yang-Mills", Astèrisque 71-72, 219-229 (1980).

[GP3] L.Gruson-Ch.Peskine: "Section plane d'une courbe gauche: postulation" in "Enumerative geometry, Proceedings Nice 1981", Progress in Math. 24, 33-35 (1982) Birkhauser

[GP4] L.Gruson-Ch.Peskine: "Postulation des courbes gauches" in "Open problems, Proceedings Ravello 1982", Lect. Notes 997, 218-227 (1983) Springer-Verlag.

[Ha] G.Halphen: "Mémoire sur la classification des courbes gauches algébriques" in Oeuvres complètes, t.III, 261-455.

[rH1] R.Hartshorne:"On the classification of algebraic space curves" in "Vector bundles..., Proceedings Nice 1979" Progress in Math. 7, 84-112 (1980) Birkhauser.

[rH2] R. Hartshorne:"Stable reflexive sheaves III" Math. Ann. 279, 517-534 (1988)

[HH] R.Hartshorne- A. Hirschowitz: "Nouvelles courbes de bon genre dans l'espace projectif" Math. Ann. 280, 353-367 (1988).

[K] J.O.Kleppe: "Liaison of families of subschemes in $\mathbb{P}^n$", ce volume.

[L] O.Laudal: "A generalized trisecant lemma" in "Algebraic geometry TromsØ 1977" Lect. Notes 687, 112-149 (1978) Springer-Verlag.

[Mi] J.Migliore: "Geometric invariants for liaison of space curves" J. of Algebra,vol. 99, No. 2, 548-572 (1986)

[Mi2] J.Migliore: "On linking double lines" Trans. of the American Math. Soc., vol 294, 177-185 (1986).

[S1] R.Strano: "A characterization of the curves complete intersection in $\mathbb{P}^3$" à paraître Proc. A.M.S.

[S2] R.Strano: "Sulle sezioni iperpiane delle curve" Preprint (1987).

SOME RESULTS ON THE CODIMENSION-TWO CHOW GROUP OF THE MODULI SPACE OF STABLE CURVES

Carel F. Faber *
Mathematisch Instituut, Universiteit van Amsterdam
Roetersstraat 15, 1018 WB Amsterdam
The Netherlands

§0. Introduction.

In this note we study the Chow group $A^2(\overline{\mathcal{M}}_g)$ of codimension-two cycles of the moduli space $\overline{\mathcal{M}}_g$ of stable curves of genus g. We work over $\mathbf{C}$ and we take the coefficients of the cycles to be $\mathbf{Q}$.

In §1 we determine generators for the subspace of $A^2(\overline{\mathcal{M}}_g)$ of cycles coming from the boundary $\overline{\mathcal{M}}_g - \mathcal{M}_g$. We use theorems of Harer ([Harer 1,2]) to achieve this.

In §2 we prove the following theorem. Put $h := [\frac{1}{2}g]$. Note that it is a consequence of Harer's results that the standard divisor classes $\lambda, \delta_0, \delta_1, \dots, \delta_h$ form a basis of the $\mathbf{Q}$-vectorspace $A^1(\overline{\mathcal{M}}_g)$.

Theorem (2.1). *For $g = 3$ and $g \geq 5$ the products of degree two of the divisor classes in a basis of $A^1(\overline{\mathcal{M}}_g)$ are linearly independent in $A^2(\overline{\mathcal{M}}_g)$, i.e., the $\frac{1}{2}(h+2)(h+3)$ products*

$$\lambda^2, \lambda\delta_0, \dots, \lambda\delta_h, \delta_0^2, \delta_0\delta_1, \dots, \delta_{h-1}\delta_h, \delta_h^2$$

of divisor classes are linearly independent in $A^2(\overline{\mathcal{M}}_g)$.

Corollary (2.2). *For $g = 3$ and $g \geq 5$ any two irreducible closed subvarieties of $\overline{\mathcal{M}}_g$ of codimension one have non-empty intersection.*

See §2 for the special cases $g = 2$ and $g = 4$.

This article is a slightly modified version of Chapter IV in our thesis [Faber]. Here is a brief summary of the other results in it. Throughout we use the notions and results of [M-Enum]. In this paper Mumford constructs the Chow ring of $\overline{\mathcal{M}}_g$, defines a set of so-called 'tautological classes' in it, calculates the classes of several geometrically defined loci in terms of them, and finally computes the Chow ring of $\overline{\mathcal{M}}_2$. In our thesis we determine completely the Chow ring of $\overline{\mathcal{M}}_3$. The result is as follows. See [M-Enum] for the definition of κ_2.

Theorem. *As a ring, $A^*(\overline{\mathcal{M}}_3) \cong \mathbf{Q}[\lambda, \delta_0, \delta_1, \kappa_2]/I$, where I is generated by three relations in codimension 3 and six relations in codimension 4. The dimensions of the Chow groups are 1, 3, 7, 10, 7, 3, 1 respectively. The pairing $A^k(\overline{\mathcal{M}}_3) \times A^{6-k}(\overline{\mathcal{M}}_3) \to \mathbf{Q}$ is perfect.*

We exhibit explicit bases for each Chow group. The first step in deriving the above result is the proof that $A^*(\mathcal{M}_3) \cong \mathbf{Q}[\lambda]/(\lambda^2)$.

---

* Supported by the Netherlands organization for scientific research (NWO)

We also have some partial results with respect to the Chow ring of $\overline{\mathcal{M}}_4$. We prove that the Chow ring of $\mathcal{M}_4$ equals $\mathbf{Q}[\lambda]/(\lambda^3)$ and we prove that the dimension of $A^2(\overline{\mathcal{M}}_4)$ equals 13. Finally, using the results obtained with respect to the Chow ring of $\overline{\mathcal{M}}_3$, we determine a.o. the Chow ring of the 'universal curve' $\overline{\mathcal{M}}_{2,1}$ of genus two.

§1. Cycles from the boundary.

To find generators for $A^2(\overline{\mathcal{M}}_g)$ one splits up the problem in three parts, then uses the standard exact sequence

$$A_k(Y) \to A_k(X) \to A_k(U) \to 0$$

for $Y \subset X$ a closed subvariety with complement $U = X - Y$ (see [Fulton, §1.8]).

The first part is to find generators for $A^2(\mathcal{M}_g)$. For small g this sometimes can be done (see [M-Enum] and [Faber]), but for general g we do not know a method. It is generally believed that for g big enough $A^2(\mathcal{M}_g)$ is generated by the tautological classes λ^2 and κ_2, but it seems that at present we are far from proving results in this direction.

The second part is to find generators for the codimension-one Chow groups of the interiors of the divisors Δ_i in the boundary. This can be done using Harer's theorems [Harer 1,2] concerning the Picard groups of the moduli spaces $\mathcal{M}_{g,h}$ of h-pointed curves of genus g.

We start, however, with the third part. One simply has to write down all possible types of stable curves of genus g with two nodes.

Proposition (1.1). *There exist* $[\frac{1}{4}(g^2-1)]+g$ *types of stable curves of genus* g *with two nodes.*

Proof. There is a unique type of irreducible curves: take a curve of genus $g-2$ and identify two pairs of points. This gives irreducible curves with two nodes of type δ_0.

To get curves with two irreducible components and with two nodes of type δ_0, take a two-pointed curve of genus i and a two-pointed curve of genus $g-i-1$, and glue them together at two points. We need $0 < i \leq g-i-1$ for this, so this gives

$$[\tfrac{1}{2}(g-1)]$$

types.

The other types of curves with two components have one node of type δ_0 and one node of type δ_i with $i > 0$. The genus of the smooth component varies between 1 and $g-1$, so this gives

$$g-1$$

types.

Finally we have to count the number of types of curves with three irreducible components. Denote the genus of the middle component by j and the genera of the outer components by i_1 and i_2, then we have to count the number of pairs (i_1, j) with

$$1 \leq j \leq g-2 \text{ and } 1 \leq i_1 \leq i_2 = g-j-i_1 .$$

This number equals

$$\sum_{j=1}^{g-2}[\tfrac{1}{2}(g-j)] = \sum_{j=2}^{g-1}[\tfrac{1}{2}j] = [\tfrac{1}{4}(g-1)^2]\,.$$

Summing up the four contributions, we get

$$[\tfrac{1}{4}(g^2-1)]+g\,.$$ ◊

Next we describe the codimension-one Chow groups (Picard groups) of the interiors of the divisors Δ_i. We will use Harer's theorems [Harer 1,2], saying that the codimension-one Chow group of the moduli space $\mathcal{M}_{g,h}$ of h-pointed curves of genus g is generated by $h+1$ classes for $g \geq 3$.

Lemma (1.2). *$A^1(Int\ \Delta_0)$ is generated by two classes for $g \geq 4$.*

Proof. Note that $Int\ \Delta_0$ is the quotient of $\mathcal{M}_{g-1,2}$ under the action of $\mathbf{Z}/2\mathbf{Z}$ exchanging the points. Following the notation of [A-C, Prop. 1], the Picard group ($\otimes\,\mathbf{Q}$) of $\mathcal{M}_{g-1,2}$ is generated for $g \geq 4$ by the classes λ, ψ_1 and ψ_2; therefore $A^1(Int\ \Delta_0)$ is generated by the two classes λ and $\psi_1+\psi_2$. ◊

Lemma (1.3). *$A^1(Int\ \Delta_1)$ is generated by two classes for $g \geq 4$.*

Proof. Describing the Picard group of $Int\ \Delta_1$ is easy; it's just the Picard group of $\mathcal{M}_{g-1,1}$, since $\mathcal{M}_{1,1} \cong \mathbf{A}^1$. So for $g \geq 4$ we have that $A^1(Int\ \Delta_1)$ is generated by two classes. ◊

Lemma (1.4). *$A^1(Int\ \Delta_2)$ is generated by three classes for $g \geq 5$.*

Proof. It's also not too difficult to find generators for the Picard group of $Int\ \Delta_2$. Namely, $Int\ \Delta_2 \cong \mathcal{M}_{2,1} \times \mathcal{M}_{g-2,1}$, and $\mathcal{M}_{2,1}$ is easily described (see e.g. [Faber, Lemma (I.1.11)]). In the Chow ring of $\mathcal{M}_{2,1}$ there is one non-trivial class coming from the divisor parametrizing the pairs (C,p) where the point p is a Weierstrass point. The complement of this divisor is a quotient of an open set in $\mathbf{A}^4$. Therefore the Picard group of $Int\ \Delta_2 \cong \mathcal{M}_{2,1} \times \mathcal{M}_{g-2,1}$ is just the direct sum of the pull-backs of the Picard groups of the two factors. So for $g \geq 5$ we have that $A^1(Int\ \Delta_2)$ is generated by $1+2=3$ classes. ◊

Finally we deal with the general case: the Picard group of $Int\ \Delta_i$ with $i \geq 3$. For $i < \frac{1}{2}g$ this locus is a product too:

$$Int\ \Delta_i \cong \mathcal{M}_{i,1} \times \mathcal{M}_{g-i,1}$$

while for $i = \frac{1}{2}g$ it is the symmetric square of $\mathcal{M}_{i,1}$.

However, in general the Picard group of a product can be larger than the direct sum of the Picard groups of the two factors. We will now show that in this situation the Picard group *is* the direct sum of the Picard groups of the two factors.

Lemma (1.5). *$A^1(Int\ \Delta_i)$ is generated by four classes for $3 \leq i < \frac{1}{2}g$, and it is generated by two classes for $i = \frac{1}{2}g \geq 3$.*

Proof. As proved in [A-C, Appendix] it is a consequence of a theorem of Harer that

$$H^1(\mathrm{M}) = 0$$

where M is a resolution of singularities of $\overline{\mathcal{M}}_{g,h}$, for $g \geq 3$. Applying [Hartshorne, Ex.(III, 12.6)] we get

$$\mathrm{Pic}(\mathrm{M} \times T) = \mathrm{Pic}\ \mathrm{M} \times \mathrm{Pic}\ T$$

for T a connected scheme. Therefore

$$A^1((\mathcal{M}_{g,h})^\circ \times T) = A^1((\mathcal{M}_{g,h})^\circ) \oplus A^1(T).$$

Since in $\mathcal{M}_{i,1}$ with $i \geq 3$ the locus of curves with automorphisms has codimension 2 or more, this means in particular

$$A^1(\mathcal{M}_{i,1} \times \mathcal{M}_{g-i,1}) = A^1(\mathcal{M}_{i,1}) \oplus A^1(\mathcal{M}_{g-i,1}).$$

Therefore

$$A^1(Int\ \Delta_i)$$

is generated by $2+2=4$ classes for $3 \leq i < \frac{1}{2}g$, and for $i = \frac{1}{2}g$ it is generated by 2 classes. ◊

Proposition (1.6). *For $g \geq 4$ the codimension-one Chow group of the union of the interiors of the divisors Δ_i is generated by $2g-3$ classes.*

Proof. For $g \geq 5$ this follows by summing up the contributions coming from the various components. For $g = 4$ we find 2 classes in $Int\ \Delta_0$, 2 classes in $Int\ \Delta_1$ and 1 class in $Int\ \Delta_2$, since this last locus is the symmetric square of $\mathcal{M}_{2,1}$. ◊

Note that for $g = 3$ we found in [Faber, Lemmas (I.1.11), (I.1.12)] only two classes in this way: one from $Int\ \Delta_0$ and one from $Int\ \Delta_1$.

Summing up the two contributions we find, for $g \geq 4$, for the number of generators of the boundary part of $A^2(\overline{\mathcal{M}}_g)$ the following number:

$$[\tfrac{1}{4}(g^2-1)] + 3g - 3.$$

§2. An independence result.

In this section we will prove that for $g \geq 5$ the products in $A^2(\overline{\mathcal{M}}_g)$ of the divisor classes in a basis of $A^1(\overline{\mathcal{M}}_g)$ are linearly independent. The same holds for the case $g = 3$, see [Faber, §I.2]. For $g =$ 2 the divisor classes λ and δ_1 span the Picard group, and in degree 2 there is the unique relation

$$(\lambda + \delta_1)\delta_1 = 0$$

(see [M-Enum, §10]). Finally, in $A^2(\overline{\mathcal{M}}_4)$ we have the relation

$$(10\lambda - \delta_0 - 2\delta_1)\delta_2 = 0$$

which follows from the relation $10\lambda - \delta_0 - 2\delta_1 = 0$ in $A^1(\overline{\mathcal{M}}_2)$; we will prove that there are no other relations in $A^2(\overline{\mathcal{M}}_4)$. Let $h := [\frac{1}{2}g]$.

Theorem (2.1). *For $g \geq 5$ the products of degree two of the divisor classes in a basis of $A^1(\overline{\mathcal{M}}_g)$ are linearly independent in $A^2(\overline{\mathcal{M}}_g)$, i.e., the $\frac{1}{2}(h+2)(h+3)$ products*

$$\lambda^2, \lambda\delta_0, \ldots, \lambda\delta_h, \delta_0^2, \delta_0\delta_1, \ldots, \delta_{h-1}\delta_h, \delta_h^2$$

of divisor classes are linearly independent in $A^2(\overline{\mathcal{M}}_g)$.

Corollary (2.2). *For $g \geq 5$ any two irreducible closed subvarieties of $\overline{\mathcal{M}}_g$ of codimension one have non-empty intersection.*

Proof of Corollary (2.2). If two subvarieties of codimension one would have empty intersection, the product of their classes vanishes in $A^2(\overline{\mathcal{M}}_g)$, which gives a relation between the products of the standard divisor classes. This contradicts Theorem (2.1). ◊

Proof of Theorem (2.1). The idea of the proof is to construct lots of test surfaces, all consisting of singular curves, and to evaluate the degree-two products of divisor classes on them. Choosing enough of these surfaces will exclude all possible relations; choosing them cleverly will keep the necessary computations to a minimum. In the last steps of the proof we will use induction on the genus. We will constantly use the methods of [H-M, §6] and [A-C, Lemma 1] to compute the pull-backs of the standard divisor classes to these test surfaces.

1) We start with a family of very simple test surfaces. Take two general curves of genera i and $g-i$ with $2 \leq i \leq h$. Take a point on both curves, and identify these two points. This gives a two-dimensional family of curves of type δ_i by varying the two points. Clearly, on this family

$$\lambda = 0, \quad \delta_j = 0 \text{ for } j \neq i.$$

Furthermore

$$\delta_i = -(K_1 + K_2)$$

(where K_l denotes the pull-back to the test surface of the canonical class on the l–th curve), thus

$$(\delta_i)^2 = 2K_1K_2 = 2(2i-2)(2(g-i)-2) \neq 0.$$

This shows that in $A^2(\overline{\mathcal{M}}_g)$ the products

$$(\delta_i)^2, \quad 2 \leq i \leq h$$

are independent, i.e., a relation in degree 2 between the products of the standard divisor classes does not contain terms $(\delta_i)^2$, $2 \leq i \leq h$.

2) The fibers in this test surface are of the form $C/(p\sim q)$ where C is a fixed general curve of genus $g-1$ and where we vary p and q. We computed the products of divisor classes for this test surface (it is the first test surface in [Faber, §I.2]):

$$\delta_0^2 = 8(g-1)(g-2),$$
$$\delta_1^2 = 4 - 2g,$$

all other products vanish.

3) Next take a general curve C of genus $g-2$, and attach two smooth elliptic tails to it at two varying points p and q. When $p = q$ we get curves with a node of type δ_2. The base of the family is the surface $C \times C$; the divisor classes are here:

$$\lambda = \delta_0 = \delta_j = 0 \; (j > 2),$$

$\delta_2 = \Delta$,

$\delta_1 = -(p_1^* K_C + p_2^* K_C + 2\Delta)$.

Therefore

$\delta_1^2 = 8(g-2)(g-3)$,

$\delta_2^2 = 6 - 2g$,

all other products vanish.

Combining (1), (2) and (3) we see that *all* products $(\delta_i)^2$ are independent. Therefore in the rest of the proof we will not care about these products anymore.

The next step is to prove that the 10 products of the divisor classes λ, δ_0, δ_1 and δ_2 are independent. We construct seven test surfaces to achieve this goal. Having done this, we will finish the proof by induction on the genus.

4) This surface consists of curves of type δ_1 . Take a general curve C of genus $g-1$ and attach a varying elliptic tail to a varying point of C . This gives the following values:

$\lambda\delta_1 = 4 - 2g$,

$\delta_0\delta_1 = 48 - 24g$,

$(\delta_1^2 = 4g - 8)$.

5) The next two test surfaces have fibers of the form

C , g − 2

E , 1 F , 1

First vary both elliptic curves E and F in simple pencils. One checks easily that this gives

$\lambda^2 = 2$, $\lambda\delta_0 = 24$,

$\lambda\delta_1 = -2$, $\lambda\delta_2 = -1$,

$\delta_0\delta_1 = -24$, $\delta_0\delta_2 = -12$,

$\delta_1\delta_2 = 1$.

6) Now vary E in a simple pencil and vary the point on C . This gives:

$\lambda\delta_2 = 6 - 2g$,

$\delta_0\delta_2 = 72 - 24g$,

$\delta_1\delta_2 = 2g - 6$.

7) The next three test surfaces have fibers of the form

C , g − 2

E , 1

First vary one point on C and one point on E . Denote by p the fixed point on C . Then

$\delta_0 = -2p_1^*[0_E] - p_2^*(K_C + 2p)$,

$\delta_1 = p_1^*[0_E]$,

$\delta_2 = p_2^*(p)$.

Thus

$\delta_0\delta_1 = 4 - 2g$,

$\delta_0\delta_2 = -2$,

$\delta_1\delta_2 = 1$.

8) Next vary one point on C and vary E in a simple pencil. Denote by x the class of a point on $\mathbf{P}^1$. Then

$$\lambda = p_1^*x ,$$
$$\delta_0 = p_1^*(10x) - p_2^*(K_C + 2p) ,$$
$$\delta_1 = 0 ,$$
$$\delta_2 = p_2^*(p) .$$

Thus

$$\lambda^2 = 0 , \qquad \lambda\delta_0 = 4 - 2g ,$$
$$\lambda\delta_2 = 1 , \qquad \delta_0\delta_2 = 10 .$$

9) Finally we vary both the point on E and the j–invariant. The basis will be a surface S with an elliptic fibering: we take the blow-up of $\mathbf{P}^2$ in the 9 points of intersection of two general cubic curves. Denote by H the pull-back of the hyperplane section, by Σ the sum of the 9 exceptional divisors and by E_0 one of them (the zero-section). Then one computes:

$$\lambda = 3H - \Sigma ,$$
$$\delta_0 = 30H - 10\Sigma - 2E_0 ,$$
$$\delta_1 = E_0 ,$$
$$\delta_2 = 0 .$$

Therefore

$$\lambda^2 = 0 , \qquad \lambda\delta_0 = -2 ,$$
$$\lambda\delta_1 = 1 , \qquad \delta_0\delta_1 = 12 .$$

We summarize the results obtained so far in a matrix.

	λ^2	$\lambda\delta_0$	$\lambda\delta_1$	$\lambda\delta_2$	$\delta_0\delta_1$	$\delta_0\delta_2$	$\delta_1\delta_2$
4)	0	0	1	0	12	0	0
5)	2	24	−2	−1	−24	−12	1
6)	0	0	0	1	0	12	−1
7)	0	0	0	0	4–2g	−2	1
8)	0	4–2g	0	1	0	10	0
9)	0	−2	1	0	12	0	0

This matrix has rank 6, and the one relation which still can exist is

$$(10\lambda - \delta_0 - 2\delta_1)\delta_2 = 0 .$$

Substituting $g = 4$ gives the result announced in the beginning of this section: in $A^2(\overline{\mathcal{M}}_4)$ the products of divisors span a 9–dimensional vector space given by the relation above.

To exclude this relation for $g \geq 5$ we construct a test surface inside Δ_2 . The basis of the test surface will be the universal curve over a pencil of curves of genus $g - 2$ as described in [Arbarello-Cornalba], and we attach a fixed one-pointed curve of genus 2 to this family of one-pointed curves of genus $g - 2$.

The pencil we take from [A–C] is called Λ_k there, with $k = g - 2 \geq 3$. It is constructed by blowing-up the $2k - 2$ base points of a Lefschetz pencil of hyperplane sections of a smooth K–3 surface of degree $2k - 2$ in $\mathbf{P}^k$. Define on the "universal curve" the divisor classes

$G :=$ the pull-back of a point on $\mathbf{P}^1$,

$\Sigma :=$ the sum of the $2k-2$ 'exceptional divisors.

Then one computes (cf. [A–C]):

$$\lambda = (k+1)(G-\Sigma),$$
$$\delta_0 = (18+6k)(G-\Sigma),$$
$$\delta_2 = -2G+\Sigma.$$

Therefore

$$\lambda\delta_2 = (k+1)(2-2k),$$
$$\delta_0\delta_2 = (18+6k)(2-2k).$$

If the one relation which still can exist:

$$(10\lambda - \delta_0 - 2\delta_1)\delta_2 = 0$$

holds, then

$$18+6k = 10(k+1) \Rightarrow k=2,$$

in contradiction with our assumption. Therefore for $g \geq 5$ the 10 products of the divisor classes λ, δ_0, δ_1, δ_2 are independent in $A^2(\bar{\mathcal{M}}_g)$.

Let now $g \geq 6$. We will prove the independence of the products with δ_3 : $\lambda\delta_3$, $\delta_0\delta_3$, $\delta_1\delta_3$, $\delta_2\delta_3$ (remember that we already dealt with δ_3^2).

10) Consider inside Δ_3 a test surface of curves consisting of a fixed one-pointed curve of genus $g-3$ attached to a varying point on a varying curve of genus 3. We take as basis of this family the universal curve belonging to the pencil Λ_3 in [A–C]. Then

$$\lambda = 4(G-\Sigma),$$
$$\delta_0 = 36(G-\Sigma),$$
$$\delta_3 = -2G+\Sigma.$$

Therefore

$$\lambda\delta_3 = -8G^2 - 4\Sigma^2 = -16,$$
$$\delta_0\delta_3 = -144.$$

11) The next two test surfaces contain curves of the following type

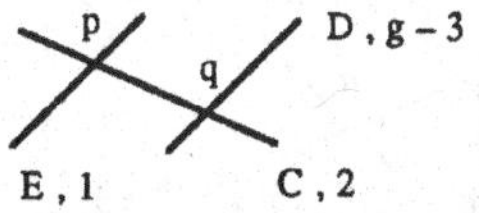

For the first surface, vary E in a simple pencil and vary the point p on C. Then

$$\lambda = p_1^*x,$$
$$\delta_0 = 12p_1^*x,$$
$$\delta_1 = -p_1^*x - p_2^*(K_C+q),$$
$$\delta_2 = p_2^*(q),$$
$$\delta_3 = -p_2^*(q).$$

Therefore

$$\lambda\delta_3 = -1, \qquad \delta_0\delta_3 = -12,$$
$$\delta_1\delta_3 = 1, \qquad \delta_2\delta_3 = 0.$$

12) For the second surface, vary both points on C. This gives

$$\lambda = \delta_0 = 0 ,$$
$$\delta_1 = -p_1^* K_C - \Delta ,$$
$$\delta_2 = \Delta ,$$
$$\delta_3 = -p_2^* K_C - \Delta .$$

Therefore

$$\delta_1\delta_3 = (p_1^* K_C)(p_2^* K_C) + \Delta^2 + \Delta(p_1^* K_C + p_2^* K_C)$$
$$= 2{\cdot}2 - 2 + 2 + 2 = 6 ,$$
$$\delta_2\delta_3 = -(2 + (-2)) = 0 .$$

So, using (10), (11) and (12), we prove the independence of $\delta_1\delta_3$, $\lambda\delta_3$ and $\delta_0\delta_3$.

Finally we consider a test surface with curves of type

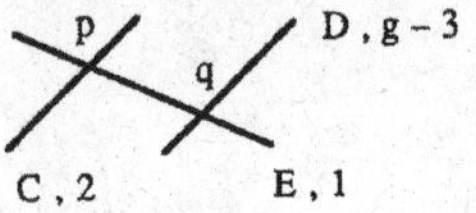

Vary the point on C and the point on D . Then

$$\delta_2 = -p_1^* K_C ,$$
$$\delta_3 = -p_2^* K_D ,$$

thus

$$\delta_2\delta_3 \neq 0 .$$

So we proved that for $g \geq 6$ the 15 products of the divisors $\lambda , \ldots , \delta_3$ are independent in $A^2(\bar{\mathcal{M}}_g)$. This will serve as the first step in an induction (on the genus) which will finish the proof of the theorem. The induction hypothesis is that for $g \geq 2k$ the products (of degree 2) of the divisors $\lambda , \ldots , \delta_k$ are independent.

Proof of the induction step: assuming the induction hypothesis we will prove that for $g \geq 2k + 2$ the products of the divisors $\lambda , \ldots , \delta_{k+1}$ are independent. This comes down to proving the independence of the products

$$\lambda\delta_{k+1} , \ldots , \delta_k\delta_{k+1} .$$

13) We start with several test surfaces with curves of type

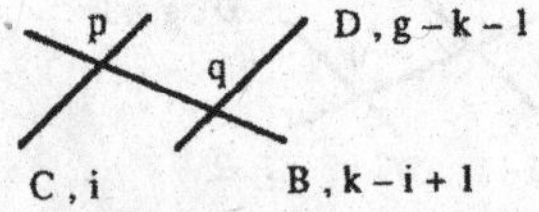

Assume $2 \leq i \leq k$. Vary a point on C and a point on D . Then

$$\delta_i = -p_1^* K_C ,$$
$$\delta_{k+1} = -p_2^* K_D ,$$

therefore

$$\delta_i\delta_{k+1} \neq 0 .$$

This leads to the independence of

$$\delta_i\delta_{k+1}$$

for $2 \leq i \leq k$.

14) Next we consider two test surfaces with curves of type

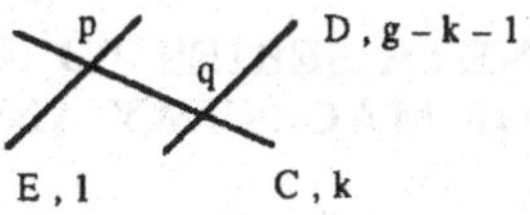

First, vary both points on C. Then

$$\delta_1 = -p_1^* K_C - \Delta ,$$
$$\delta_k = \Delta ,$$
$$\delta_{k+1} = -p_2^* K_C - \Delta .$$

Thus, on this test surface:

$$\delta_1 \delta_{k+1} = (2k-2)(2k-1) \neq 0 .$$

As a consequence we get the independence of $\delta_1\delta_{k+1}$. (As explained before, this means that $\delta_1\delta_{k+1}$ is independent of the other products; i.e., in any linear relation in $A^2(\bar{\mathcal{M}}_g)$ between the products of the divisor classes the coefficient of $\delta_1\delta_{k+1}$ vanishes. In fact we already proved that in such a relation the coefficients of all products of the divisors $\lambda , \ldots , \delta_k$ and of the products $\delta_i\delta_{k+1}$ with $1 \le i \le k+1$ vanish.)

15) Secondly, vary E in a simple pencil and vary the point on D. As one checks easily, this leaves as the only relation possible :

$$\delta_0\delta_{k+1} = 12\lambda\delta_{k+1} .$$

16) Finally we exclude this relation by taking a surface inside Δ_{k+1} based on the universal curve of the pencil Λ_{k+1} from [A–C]. Namely, this leaves as the only possible relation:

$$(k+2)\delta_0\delta_{k+1} = (24+6k)\lambda\delta_{k+1}$$

and

$$12(k+2) = (24+6k) \Leftrightarrow k = 0 .$$

Therefore we conclude that

$$\delta_0\delta_{k+1} \text{ and } \lambda\delta_{k+1}$$

are also independent.

This finishes the proof of the induction step and therefore also the proof of Theorem (2.1). ◊

References.

[A-C] E. Arbarello, M. Cornalba - The Picard groups of the moduli spaces of curves, Topology **26**, 153-171 (1987).

[Faber] C. Faber - *Chow rings of moduli spaces of curves*, thesis, Universiteit van Amsterdam, 1988.

[Fulton] W. Fulton - *Intersection Theory*, Ergebnisse, Springer-Verlag, 1984.

[Harer 1] J. Harer - The second homology group of the mapping class group of an orientable surface, Invent. Math. **72**, 221-239 (1983).

[Harer 2] J. Harer - The cohomology of the moduli space of curves, C.I.M.E. notes, Montecatini, 1985.

[Hartshorne] R. Hartshorne, *Algebraic Geometry*, Graduate Texts in Math., Springer-Verlag, 1977.

[H-M] J. Harris, D. Mumford - On the Kodaira dimension of the moduli space of curves, Invent. Math. **67**, 23-88 (1982).

[M-Enum] D. Mumford - Towards an enumerative geometry of the moduli space of curves, in *Arithmetic and Geometry* (dedicated to I. Shafarevich), Vol. II, Birkhaüser, 1983.

RESTRICTIONS OF LINEAR SERIES TO HYPERPLANES, AND SOME RESULTS OF MACAULAY AND GOTZMANN

Mark Green*
University of California, Los Angeles

A very beautiful line of algebraic inquiry, going back to Macaulay, deals with the following question: for $W \subseteq H^0(\mathcal{O}_{\mathbf{P}^r}(d))$ a linear subspace of a given dimension, what is the smallest possible dimension of the ideal it generates in the homogeneous coordinate ring in degree $d+1$? This question was answered by Macaulay ([M], see also [S], [B]) . A similar question, which I first learned of from F. Oort, is: what is the smallest possible dimension of the restriction W_H of W to a general hyperplane H. This is answered by Theorem 1. In both cases, monomial ideals achieve the bounds (see [S]).

G. Gotzmann [Go] carried Macaulay's work further by showing that if W achieves Macaulay's bound for the worst possible behavior, then the dimensions of the ideal it generates are determined in all degrees $\geq d$, and at each step it achieves Macaulay's bound. In [G1], I made use of Gotzmann's result to prove a Hodge-theoretic result about the Noether-Lefschetz locus of surfaces in $\mathbf{P}^3$ whose Picard group is not generated by the hyperplane section. I would expect there to be other applications of the line of thought of Macaulay and Gotzmann in this area, as there has been in others, e.g. [B], [B-M], [I], [T]. So it seems worthwhile to obtain as thorough an understanding of Gotzmann's work as possible.

The essential step in Gotzmann's proof is a bound for the regularity of an ideal sheaf in terms of its Hilbert polynomial. We give a somewhat different proof of this result, and indeed are able to get a vanishing theorem for the cohomology of the ideal sheaf that improves his result (although not his bound for the regularity.)

What I find most interesting is the possible interweaving of the algebraic ideas of Macaulay and Gotzmann with geometric arguments. An example of one way this could be done is given in Theorems 3 and 4. In the course of writing this paper, I have come to realize the great strength of the results of [B], which contains a very significant and useful improvement of Gotzmann's results. I hope in a later paper to explore some applications of Bayer's techniques.

I wish to thank David Eisenbud, Dave Bayer, and Tony Iarrobino for introducing me to the work of Macaulay and Gotzmann, and the organizers of the Trento conference for providing me the opportunity to write this paper.

Every positive integer c can be written uniquely in the form

$$c = \binom{k_d}{d} + \binom{k_{d-1}}{d-1} + \cdots + \binom{k_1}{1}$$

* Research partially supported by N.S.F. Grant DMS 88-02020

where

$$k_d > k_{d-1} > \cdots k_1 \geq 0$$

. We will refer to this as the d'th *Macaulay representation* of c, and $k_d, k_{d-1}, \ldots, k_1$ as the d'th *Macaulay coefficients* of c. The main property of the Macaulay representations, other than uniqueness, is that the lexicographic order coincides with the usual order.

For $c \neq 0$, if $\delta = \min\{m \mid k_m \geq m\}$, we have the alternate presentation

$$c = \binom{k_d}{d} + \binom{k_{d-1}}{d-1} + \cdots \binom{k_\delta}{\delta}$$

with

$$k_d > k_{d-1} > \cdots k_\delta \geq \delta > 0.$$

If c has d'th Macaulay coefficients $k_d, k_{d-1}, \ldots, k_1$, then let

$$c_{<d>} = \binom{k_d - 1}{d} + \binom{k_{d-1} - 1}{d-1} + \cdots + \binom{k_1 - 1}{1},$$

where we adopt the convention $\binom{m}{n} = 0$ if $m < n$. This has the two elementary properties that

$$c_{<d>} \leq c'_{<d>} \ if \ c \leq c'$$

and

$$(c-1)_{<d>} < c_{<d>} \ if \ k_\delta \neq \delta$$

Theorem 1. *Let $W \subseteq H^0(\mathcal{O}_{\mathbf{P}^r}(d))$ be a linear system with* $\operatorname{codim}(W, H^0(\mathcal{O}_{\mathbf{P}^r}(d))) = c$. *Let $W_H \subseteq H^0(\mathcal{O}_H(d))$ be the restriction of W to a general hyperplane H, and c_H the codimension of W_H. Then*

$$c_H \leq c_{<d>}.$$

Proof: Let c_H have d'th Macaulay coefficients $l_d, l_{d-1}, \ldots, l_1$. The inequality of the theorem is equivalent to

$$c \geq \binom{l_d + 1}{d} + \binom{l_{d-1} + 1}{d-1} + \cdots + \binom{l_\delta + 1}{\delta},$$

where $\delta = \min\{m \mid l_m \geq m\}$ (Note that δ is defined unless $c = 0$, where the result is obvious). If the theorem fails, then

$$c - c_H < \binom{l_d}{d-1} + \binom{l_{d-1}}{d-2} + \cdots + \binom{l_\delta}{\delta - 1}.$$

We will denote by $W(-H)$ the subspace of all polynomials P of degree $d-1$ for which PH belongs to W. We have the basic exact sequence

$$0 \to W(-H) \to W \to W_H \to 0,$$

which has as a consequence that

$$c = c_H + \operatorname{codim}(W(-H)).$$

If H, H^* are general hyperplanes, we have from the restriction sequence

$$0 \to W_H(-(H \cap H^*)) \to W_H \to W_{H\cap H^*} \to 0$$

that

$$\operatorname{codim} W_H = \operatorname{codim} W_{H\cap H^*} + \operatorname{codim} W_H(-(H \cap H^*)).$$

Since

$$W(-H^*)_H \subseteq W_H(-(H \cap H^*)),$$

we see that

$$\operatorname{codim} W_H \leq \operatorname{codim} W_{H\cap H^*} + \operatorname{codim} W(-H^*)_H.$$

By induction on dimension and on degree, we may assume that the estimate of the theorem holds for restrictions to a general hyperplane for W_H and $W(-H^*)$. For dimension 1 the theorem is easy, and likewise for degree 1 (This is where we use the fact that H is a general hyperplane).

The dimension of W_H is c_H and the dimension of $W(-H^*)$ is $c - c_H$. Thus

$$c_H \leq (c_H)_{<d>} + (c - c_H)_{<d-1>}.$$

There are now two cases to consider. If $\delta = 1$, then

$$c - c_H \leq \binom{l_d}{d-1} + \binom{l_{d-1}}{d-2} + \cdots \binom{l_2}{1}.$$

Now

$$(c - c_H)_{<d-1>} \leq \binom{l_d - 1}{d-1} + \cdots + \binom{l_2 - 1}{1}$$

and

$$(c_H)_{<d>} = \binom{l_d - 1}{d} + \cdots + \binom{l_1 - 1}{1}.$$

So adding these

$$c_H \leq \binom{l_d}{d} + \cdots + \binom{l_2}{2} + \binom{l_1 - 1}{1} < c_H$$

which is a contradiction.

If on the other hand $\delta > 1$, then

$$c - c_H < \binom{l_d}{d-1} + \binom{l_{d-1}}{d-2} + \cdots + \binom{l_\delta}{\delta - 1}.$$

Now

$$(c - c_h)_{<d-1>} < \binom{l_d - 1}{d-1} + \cdots + \binom{l_\delta - 1}{\delta - 1}.$$

The reason we get a strict inequality here is that we are taking $_{<d-1>}$ of both sides of a strict inequality, and $l_\delta - 1 > \delta - 1$. Now adding the two inequalities gives

$$c_H < \binom{l_d}{d} + \cdots + \binom{l_\delta}{\delta} = c_H$$

which is a contradiction.

From this theorem about restriction to a general hyperplane it is possible to recover the classical theorem of Macaulay. Let W, c be as above. Let W_1 denote the image of the multiplication map $W \otimes H^0(\mathcal{O}_{\mathbf{P}^r}(1)) \to H^0(\mathcal{O}_{\mathbf{P}^r}(d+1))$ and c_1 its codimension. If c has d'th Macaulay coefficients $k_d, k_{d-1}, \ldots, k_1$, let

$$c^{<d>} = \binom{k_d + 1}{d+1} + \cdots + \binom{k_1 + 1}{2}.$$

We remark the elementary properties

$$c \leq c' \text{ implies } c^{<d>} \leq c'^{<d>}$$

and

$$(c+1)^{<d>} = \begin{cases} c^{<d>} + k_1 + 1 & \text{if } \delta = 1 \\ c^{<d>} + 1 & \text{if } \delta > 1 \end{cases}$$

where as before $\delta = \min\{m \mid k_m \geq m\}$. We then have:

Theorem 2 (Macaulay). $c_1 \leq c^{<d>}$.

Proof: From the restriction sequence for W_1 we have

$$\text{codim } W_1 = \text{codim } W_1(-H) + \text{codim } (W_1)_H.$$

Since $W \subseteq W_1(-H)$, it follows that

$$c_1 \leq c + \text{codim } (W_1)_H.$$

If c_1 has $(d+1)st$ Macaulay coefficients $l_{d+1}, l_d, \ldots, l_1$, then

$$\text{codim } (W_1)_H \leq \binom{l_{d+1} - 1}{d+1} + \cdots + \binom{l_1 - 1}{1}.$$

It follows that

$$c \geq \binom{l_{d+1} - 1}{d} + \cdots + \binom{l_2 - 1}{1} + \binom{l_1 - 1}{0}.$$

If $\delta > 1$, then $l_1 = 0$ so

$$c^{<d>} \geq \binom{l_{d+1}}{d+1} + \cdots + \binom{l_2}{2} = c_1$$

and we are done. If $\delta = 1$, then by the second elementary property of ${}^{<d>}$,

$$c^{<d>} \geq \binom{l_{d+1}}{d+1} + \cdots + \binom{l_3}{3} + \binom{l_2}{2} + l_2.$$

Since $l_2 > l_1$,we get that

$$c^{<d>} > c_1$$

which completes the proof.

We now consider the following slightly different situation: let $I_\bullet$ be a graded ideal, with *Hilbert polynomial* $P(k)$, i.e.

$$P(k) = \operatorname{codim}(I_k, H^0(\mathcal{O}_{\mathbf{P}^r}(k))) \quad \text{for } k >> 0.$$

If $\mathcal{I}$ is the ideal sheaf corresponding to $I_\bullet$, and

$$\mathcal{F} = \mathcal{O}_{\mathbf{P}^r}/\mathcal{I},$$

then

$$P(k) = \chi(\mathcal{F}(k)).$$

It is convenient to recall the standard properties of regularity (see [G2]). A coherent analytic sheaf $\mathcal{S}$ is said to be *m-regular* if $H^q(\mathcal{S}(m-q)) = 0$ for all $q > 0$. If $\mathcal{S}$ is m-regular, then it is $(m+1)$-regular; often the lowest value of m for which $\mathcal{S}$ is m-regular is called the *regularity* of $\mathcal{S}$. For a coherent analytic sheaf $\mathcal{S}$ on $\mathbf{P}^r$, m-regularity is equivalent to the minimal free resolution of $\mathcal{S}$ having the form

$$\cdots \to \oplus\mathcal{O}_{\mathbf{P}^r}(-a_{1i}) \to \oplus\mathcal{O}_{\mathbf{P}^r}(-a_{0i}) \to \mathcal{S} \to 0$$

where

$$a_{pi} \leq m + p \quad \text{for all } p \geq 0.$$

Gotzmann's Regularity Theorem. *Any graded ideal $I_\bullet$ has Hilbert polynomial of the form*

$$P(k) = \binom{k+a_1}{a_1} + \binom{k+a_2-1}{a_2} + \cdots + \binom{k+a_s-(s-1)}{a_s}$$

where

$$a_1 \geq a_2 \geq \cdots \geq a_s \geq 0.$$

Furthermore, the associated ideal sheaf $\mathcal{I}$ is s-regular.

Remark: We will prove a somewhat stronger result. Let

$$s_q = \operatorname{card}\{i \mid a_i \geq q-1\}.$$

Note that $s_0 = s$. Then in fact for each $q > 0$,

$$H^q(\mathcal{I}(k-q)) = 0 \quad \text{for } k \geq s_q.$$

Proof: Let H be a general hyperplane, and $\mathcal{I}' = \mathcal{I} \otimes \mathcal{O}_H$. We have the exact sequence

$$0 \to \mathcal{I}(k-1) \to \mathcal{I}(k) \to \mathcal{I}'(k) \to 0$$

arising from restriction. If $P_H(k)$ is the Hilbert polynomial of $\mathcal{I}'$ viewed as a coherent sheaf on H, then we have by this sequence that

$$P_H(k) = P(k) - P(k-1).$$

By induction on dimension of the ambient projective space, the case of dimension 0 being obvious, we may assume that $P_H(k)$ has the form

$$P_H(k) = \binom{k+b_1}{b_1} + \binom{k+b_2-1}{b_2} + \cdots + \binom{k+b_t-(t-1)}{b_t}$$

where

$$b_1 \geq b_2 \geq \cdots b_t \geq 0.$$

We may further assume that if

$$t_q = \operatorname{card}\{i \mid b_i \geq q-1\},$$

then for every $q > 0$,

$$H^q(\mathcal{I}'(k-q)) = 0 \quad \text{for } k \geq t_q.$$

We immediately conclude that

$$P(k) = \binom{k+a_1}{a_1} + \binom{k+a_2-1}{a_2} + \cdots + \binom{k+a_s-(s-1)}{a_s} + e$$

where e is an unknown constant and $a_i = b_i + 1$. Thus $s_{q+1} = t_q$ for $q \geq 0$. We immediately see by the restriction sequence for $\mathcal{I}'$, together with Theorem B, that for all $q > 1$,

$$H^q(\mathcal{I}(k-q)) = 0 \quad \text{for } k \geq s_q.$$

It remains to show that this also holds for $q = 1$, and that $e \geq 0$.

Let $f_d = \operatorname{codim}(H^0(\mathcal{I}(d)), H^0(\mathcal{O}_{\mathbf{P}^r}(d)))$, and $f_{d,H}$ the analogous numbers for $\mathcal{I}'$. If $e < 0$, then for $d >> 0$,

$$f_d < \binom{k+a_1}{a_1} + \binom{k+a_2-1}{a_2} + \cdots + \binom{k+a_s-(s-1)}{a_s}.$$

By our result on codimensions for restriction to a general hyperplane, for $d >> 0$,

$$f_{d,H} < \binom{k+b_1}{b_1} + \binom{k+b_2-1}{b_2} + \cdots + \binom{k+b_t-(t-1)}{b_t} = P_H(k),$$

which is a contradiction. So $e \geq 0$. Setting $a_{t+1} = a_{t+2} = \cdots = a_{t+e} = 0$, we get that $P(k)$ has the desired formula, where $s = t + e$.

By the vanishing of cohomology that we have so far, we have that

$$f_d \leq P(d), \quad \text{for } d \geq s_2 - 2,$$

with equality holding if and only if $H^1(\mathcal{I}(d)) = 0$. For $d = s - 1$, we may write

$$P(d) = \binom{d+a_1}{d} + \binom{d+a_2-1}{d-1} + \cdots + \binom{d+a_s-(s-1)}{d-(s-1)}.$$

By Macaulay's theorem, if $f_{s-1} < P(s-1)$, then it remains behind forever, contradicting the fact that $f_k = P(k)$ for $k >> 0$. Thus $H^1(\mathcal{I}(s-1)) = 0$, which is the last thing we need to conclude $\mathcal{I}$ is s-regular. This completes the proof of Gotzmann's Regularity Theorem.

Gotzmann's Persistence Theorem. *In the situation of Theorem 2, if c has Macaulay coefficients of degree d given by $k_d, k_{d-1}, \ldots, k_1$, and if $c_1 = c^{<d>}$, then*

$$c_\nu = \binom{k_d+\nu}{d+\nu} + \binom{k_{d-1}+\nu}{d-1+\nu} + \cdots + \binom{k_1+\nu}{1+\nu}$$

for all $\nu \geq 0$.

Proof: By the exact sequence

$$0 \to W_1(-H) \to W_1 \to W_{1,H} \to 0,$$

and the fact that

$$W \subseteq W_1(-H),$$

we have that

$$c_{1,H} \geq c_1 - c.$$

Now

$$(c_{<d>})^{<d>} \geq c_H^{<d>} \geq c_{1,H} \geq c_1 - c = (c_{<d>})^{<d>}.$$

The above string of inequalities must all be equalities, and thus $c_{1,H}$ has the predicted value. By induction on the dimension of the ambient projective space, we may thus assume that $c_{\nu,H}$ has the predicted value for all $\nu \geq 0$. The Hilbert polynomial of the homogeneous ideal generated by W_H is therefore

$$P_H(k) = \binom{k+k_d-d-1}{k_d-d} + \binom{k+k_{d-1}-(d-1)-1}{k_{d-1}-(d-1)} + \cdots + \binom{k+k_1-1-1}{k_1-1}.$$

If we let $\mathcal{I}$ denote the ideal sheaf of the homogeneous ideal generated by W, and $\mathcal{I}'$ the ideal sheaf of the homogeneous ideal generated by W_H, then we see by the Gotzmann Regularity Theorem that $\mathcal{I}'$ is d-regular. If $P(k)$ is the Hilbert polynomial of the homogeneous ideal generated by W, then we know by the vanishing of the higher cohomology of twists of $\mathcal{I}$ that occurred in the proof of Gotzmann's Regularity Theorem that

$$c \leq P(d).$$

Once again, if strict inequality occurs, then by Macaulay's Theorem, the codimension of the ideal can never catch up with the Hilbert polynomial. Thus

$$c = P(d).$$

Since we know the polynomial $P_H(k)$, this one additional bit of information shows that

$$P(k) = \binom{k+k_d-d}{k_d-d} + \binom{k+k_{d-1}-(d-1)}{k_{d-1}-(d-1)} + \cdots + \binom{k+k_1-1}{k_1-1}$$

and that $\mathcal{I}$ is d-regular. Thus

$$c_\nu = P(d+\nu)$$

for $\nu \geq 0$, which proves the Persistence Theorem.

There are a number of interesting questions in this area. In the situation of Theorem 1, the lowest c for which the restriction of W to a general hyperplane may fail to be the complete linear system of all polynomials of degree d is $c = d+1$. In this case, W can be the ideal of a line, for then the restriction of W to a general hyperplane is the ideal of a point. This is the only possibility.

Proposition. *Let $W \subset H^0(\mathcal{O}_{\mathbf{P}^r}(d))$ be a linear subspace with*

$$c = d+1, \quad c_H = 1.$$

Then

$$W = I_d(L)$$

for some line L.

Proof: It is possible to give a geometric proof of this, but in order to illustrate the power of Gotzmann's results, we'll give a proof in the spirit of his approach. By Theorem 2, we know that $c_1 \leq d+2$. If equality holds, then by Gotzmann's Persistence Theorem, $c_k = d+k+1$ for all $k \geq 0$. Thus $P(k) = k+1$. Now by Gotzmann's Regularity Theorem, the ideal sheaf $\mathcal{I}$ associated to the ideal generated by W is 1-regular. Thus $\operatorname{codim}(H^0(\mathcal{I}(1))) = 2$. Thus W contains the $I_d(L)$ for some line L, and equals it since both spaces have the same dimension.

If $c_1 \leq d+1$, then by Theorem 2 we have $c_k \geq c_{k+1}$ for all $k \geq 1$. Thus $P(k)$ is a constant $a \leq d+1$. By Gotzmann's Regularity Theorem, it follows that $\mathcal{I}$ is $(d+1)$-regular. So $H^1(\mathcal{I}(d)) = 0$. Now let $\mathcal{R}$ be defined by the exact sequence of sheaves

$$0 \to \mathcal{R} \to W \otimes \mathcal{O}_{\mathbf{P}^r} \to \mathcal{I}(d) \to 0.$$

Let

$$\mathcal{R}' = \mathcal{R} \otimes \mathcal{O}_H, \quad \mathcal{I}' = \mathcal{I} \otimes \mathcal{O}_H,$$

where H is a general hyperplane. We have exact sequences

$$0 \to \mathcal{R}' \to W \otimes \mathcal{O}_H \to \mathcal{I}'(d) \to 0$$

and

$$0 \to \mathcal{R}(-1) \to \mathcal{R} \to \mathcal{R}' \to 0.$$

Because $c_H = 1$ and $\mathcal{I}' = \mathcal{O}_H$, we see that $h^1(\mathcal{R}') = 1$. Since $H^1(\mathcal{I}(d)) = 0$, we have that $H^2(\mathcal{R}) = 0$. Thus $h^1(\mathcal{I}(d-1)) = h^2(\mathcal{R}(-1)) = 1$. So $\operatorname{codim}(H^0(\mathcal{I}(d-1))) = d$. Now by Gotzmann's Persistence Theorem, the ideal generated by $H^0(\mathcal{I}(d-1))$ has Hilbert polynomial $P(k) = k+1$, and is thus the ideal of a line. Thus W contains $I_d(L)$ for some line L, with equality once again by dimension.

A more geometric argument gives a generalization of this result.

Theorem 3. *Let $W \subset H^0(\mathcal{O}_{\mathbf{P}^r}(d))$ be a linear subspace with*

$$c = \binom{d+m}{d}, \quad c_H = \binom{d+m-1}{d},$$

where $d \geq 1$ and $m \geq 1$. Then

$$W = I_d(P)$$

for some m-dimensional linear space P.

Proof: We may do induction on d, since the case $d = 1$ is obvious. From the exact sequence

$$0 \to W(-H) \to W \to W_H \to 0,$$

we conclude that for a general hyperplane H, $\operatorname{codim}(W(-H)) = \binom{d+m-1}{d-1}$. If H' is another general hyperplane, we have the inclusion $W(-H)_{H'} \subseteq W_{H'}(-(H \cap H'))$, and thus $\operatorname{codim}(W_{H'}(-(H \cap H'))) \leq \binom{d+m-2}{d-1}$. From the exact sequence

$$0 \to W_{H'}(-(H \cap H')) \to W_{H'} \to W_{H \cap H'} \to 0,$$

we see that $\operatorname{codim}(W_{H \cap H'}) \geq \binom{d+m-2}{d}$. Hence by Theorem 1, the last two inequalities are equalities, and $W(-H)_{H'} = W_{H'}(-(H \cap H'))$. The hypotheses of the theorem now apply to $W(-H)$, so by induction on degree,

$$W(-H) = I_{d-1}(P_H)$$

where P_H is a linear space of dimension m depending on H. Now

$$W_{H'}(-(H \cap H')) = I_{d-1}(P_H \cap H').$$

We therefore see that for H_1, H_2 two general hyperplanes, $P_H \cap H_1$ and $P_H \cap H_2$ are constant for H a general element of the pencil of hyperplanes spanned by H_1 and H_2. Thus P_H is constant on this pencil, and hence is constant. This proves the theorem.

As a final illustration, we will characterize ideals of plane curves by their properties under hyperplane restriction. Notice that

$$\binom{d+1}{d} + \binom{d}{d-1} \cdots + \binom{d-(k-2)}{d-(k-1)} = kd + 1 - \binom{k-1}{2}.$$

Theorem 4. *Let $W \subset H^0(\mathcal{O}_{\mathbf{P}^r}(d))$ be a linear subspace with*

$$c = kd + 1 - \binom{k-1}{2}, \quad c_H = k.$$

If $d \geq k \geq 0$, then

$$W = I_d(C),$$

where C is a plane curve of degree k.

Proof: We will do an induction on d, starting with the case $d = k$. For H a general hyperplane, the same argument as in Theorem 3 shows that $\mathrm{codim}(W(-H)) = \binom{k+1}{2}$. If H' is a second general hyperplane, then we see by Theorem 1 that $\mathrm{codim}(W_{H \cap H'}) = 0$ and therefore $\mathrm{codim}(W_{H'}(-(H \cap H'))) = k$. By Theorem 3, we have that $W(-H) = I_{k-1}(P_H)$, where P_H is a 2-plane depending on H. By the same argument as in Theorem 3, we see that in fact P_H is a 2-plane P independent of H. Thus $I_k(P) \subseteq W$, and hence by dimension W is spanned by one extra generator F. Thus in this case $W = I_k(C)$, where $C = P \cap \{F = 0\}$.

Now, for higher d, the same sort of dimension count we have already done shows that for a general hyperplane H, $W(-H)$ satisfies the hypotheses of the theorem. Thus $W(-H) = I_{d-1}(C_H)$ for some plane curve C_H of degree k depending on H. The same argument as in Theorem 3 shows that in fact C_H is independent of H, and thus that $W = I_d(C)$ for a plane curve C of degree k.

BIBLIOGRAPHY

[B] D.Bayer, "The division algorithm and the Hilbert scheme," Thesis, Harvard University.

[B-M] E.Bierstone and P.Milman, "The local geometry of analytic mappings," preprint.

[Go] G. Gotzmann, "Eine Bedingung für die Flachheit und das Hilbertpolynom eines graduierten Ringes," Math. Z. 158 (1978), 61-70.

[G1] M.Green, "Components of maximal dimension in the Noether-Lefschetz locus," J. Diff. Geom., to appear.

[G2] M.Green, "Koszul cohomology and geometry," preprint.

[I] A.Iarrobino, "Hilbert schemes of points: Overview of last ten years," Algebraic Geometry, Bowdoin 1985, Proc. Symp. in Pure Math. vol 46, 297-320.

[M] F.S.Macaulay, "Some properties of enumeration in the theory of modular systems," Proc. Lond. Math. Soc. 26 (1927), 531-555.

[S] R.Stanley, "Hilbert functions of graded algebras," Adv. in Math. 28 (1978), 57-83.

[T] B.Teissier, "Variétès toriques et polytopes," Séminaire Bourbaki 565 (1980-1), 71-82.

La Rationalité des schémas de Hilbert de courbes gauches rationnelles suivant Katsylo

André Hirschowitz

En matière de rationalité un progrès significatif a été accompli par Katsylo [K], qui a démontré que le quotient par le groupe GL(2, k) du schéma des formes homogènes de degré d non identiquement nulles à coefficients dans k est rationnel. Un corollaire immédiat de ce résultat est le

Théorème : les schémas de Hilbert de courbes rationnelles dans les espaces projectifs de dimension $n \geqslant 3$ (et même $n \geqslant 2$) sont rationnels.

Le théorème de Katsylo correspond au cas n=0. Ce corollaire, qu'on attendait depuis longtemps (cf. [EV]) n'est pas mentionné par Katsylo, ni par Bogomolov [Bo], ni même par Dolgachev dans son exposé de synthèse [D]. L'objet de la présente note est d'abord de réparer cet oubli, et ensuite de montrer qu'il suffit, pour prouver ce résultat d'utiliser certaines idées très simples introduites par Katsylo et Bogomolov, dont l'utilisation ne semble pas définitivement limitée au cas des courbes. Nous ne traitons que le cas $n \geqslant 3$, le cas n=2 demandant un examen un peu plus attentif.

Je remercie chaleureusement Fedor Bogomolov qui m'a expliqué la démonstration de Katsylo.

1.- LE CAS DU DEGRE IMPAIR

On traite ici les schémas de Hilbert de courbes rationnelles de degré d impair. La démonstration proposée reprend le mécanisme utilisé par Ballico [Ba] pour prouver la rationalité de certaines variétés de courbes rationnelles, mécanisme qu'on peut décrire comme suit :

Critère : Soit $A \xrightarrow{f} B \xleftarrow{g} C$ un diagramme rationnel de variétés, où A et C sont birationnelles à des produits de B par des variétés rationnelles. Si la dimension de A majore celle de C et si C est rationnelle alors A l'est aussi.

On va montrer comment ce critère permet de traiter le cas qui nous occupe.

Notons S_d^n la variété des morphismes de degré d de $\mathbb{P}^1$ vers $\mathbb{P}^n$. Le groupe PGL(2) des automorphismes de $\mathbb{P}^1$ agit de façon équivariante sur S_d^n et sur $S_d^n \times \mathbb{P}^1$. Sur l'ouvert des plongements, le morphisme naturel de S_d^n vers H_d^n fait de H_d^n le quotient correspondant, tandis que parallèlement, la courbe tautologique C_d^n sur H_d^n est le quotient de l'ouvert correspondant de $S_d^n \times \mathbb{P}^1$.

Nous avons besoin de la construction analogue pour $n = 2$. Dans ce cas, il faut considérer l'ouvert des immersions à croisements normaux. Sur cet ouvert, H_d^2 est encore le quotient de S_d^2 par PGL(2). La courbe tautologique est singulière ; c'est sa normalisée qu'on note C_d^2 et qui est le quotient de l'ouvert adéquat de $S_d^2 \times \mathbb{P}^1$ par l'action de PGL(2). On prend donc pour A la variété H_d^{n-1}. Pour f, on prend la composition avec une projection (rationnelle linéaire) de $\mathbb{P}^n$ vers $\mathbb{P}^{n-1}$, disons de centre O. Enfin pour C on prend la variété S_d^{n-1}, qui est évidemment rationnelle de dimension $(n-1)(d+1)-1$. Comme l'indique Ballico, et c'est ici qu'intervient l'hypothèse sur d, S_d^{n-1} est birationnel au produit de H_d^{n-1} par PGL(2). Comme pour $d \geqslant 2$, la dimension de H_d^n majore celle de S_d^{n-1}, il ne reste qu'à voir que f identifie birationellement H_d^n au produit de H_d^{n-1} par un espace numérique:

soit $\pi : C_d^{n-1} \longrightarrow H_d^{n-1}$ la projection et $\mathcal{O}_C(1)$ l'image inverse sur C_d^{n-1} du fibré hyperplan sur $\mathbb{P}^{n-1}$. Comme $\mathbb{P}^n - \{O\}$ s'identifie au fibré géométrique $\mathbf{V}(\mathcal{O}_{\mathbb{P}^{n-1}}(-1))$, le morphisme f se factorise à travers le fibré géométrique $V = \mathbf{V}((\pi_* \mathcal{O}_C(1))^\vee)$. Le morphisme correspondant $\hat{f} : H_d^n \longrightarrow V$ est un plongement ouvert. Pour s'en convaincre, on contruit un morphisme inverse d'un ouvert de V vers H_d^n, comme suit. Soit C_V le produit fibré de C_d^{n-1} avec V sur H_d^{n-1}. Sur C_V on dispose d'un morphisme vers $\mathbb{P}^{n-1}$ et d'une section tautologique de $\mathcal{O}_{C_V}(1)$. Il y correspond un morphisme de C_V vers $\mathbb{P}^n - \{O\}$ qui induit l'inverse cherché de V vers H_d^n.

Remarque : Insistons sur le fait que cette méthode repose sur le fait que la courbe tautologique est birationnellement triviale. On traiterait de même les images de degré d de $\mathbb{P}^n$ dans $\mathbb{P}^N$, chaque fois que d et $n + 1$ sont premiers entre-eux. Par exemple la variété des surfaces de Veronese dans $\mathbb{P}^n$ est rationnelle.

II. - LE CAS DU DEGRE PAIR.

Pour traiter le cas du degré pair, on applique la méthode sans nom, recommandée par Bogomolov (cf. [D], § 4). Précisons d'abord le vocabulaire. Soit G un groupe algébrique réductif et V une représentation linéaire de G. On note V/G le quotient d'un ouvert adéquat de V par G (pour un ouvert canonique voir [DR]). On dit que la représentation V est presque libre si le stabilisateur du point général de V est réduit à l'identité. Enfin on dit que la variété X est n-rationnelle si $X \times \mathbf{A}^n$ est rationnelle.

On retiendra de la méthode sans nom le résultat suivant qu'on démontre en projetant $(V \times W)/G$ successivement sur V/G et sur W/G :

Proposition [BK] lemme 1.3. Soit G un groupe algébrique admettant une représentation linéaire V presque libre de dimension r telle que le quotient V/G soit rationnel. Alors pour toute représentation linéaire presque libre W de G le quotient W/G est r-rationnel.

On voudrait appliquer ce résultat aux représentations de $GL(H)$ dans $S^d H^\vee$ où H est de dimension deux, ou dans $S^d H^\vee \otimes M$, le groupe n'agissant pas sur M. Mais contrairement à ce qu'écrit Dolgachev [D], l'action de $GL(H)$ sur $S^d H$ n'est pas presque libre. L'astuce de Bogomolov-Katsylo consiste à changer de groupe. Le groupe $SL(H)$ opère aussi sur $S^d H^\vee \otimes M$ et, comme d est pair, le centre $\{-1, +1\}$ opère trivialement. On a donc une action de $PGL(H)$ sur $S^d H \otimes M$. On complète cette action par celle de $\mathbb{G}_m$ de sorte que le quotient de $S^d H^\vee \otimes M$ par $GL(H)$ est le même que par $PGL(H) \times \mathbb{G}_m$ et ce qu'on a gagné est que l'action de $PGL(H) \times \mathbb{G}_m$ sur $S^d H^\vee$, et a-fortiori $S^d H^\vee \otimes M$, est presque libre pour d pair, $d \geq 6$. Cela résulte du fait que dans un sextuplet général de $\mathbf{P}^1$, deux quadruplets différents ont des birapports différents. On note désormais G le groupe $PGL(H) \times \mathbb{G}_m$.

Il nous faut alors trouver une représentation presque libre de G à quotient rationnel. On prend pour ça l'espace $V = S^2 H^\vee \otimes M$ avec M de dimension trois. Les vecteurs généraux de V induisent un plongement de $H/\{\pm 1\}$ dans M de sorte que l'action est presque libre. Quant au quotient c'est la variété des coniques du plan projectif associé à M, variété qui est bien rationnelle.

Pour conclure on veut montrer que le quotient de $R = S^d H^\vee \otimes N$, avec N de dimension au moins 4, est rationnel. On écrit $N = D \oplus D'$ avec D et D' de dimension au moins deux et on pose $W = S^d H^\vee \otimes D$. Alors R/G est birationnel à $W/G \times \mathbf{A}^m$ où m est la dimension de $S^d H^\vee \otimes D'$ donc $m \geq 2d + 2 \geq 9$. D'après ce qu'on a vu, $W/G \times \mathbf{A}^9$ est rationnel donc aussi R/G.

Remarque : Cette méthode permet de traiter plus généralement la variété des images de degré $p(n+1)$ de $\mathbf{P}^n$ dans $\mathbf{P}^N$ pour p ou N suffisamment grand si on connait une représentation presque libre à quotient rationnel de $PGL(n+1) \times \mathbb{G}_m$. Cependant la démarche la plus raisonnable consiste à passer par le théorème de Katsylo et cette note est donc surtout un encouragement à étudier [K] et [Bo].

BIBLIOGRAPHIE

[Ba] E. BALLICO : On the rationality of the variety of smooth rational space curves with fixed degree and normal bundle. Proc. Amer. Math. Soc. 91 (1984) 510-512.

[Bo] F.A. BOGOMOLOV : Rationality of the moduli of hyperelliptic curves of arbitrary genus. Canadian Math. Soc. Conference Proceedings 6 (1986) 17-37.

[BK] F.A. BOGOMOLOV - P. KATSYLO : Rationality of certain quotient varieties. Math. Sbornik 126(168) (1985) 584-589. Math. of the USSR Sbornik 54 (1986) 571-576.

[DR] J. DIXMIER - M. RAYNAUD : Sur le quotient d'une variété algébrique par un groupe algébrique. Mathematical Analysis and Applications, Part A, 327-344, Adv. in Math. Suppl. Stud., 7a, Academic Press, New-York - London, 1981.

[D] I. DOLGACHEV : Rationality of fields of invariants. Algebraic Geometry Bowdoin 1985. Proc. of Symp. in Pure Math. 46, 2 (1987) 3-16.

[EV] D. EISENBUD - A. VAN DE VEN : On the Variety of Smooth rational space curves with given degree and normal bundle. Invent. Math. 67 (1982) 89-100.

[K] P. KATSYLO : Rationality of the moduli spaces of hyperelliptic curves. Izv. Akad. Nauk SSSR Ser. Mat. 48 (1984) 705-710.

Author's Address: Mathématiques
Université de Nice
Parc Valrose
F-06034 Nice Cedex
(France)

Cohérence et dualité sur le gros site de Zariski

par André Hirschowitz

« Citez toujours Grothendieck.
Si vous ne savez pas pourquoi,
Lui, sans doute, saurait. »
(d'après un proverbe bantou)

O. INTRODUCTION

Ce travail est une réflexion sur les images directes de faisceaux cohérents, essentiellement dans le cas propre et plat. On a toujours su que la formation de ces images directes ne commute pas aux changements de base et on s'est accomodé tant bien que mal de cet état des choses. L'arrangement le plus communément admis consiste à considérer les images directes dans une catégorie dérivée. Cela comporte, outre l'inconvénient de la lourdeur, celui qu'on ne peut plus guère considérer une image directe $R^i\pi_* \mathcal{F}$ toute seule.

On propose ci-dessous le point de vue qui nous semble le mieux adapté à l'étude des images directes, notamment en ce qu'il évite les deux écueils précédents. Le gros site de Zariski est une topologie (au sens de Grothendieck) à peine plus fine que celle de Zariski ([SGA IV]). En considérant les images directes comme des faisceaux dans cette topologie, on intègre dans la nouvelle notion la façon dont la formation de l'ancienne se comportait par changement de base. Ces nouvelles images directes sont des cohomologies de complexes qui ont leur existence propre sans référence à aucun complexe.

On appelle U-modules les faisceaux de modules sur le gros site de Zariski. On introduit la catégorie des U-modules U-cohérents, c'est la plus petite catégorie abélienne contenant les anciens faisceaux cohérents. Bien entendu, les images directes de faisceaux propres et plats sont U-cohérentes (l'hypothèse de platitude est sans doute superflue). Notre résultat principal assure que la U-dualité, c'est-à-dire le foncteur $\mathcal{H}om(.\,,\mathcal{O})$, est une involution exacte de la catégorie des U-modules U-cohérents, qui prolonge donc la dualité ordinaire des faisceaux localement libres. La U-dualité permet de donner au théorème de dualité relative une forme relativement peu barbare, du moins dans le seul cas qui me soit familier, celui des morphismes projectifs lisses cf. § 8.

Le lecteur hésitant se demande comment on peut être conduit à écrire des articles aussi théoriques quand on connait tant de questions concrètes sur les groupes de points dans $\mathbb{P}^2$: j'ai compris il y a longtemps que la démonstration du théorème de liaison-postulation relative (cf. [BH] 8.4) serait simplifiée ou du moins clarifiée par l'introduction d'une nouvelle notion de fonction constructible permettant notamment de définir le rang des images directes "même au-dessus des nilpotents". Ce projet était en sommeil lorsque j'ai reçu le manuscrit [GL] de Green et Lazarsfeld où apparaissent des idées voisines. J'ai alors entrepris d'écrire un travail sur les nouvelles fonctions constructibles [H1] avec son prolongement [H2]. C'est au cours de ces rédactions que j'ai trop mal supporté de ne pas bien comprendre les images directes et que j'ai trop ressenti le besoin des nouvelles images directes introduites ici.

Je remercie Mark Green ou Robert Lazarsfeld de m'avoir adressé [GL] avant sa parution. Par ailleurs, je remercie vivement O. Gabber qui a identifié la topologie que j'avais choisie comme le gros site de Zariski et la dualité que j'avais construite comme $\mathcal{H}om(.\,,\mathcal{O})$ et qui m'a indiqué la référence qui me manquait ([SGA IV]).

PLAN.

1. U-généralités.
2. U-modules cohérents et cocohérents
3. U-cohomologies
4. Structure des U-modules U-cohérents
5. U-projectifs
6. La bidualité
7. Le théorème
8. La dualité relative.

1. U-GENERALITES

1.1 U-topologie. On prend pour topologie (au sens de Grothendieck, cf. [SGA IV]) sur le schéma quasi-projectif X toute la catégorie des X-schémas quasi-projectifs avec, pour cribles, ceux engendrés par les recouvrements de Zariski. On appellera cette topologie U-topologie.

1.2 U-faisceaux. On appellera U-faisceau sur X tout faisceau sur X pour la U-topologie. Tout U-faisceau F sur X induit par restriction un faisceau noté F_X dans la topologie de Zariski. Inversement, tout faisceau de modules cohérents dans la topologie de Zariski induit par changement de base un U-faisceau sur X.

1.3 Le U-faisceau structural. Le faisceau structural $\mathcal{O}_X$ induit par changement de base un U-faisceau encore noté $\mathcal{O}_X$. On a donc $H^0(S, \mathcal{O}_X) = H^0(S, \mathcal{O}_S)$. On vérifie que, dans la catégorie des U-faisceaux, $\mathcal{O}_X$ est un anneau (cf. [EGA] 0.8.2).

1.4 U-modules. On appellera U-module sur X tout U-faisceau de modules sur le U-faisceau d'anneaux $\mathcal{O}_X$. Les U-modules constituent une catégorie abélienne dont les faisceaux de modules cohérents forment une sous-catégorie pleine (on vérifie facilement que $\mathcal{H}om(\mathcal{O},F)$ s'identifie à F) mais non abélienne.

1.5 U-cohérence. On dira qu'un U-module est (localement) U-cohérent s'il est (localement) dans la plus petite catégorie abélienne de U-modules contenant les faisceaux localement libres de type fini. Les faisceaux cohérents sont U-cohérents parce que les changements de base sont exacts à droite. On démontrera plus bas (5.6) que les U-modules localement U-cohérents sont U-cohérents.

1.6 Exactitude. Plus précisément le foncteur de changement de base qui à un module cohérent associe le U-module correspondant est exact à droite (mais pas à gauche). Ses foncteurs dérivés à gauche sont les foncteurs $S \longmapsto H^0(S, \mathcal{T}or_i^{\mathcal{O}_X}(\mathcal{O}_S, .))$.

2. U-MODULES COHERENTS, U-DUALITE ET U-MODULES COCOHERENTS

2.1 Définitions. Les U-modules cohérents sont donc par définition ceux induits par des modules cohérents. Si F est un U-module sur X, on définit le U-dual F^U de F comme le U-module $\mathcal{H}om(F,\mathcal{O})$: il s'agit donc du U-faisceau des morphismes de U-modules. Si F est cohérent, on voit à l'aide d'une présentation locale libre : $L_1 \to L_0 \to F \to 0$ que F^U admet une présentation locale $0 \to F^U \to L_0^{\vee} \to L_1^{\vee}$. On dira qu'un U-module est cocohérent si c'est le U-dual d'un U-module cohérent.

2.2 Le vectoriel $V_\varepsilon(F)$. Soit F un module cohérent sur X. On connait déjà le "fibré" vectoriel V(F) associé à F(cf. [EGA] II.1.7). On note $V_\varepsilon(F)$ le premier voisinage infinitésimal de la section nulle dans V(F), c'est-à-dire le sous-schéma défini par le carré de l'idéal de la section nulle. Il est amusant de constater que les opérations structurales qui font de V(F) un vectoriel sur X se restreignent à $V_\varepsilon(F)$ pour en faire un nouveau vectoriel sur X. Ces vectoriels sont munis d'une section linéaire tautologique τ ou τ_F de F^U définie comme suit sur V(F). Les sections de F sur V(F) forment le produit tensoriel $F \otimes SF$ où SF est l'algèbre symétrique des fonctions sur V(F). La forme linéaire tautologique est définie par le produit $F \otimes SF \longrightarrow SF$.

2.3 Proposition. Si F et G sont deux U-modules cohérents sur X, alors $\mathcal{H}om(F^U, G)$ est égal à $F \otimes G$.

Preuve. Définissons d'abord le morphisme naturel a de $\mathcal{H}om(F^U, G)$ vers $F \otimes G$. Il associe à f : $F^U \longrightarrow G$ l'image par f de τ_F dans G, qui est bien une section de $F \otimes G$.
Inversement la contraction définit un morphisme naturel c de $F \otimes G$ vers $\mathcal{H}om(F^U, G)$. On vérifie facilement que $a \circ c$ est l'identité. Il nous reste donc à voir que a est injectif. Pour cela soit f un morphisme de F^U vers G. Alors l'image de la section tautologique sur V(F) par f est a priori dans $G \otimes S(F)$. Mais du fait que τ est linéaire pour la structure vectorielle sur V(F), on vérifie que $f(\tau)$ est en fait dans $G \otimes F$. Et comme la restriction à $V_\varepsilon(F)$ induit l'identité sur $G \otimes F$, on a égalité entre $f(\tau)$ et a(f). On en déduit que si a(f) est nul alors f l'est parce que toute section de F^U se déduit de τ par un changement de base.

2.4 Corollaire. Tout faisceau cohérent s'identifie à son U-bidual.

2.5 Corollaire. Si F et G sont deux U-modules cohérents sur X, alors $\mathcal{H}om(F^U, G^U)$ est naturellement isomorphe à $\mathcal{H}om(G,F)$.

Preuve. La transposition définit un morphisme de $\mathcal{H}om(G,F)$ vers $\mathcal{H}om(F^U, G^U)$. Pour montrer que c'est un isomorphisme, on peut raisonner dans des ouverts où G admet une présentation libre $M \longrightarrow L \longrightarrow G \longrightarrow 0$. On a alors la présentation duale $0 \longrightarrow G^U \longrightarrow L^{\vee} \longrightarrow M^{\vee}$ pour G^U. Compte tenu de 2.3, cette présentation fait apparaître $\mathcal{H}om(F^U, G^U)$ comme le noyau de $F\otimes L^{\vee} \longrightarrow F\otimes M^{\vee}$, noyau que la présentation de G identifie à $\mathcal{H}om(G,F)$.

2.6 Corollaire. Si A,B,C sont trois U-modules cohérents et $A \longrightarrow B$ un épimorphisme, alors $\mathcal{H}om(A^U, C) \longrightarrow \mathcal{H}om(B^U, C)$ est surjectif.

Preuve immédiate.

2.7 Corollaire. Si N et N' sont deux U-modules cocohérents et $0 \to N \to L^1 \to L^2$, $0 \to N' \to L'^1 \to L'^2$ des présentations libres, alors tout morphisme de N dans N' provient d'un diagramme commutatif :

$$\begin{array}{ccc} L^1 & \longrightarrow & L^2 \\ \downarrow & & \downarrow \\ L'^1 & \longrightarrow & L'^2 \end{array}$$

Preuve. Soit F un U-module cohérent tel que N soit égal à F^U. Alors pour tout G cohérent, on a

$$\begin{aligned} \mathcal{H}om(F,G) &= \mathcal{H}om(G^U, N) \\ &= \ker(\mathcal{H}om(G^U, L^1) \longrightarrow \mathcal{H}om(G^U, L^2)) \\ &= \ker(\mathcal{H}om(L^{1\vee}, G) \longrightarrow \mathcal{H}om(L^{2\vee}, G)). \end{aligned}$$

Cela signifie que F est le conoyau de $L^{2\vee} \to L^{1\vee}$ dans la catégorie des modules, donc aussi dans celle des U-modules. Ainsi les suites $L^{2\vee} \to L^{1\vee} \to F \to 0$ et $L'^{2\vee} \to L'^{1\vee} \to F' \to 0$, avec $N' = F'^U$ sont exactes.

Il reste pour conclure à étendre à ces résolutions le morphisme transposé $F' \to F$, puis à transposer le diagramme commutatif obtenu.

2.8 Variante. Dans la situation du corollaire précédent, tout diagramme commutatif

$$\begin{array}{ccccc} 0 & \longrightarrow & N & \longrightarrow & L^1 \\ & & \downarrow & & \downarrow \\ 0 & \longrightarrow & N' & \longrightarrow & L'^1 \end{array}$$

provient d'un diagramme commutatif

$$\begin{array}{ccccccc} 0 & \longrightarrow & N & \longrightarrow & L^1 & \longrightarrow & L^2 \\ & & \downarrow & & \downarrow & & \downarrow \\ 0 & \longrightarrow & N' & \longrightarrow & L'^1 & \longrightarrow & L'^2 \end{array}$$

Preuve. On se ramène au cas cohérent dual comme dans la preuve précédente.

3. U-COHOMOLOGIES

3.1 Définition. On dira qu'un U-module sur X est une U-cohomologie s'il est localement dans X la cohomologie (au sens des U-modules) d'un complexe à trois termes de faisceaux localement libres.

3.2 Exemples.

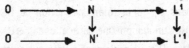

a) Soit $f : Y \to X$ un morphisme et F un module X-propre et X-plat sur Y. Alors les images directes supérieures $R^i f_* F$ peuvent être considérées comme des U-modules (U-cohérents mais non cohérents en général); et un théorème de Grothendieck (cf.[M]) assure que ce sont en fait des U-cohomologies.

b) Plus généralement, dans la situation précédente, si G est un second faisceau cohérent X-propre et X-plat sur Y, les groupes d'extension relatifs $\mathrm{Ext}^i{}_f(F,G)$ sont eux aussi des U-cohomologies (cf. [BPS], voir aussi [K],[L]).

3.3 Présentations. Une présentation (éventuellement locale) d'une U-cohomologie H est un complexe L^* à trois termes $L^0 \to L^1 \to L^2$ de modules localement libres dont H est la cohomologie (en degré un).

3.4 Déballage d'une U-cohomologie.
Soit L^* une présentation (locale) de la U-cohomologie H. On appellera déballage de H le diagramme de U-modules

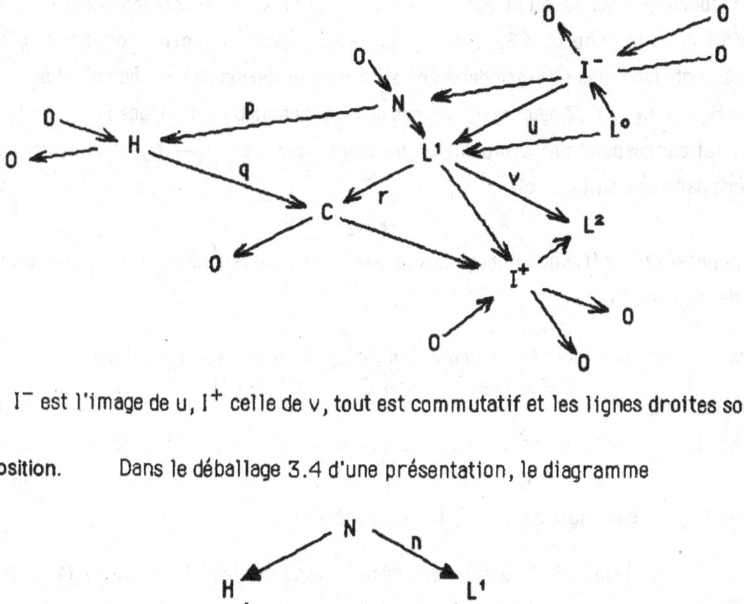

dans lequel I^- est l'image de u, I^+ celle de v, tout est commutatif et les lignes droites sont exactes.

3.5 Proposition. Dans le déballage 3.4 d'une présentation, le diagramme

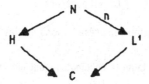

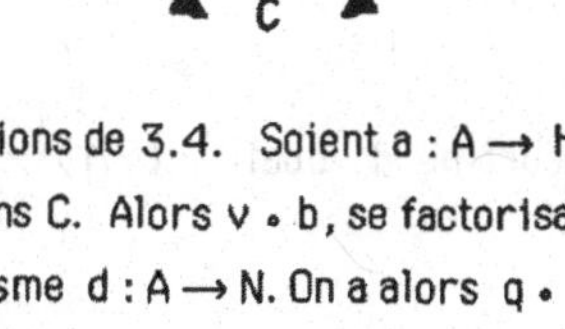

est cartésien.

Preuve. On reprend les notations de 3.4. Soient $a : A \to H$ et $b: A \to L^1$ deux morphismes de U-modules ayant même image dans C. Alors $v \circ b$, se factorisant à travers H, est nul. Par suite b se factorise à travers un morphisme $d : A \to N$. On a alors $q \circ p \circ d = r \circ b = q \circ a$. Comme q est un monomorphisme, on conclut que $a = p \circ d$. Il reste à observer que la factorisation est unique puisque n est un monomorphisme.

3.6 Proposition. Si H et H' sont deux U-cohomologies sur X munies de présentations L et L' alors tout morphisme de H vers H' provient localement d'un morphisme de L vers L'.

Preuve. On utilise les déballages (cf.3.4), avec les U-modules C et N pour H et C' et N' pour H'. Le morphisme donné $f: H \longrightarrow H'$ induit un morphisme de N vers C' qui se prolonge à L^1 (cf.2.6). Le morphisme de L^1 vers C' est nul sur I^- donc sur L^0 et se factorise par conséquent à travers C. Le morphisme de C vers C' se relève en morphisme de L^1 vers L'^1 et L^0 vers L'^0. D'après (3.5), les morphismes de N vers H' et L'^1 se factorisent à travers N' et le morphisme induit de N vers N' se prolonge (2.8) en morphisme de L^2 vers L'^2.

4. STRUCTURE DES U-MODULES U-COHERENTS

4.1 Proposition. Tout U-module U-cohérent est localement l'image d'un morphisme de U-cohomologies.

Preuve. Il nous faut montrer que la catégorie de ces images contient ses noyaux et ses conoyaux, c'est-à-dire que le (co)noyau d'un morphisme d'images de morphismes de U-cohomologies est de même nature.

On remarque d'abord que les noyau et conoyau d'un morphisme de U-cohomologies sont aussi des images. En effet compte tenu de (3.6) et grâce au cylindre d'application (cf [rH2]), tout morphisme de U-cohomologies s'insère dans une suite exacte longue de U-cohomologies.

Soient donc $f_i : H_i \longrightarrow K_i$, i=1,2 deux morphismes de U-cohomologies d'images I_i et $a : I_1 \rightarrow I_2$ un morphisme. Il lui correspond par composition un morphisme $a': H_1 \rightarrow K_2$. Comme on vient de le voir a' s'insère dans une suite exacte

$$A \longrightarrow H_1 \longrightarrow K_2 \longrightarrow B$$

de U-cohomologies et on vérifie que le noyau de a est l'image de A dans K_1 tandis que le conoyau de a est l'image de H_2 dans B.

4.2 Corollaire. Tout U-module U-cohérent F s'insère dans une suite exacte

$$0 \longrightarrow F \longrightarrow C^0 \longrightarrow C^1 \longrightarrow C^2 \longrightarrow 0$$

où les C^i sont cohérents.

Preuve. F s'injecte d'après ce qui précède dans une U-cohomologie qui elle-même s'injecte dans un faisceau cohérent C^0. Le quotient C^0/F est un nouveau U-module qui s'injecte à son tour dans un faisceau cohérent C^1 et le conoyau de $C^0 \rightarrow C^1$ est cohérent.

4.3 Corollaire. Tout U-module M admet des présentations locales $P_1 \rightarrow P_0 \rightarrow M \rightarrow 0$ avec P_1 et P_2 cocohérents.

Preuve. Résulte de même du fait que toute U-cohomologie est localement quotient d'un U-module cocohérent.

4.4 Remarque. Il résultera du théorème de dualité (7.1) que le noyau de $P_1 \rightarrow P_0$ est cocohérent.

5. U-PROJECTIFS

Dans ce paragraphe, on montre que les U-modules cocohérents sont projectifs dans le cas affine.

5.1 Proposition. Soit F cohérent sur X. On note z la section nulle de $V_\varepsilon(F): X \longrightarrow V_\varepsilon(F)$. Alors pour tout U-module U-cohérent G, $Hom(F^U,G)$ s'identifie naturellement au noyau de $z^*: G(V_\varepsilon(F)) \longrightarrow G(X)$.

Preuve. Il suffit de traiter le cas où X est affine, ce qu'on suppose dans la suite. Traitons d'abord le cas où G est cohérent. Alors $Hom(F^U,G)$ est égal à $H^0(X,F\otimes G)$ d'après (2.3). De l'autre côté $G(V_\varepsilon(F))$ est égal à $H^0(X,G\otimes(\mathcal{O}\oplus F))$ ou encore à $H^0(X;G)\oplus H^0(X,G\otimes F)$, z^* étant la projection sur le premier facteur.

Pour le cas général on utilise une présentation

$$0 \longrightarrow G \longrightarrow C^0 \longrightarrow C^1$$

avec C^0 et C^1 cohérents. On a un diagramme exact

$$\begin{array}{ccccccc}
 & 0 & & 0 & & 0 & \\
 & \downarrow & & \downarrow & & \downarrow & \\
\longrightarrow & Hom(F^U,G) & \longrightarrow & G(V_\varepsilon(F)) & \longrightarrow & G(X) & \longrightarrow 0 \\
 & \downarrow & & \downarrow & & \downarrow & \\
\longrightarrow & Hom(F^U,C^0) & \longrightarrow & C^0(V_\varepsilon(F)) & \longrightarrow & C^0(X) & \longrightarrow 0 \\
 & \downarrow & & \downarrow & & \downarrow & \\
\longrightarrow & Hom(F^U,C^1) & \longrightarrow & C^1(V_\varepsilon(F)) & \longrightarrow & C^1(X) & \longrightarrow 0
\end{array}$$

et le lemme du serpent permet de conclure.

5.2 Proposition. Si X est affine, les U-modules cocohérents sont projectifs dans la catégorie des U-modules U-cohérents.

Preuve. Soit F cohérent sur X et $A \longrightarrow B \longrightarrow 0$ un épimorphisme de U-modules U-cohérents. Il nous faut montrer que $Hom(F^U,A) \longrightarrow Hom(F^U,B)$ est surjectif.
Soient A_ε, B_ε les faisceaux induits par A et B sur $V_\varepsilon(F)$ et A_0, B_0 ceux induits sur X. On a le diagramme de suites exactes de faisceaux sur $V_\varepsilon(F)$:

$$\begin{array}{ccccccccc}
 & & & & K_\varepsilon & \longrightarrow & K_0 & & \\
 & & & & \downarrow & & \downarrow & & \\
0 & \longrightarrow & A' & \longrightarrow & A_\varepsilon & \longrightarrow & A_0 & \longrightarrow & 0 \\
 & & \downarrow{\scriptstyle u'} & & \downarrow{\scriptstyle u_\varepsilon} & & \downarrow{\scriptstyle u_0} & & \\
0 & \longrightarrow & B' & \longrightarrow & B_\varepsilon & \longrightarrow & B_0 & \longrightarrow & 0
\end{array}$$

et on veut montrer que $H^0(A') \longrightarrow H^0(B')$ est surjectif. Pour cela, comme le foncteur H^0 est exact, il suffit d'après le lemme de serpent de montrer que $H^0(K_\varepsilon) \longrightarrow H^0(K_0)$ est surjectif. Cela résulte du fait qu'on dispose d'un relèvement fonctoriel de (A_0,B_0) vers $(A_\varepsilon,B_\varepsilon)$ parce que $V_\varepsilon(F)$ se projette sur X.

5.3 Corollaire. Le U-dual d'un U-module U-cohérent est U-cohérent.

Preuve. Un faisceau U-cohérent admet donc des présentations locales par des U-modules cocohérents et on conclut par (2.4).

5.4 Remarque. Il résultera du théorème de dualité que les faisceaux cohérents sont des

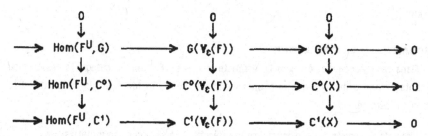

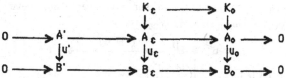

U-modules U-cohérents injectifs sur les schémas affines.

5.5 Proposition. Sur un schéma X quasi-projectif, tout U-module localement U-cohérent P est quotient d'un U-module cocohérent.

Preuve. On recouvre X par des complémentaires X_i de diviseurs H_i sur chacun desquels P est quotient d'un cocohérent F_i^U. Il nous suffit alors de prolonger chacun des morphismes $F_i^U \to P$ pour disposer du morphisme surjectif $\oplus F_i^U \to P$.

Pour prolonger un morphisme $m : F^U \to P$ à travers un diviseur H, on prolonge d'abord F puis on invoque (5.1). D'après [rH1] II.5.3, m se prolonge à valeurs dans un produit tensoriel $P \otimes \mathcal{O}(nH)$. Par suite m se prolonge de $F(nH)^U$ vers P.

5.6 Corollaire. Sur un schéma X quasi-projectif, les U-modules localement U-cohérents sont U-cohérents.

6. BIDUALITE : LE CAS AFFINE

6.0 Dans ce paragraphe, on note D le foncteur $F \longmapsto F^U$ sur la catégorie des U-modules U-cohérents et on démontre que sur un schéma affine X, le foncteur de bidualité D^2 est isomorphe à l'identité.

6.1 La catégorie des carrés. On considère la catégorie Q des carrés commutatifs

$$(q) \qquad \begin{array}{ccc} A & \xrightarrow{\ h\ } & B \\ {\scriptstyle g}\downarrow & & \downarrow{\scriptstyle d} \\ C & \xrightarrow{\ b\ } & E \end{array}$$

de faisceaux localement libres sur X.

Cette catégorie est munie d'un morphisme $\pi : Q \longrightarrow M$ dans celle des U-modules U-cohérents sur X : au carré ci-dessus, π associe naturellement le conoyau de $\ker h \longrightarrow \ker b$.

6.2 La dualité des carrés. On introduit un foncteur Δ de Q vers Q qui relève la U-dualité. Au carré ci-dessus, Δ associe naturellement le carré

$$\Delta(q) : \qquad \begin{array}{ccc} B^U \oplus E^U & \xrightarrow{\ (-h^\vee,0)\ } & A^U \\ {\scriptstyle \begin{pmatrix} id & -d^\vee \\ 0 & b^\vee \end{pmatrix}}\downarrow & & \downarrow{\scriptstyle id} \\ B^U \oplus C^U & \xrightarrow[\ (-h^\vee,-g^\vee)\ ]{} & A^U \end{array}$$

Lemme. Le foncteur Δ relève la U-dualité c'est-à-dire que les foncteurs $\pi \circ \Delta$ et $D \circ \pi$ sont isomorphes.

Preuve. Soit q le carré considéré plus haut. Alors l'image P de q par π s'insère dans le diagramme commutatif exact

$$
\begin{array}{ccccccc}
0 & \longrightarrow & F^{U} & \longrightarrow & A & \xrightarrow{h} & B \\
 & & {\scriptstyle k}\downarrow & & \downarrow{\scriptstyle g} & & \downarrow{\scriptstyle d} \\
0 & \longrightarrow & G^{U} & \longrightarrow & C & \xrightarrow{b} & E \\
 & & \downarrow & & & & \\
 & & P & & & & \\
 & & \downarrow & & & & \\
 & & 0 & & & &
\end{array}
$$

et P^{U} s'insère dans le diagramme dual

$$
\begin{array}{ccccccc}
 & & & & 0 & & \\
 & & & & \downarrow & & \\
 & & & & P^{U} & & \\
 & & & & \downarrow & & \\
E^{U} & \xrightarrow{b^{\vee}} & C^{U} & \longrightarrow & G & \longrightarrow & 0 \\
{\scriptstyle d^{\vee}}\downarrow & & \downarrow{\scriptstyle g^{\vee}} & & \downarrow & & \\
B^{U} & \xrightarrow{h^{\vee}} & A^{U} & \longrightarrow & F & \longrightarrow & 0
\end{array}
$$

On considère le carré de gauche du diagramme précédent comme un morphisme de complexes (de la ligne du haut vers la ligne du bas) et $G \longrightarrow F$ comme un morphisme induit en cohomologie.
On construit le cylindre d'application correspondant :

$$
0 \longrightarrow E^{U} \xrightarrow{\begin{pmatrix} -d^{\vee} \\ b^{\vee} \end{pmatrix}} B^{U} \oplus C^{U} \xrightarrow{(h^{\vee},\, g^{\vee})} A^{U} \longrightarrow 0 .
$$

La suite longue de cohomologie correspondante fait apparaître P^{U} comme conoyau du morphisme induit en cohomologie par la flèche verticale centrale dans le diagramme suivant

$$
\begin{array}{ccccc}
0 & \longrightarrow & B^{U} & \xrightarrow{-h^{\vee}} & A^{U} \\
{\scriptstyle 0}\downarrow & & \downarrow{\scriptstyle \begin{pmatrix} id \\ 0 \end{pmatrix}} & & \downarrow{\scriptstyle id} \\
E^{U} & \xrightarrow{\begin{pmatrix} -d^{\vee} \\ b^{\vee} \end{pmatrix}} & B^{U} \oplus C^{U} & \xrightarrow{(-h^{\vee},\, -g^{\vee})} & A^{U}
\end{array}
$$

Par suite P^{U} est bien l'image dans M du carré $\Delta(q)$.

De même un morphisme $\mu : q \longrightarrow q'$ de carrés induit un morphisme entre le diagramme

$$
\begin{array}{ccccc}
0 & \longrightarrow & B^{U} & \longrightarrow & A^{U} \\
 & & \downarrow & & \\
E^{U} & \longrightarrow & B^{U} \oplus C^{U} & \longrightarrow & A^{U} \\
 & & \downarrow & & \\
E^{U} & \longrightarrow & C^{U} & & \\
 & & \downarrow & & \\
B^{U} & \longrightarrow & A^{U} & &
\end{array}
$$

et son homologue correspondant à q'. Il y correspond un morphisme $\bar{\mu}$ entre la suite de cohomologie

$$
H \xrightarrow{u} I \longrightarrow G \xrightarrow{v} F
$$

et son homologue associée à q'.

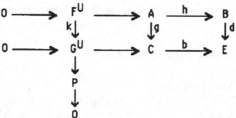

Le morphisme associé à μ par $\pi \circ \Delta$ est celui induit par $\bar{\mu}$ sur le conoyau de u tandis que celui associé par $D \circ \pi$ est celui induit par $\bar{\mu}$ sur le noyau de v; et ils sont bien égaux.

6.3 La bidualité des carrés.

Lemme. Les foncteurs $\pi \circ \Delta^2$ et π sont isomorphes.

Preuve. On calcule $\Delta^2(q)$ et on trouve

$$\begin{array}{ccc}
A \oplus A & \xrightarrow{\begin{pmatrix} h & 0 \\ 0 & 0 \end{pmatrix}} & B \oplus E \\
\Delta^2(q) : \begin{pmatrix} id & -id \\ 0 & -h \\ 0 & -g \end{pmatrix} \Big\downarrow & & \Big\downarrow \begin{pmatrix} id & 0 \\ 0 & id \end{pmatrix} \\
A \oplus B \oplus C & \xrightarrow{\begin{pmatrix} h & -id & 0 \\ 0 & d & -b \end{pmatrix}} & B \oplus E
\end{array}$$

Exhibons d'abord un morphisme naturel de $\pi(q)$ vers $\pi\, \Delta^2(q)$: on injecte F^U dans $A \oplus A$ par $\begin{bmatrix} -i \\ -i \end{bmatrix}$ où i est l'injection $F \to A$, et on injecte G^U dans $A \oplus B \oplus C$ par $\begin{bmatrix} 0 \\ 0 \\ i' \end{bmatrix}$ où i' est l'injection $G^U \longrightarrow C$. Ces deux injections sont naturelles, compatibles entre elles, et envoient F^U et G^U dans les noyaux des flèches horizontales de $\Delta^2(q)$. On a ainsi un morphisme naturel de $\pi(q)$ vers $\pi\, \Delta^2(q)$.

Pour voir que c'est un isomorphisme voyons d'abord qu'il est injectif. Si γ est une section de G^U dont l'image dans $A \oplus B \oplus C$ est de la forme $(\alpha - \alpha', -h(\alpha'), -g(\alpha'))$ avec $h(\alpha)=0$, alors α et α' sont égaux et γ est l'image de $-\alpha$, donc donne zéro dans P^U. Montrons maintenant qu'il est surjectif : soit (α,β,γ) une section (choisie après un éventuel changement de base) de $A \oplus B \oplus C$ dans le noyau de $\begin{bmatrix} h & -id & 0 \\ 0 & d & -b \end{bmatrix}$ c'est-à-dire avec $h(\alpha) = \beta$, $d\beta = b\gamma$.

Alors on observe que (α,β,γ) est la somme de l'image de $(0,\alpha)$ avec $(0,0,\gamma - g(\alpha))$.

Corollaire. Les foncteurs $D^2 \circ \pi$ et π sont isomorphes.

6.4 L'involutivité dans le cas affine. Pour conclure de l'énoncé précédent que D^2 est isomorphe à l'identité, il reste à montrer que le foncteur π est essentiellement surjectif, sur les objets et sur les flèches. C'est ce qu'on va prouver maintenant. La surjectivité sur les objets résulte du fait (5.5) que tout U-module U-cohérent admet une présentation globale par des U-modules cocohérents, lesquels admettent les résolutions libres dont on a besoin.

Voyons maintenant la surjectivité sur les flèches. Compte tenu de (5.2), tout morphisme de U-cohérents se relève en carré commutatif de U-modules cocohérents. En dualisant il nous faut donc prouver que tout carré commutatif de faisceaux cohérents se relève en carré commutatif de résolutions. Pour cela, il suffit bien sûr de traiter le cas des zéro-résolutions. Le point important est qu'on choisit la résolution en fonction du problème.

Fixons les notations. Soit :

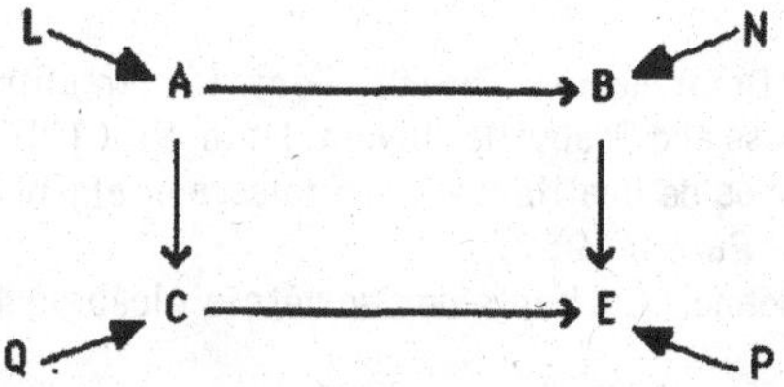

le diagramme donné avec A, B, C, E cohérents et L,N,P,Q localement libres. Choisissons d'abord une factorisation de N ⟶ E via P. Quitte à accroître N, on peut supposer, en notant F le produit fibré de B et P sur E, que N ⟶ F est surjectif. Maintenant on factorise Q ⟶ E à travers P et L⟶C à travers Q. Enfin L ⟶ B et L ⟶ P définissent un morphisme L ⟶ F qu'on factorise via N.

7. LE THEOREME DE U-DUALITE

Ici on conclut la démonstration du

7.1 Théorème. La U-dualité est l'unique (à isomorphisme près) foncteur involutif exact sur la catégorie fibrée des U-modules U-cohérents qui prolonge la dualité des faisceaux localement libres.

Preuve. L'involutivité a été vue au §6 dans le cas affine, ce qui est suffisant puisqu'on dispose d'un morphisme naturel de l'identité vers D^2. Maintenant c'est un phénomène général qu'un isomorphisme additif entre deux catégories additives qui sont abéliennes est nécessairement exact.

7.2 Corollaire. Sur un X affine, les U-modules cohérents sont injectifs.

7.3 Corollaire. Toute suite U-exacte (i.e. exacte comme suite de U-modules) de faisceaux cohérents est localement scindée.

8. DUALITE RELATIVE

Soit $p: X \longrightarrow S$ un morphisme projectif lisse de dimension relative n et soient E et F deux modules cohérents S-plats sur X. La dualité entre $Ext_p^i(E,F)$ et $Ext_p^{n-i}(F, E \otimes \omega_{X/S})$ est déjà mentionnée par Grothendieck [G] au moins dans le cas absolu où S est le Spec d'un corps. Ce cas a été redémontré et utilisé par Drezet-Le Potier [DLP].

Le cas relatif avec F localement libre est bien connu dans le langage des catégories dérivées (cf. [rH2]). L'énoncé analogue pour le cas où F est quelconque s'en déduit par récurrence sur la dimension homologique. En passant aux U-cohomologies, on obtient le :

Théorème de dualité relative : Dans les hypothèses ci-dessus, les U-modules $Ext_p^i(E,F)$ et $Ext_p^{n-i}(F, E \otimes \omega_{X/S})^U$ sont naturellement isomorphes.

Bibliographie

[BP] **C. Banica, M. Putinar, G. Schumacher,** Variation der globalen Ext in Deformationen kompakter komplexer Raüme, Math. Ann. 250 (1980), 135-155

[BH] **J. Brun, A. Hirschowitz,** Le problème de Brill-Noether pour les idéaux de $\mathbb{P}^2$, Ann. Sci. ENS Paris 20 (1987), 171-200

[DLP] **J.M. Drezet, J. Le Potier,** Fibrés stables et fibrés exceptionnels sur $\mathbb{P}^2$, Ann. Sci. ENS Paris 18 (1985), 193-244

[EGA] cf. [GD]

[GL] **M. Green, R. Lazarsfeld**, Deformation Theory, Generic Vanishing Theorems and some Conjectures of Enriques, Catanese and Beauville, Invent. Math. 90 (1987), 389-407

[G] **A. Grothendieck**, Théorèmes de dualité pour les faisceaux algébriques cohérents, Sém. Bourbaki 149, Secr. Math. IHP Paris (1957)

[GD] **A. Grothendieck, J. Dieudonné**, Eléments de Géométrie Algébrique, Publ. Math. IHES 8 (1961) et 11 (1961)

[rH1] **R. Hartshorne**, Algebraic Geometry, Graduate texts in Math., vol.52, Berlin-Heidelberg - New York: Springer 1977

[rH2] **R. Hartshorne**, Residues and Duality, Lect. Notes in Math., 20, Springer Verlag (1966)

[aH1] **A. Hirschowitz**, Nouvelles fonctions constructibles et rang des images directes. Dans ce volume

[aH2] **A. Hirschowitz**, Préjugés favorables pour les stratifications cohomologiques, (en préparation)

[K] **S.L. Kleiman**, Rudiments of a General Base Change Theory, Appendix in [LK]

[L] **H. Lange**, Universal Families of Extensions, Journal of Algebra 83 (1983), 101-112

[LK] **K. Lonsted, S.L. Kleiman**, Basics on Families of Hyperelliptic Curves, Compos. Math. 38 (1979), 83-111

[M] **D. Mumford**, Abelian Varieties, Oxford Univ. Press (1970).

[SGA IV] Théorie des topos et cohomologie étale des schémas. Dirigé par M.Artin, A. Grothendieck et J.L.Verdier. Lect. Notes in Math., 269, Springer-Verlag (1972).

Nouvelles fonctions constructibles et rang des images directes

A. Hirschowitz

Si F est un faisceau cohérent S-propre et S-plat sur $S \times X$, on lui associe la stratification de S par les sous-ensembles définis par des inégalités de la forme $h^i(F(s)) \geq j$ où h^i désigne la dimension du groupe de cohomologie H^i. De tout temps (cf. [B]) on a muni certaines de ces strates de structures schématiques. Cependant, jusqu'ici ces constructions se faisaient au coup par coup. On propose ici un texte de référence pour la structure schématique des strates cohomologiques. Le point de vue choisi consiste à considérer les $h^i(F(s))$ comme des "nouvelles fonctions semi-continues supérieurement (scs)", ces nouvelles fonctions f étant par définition des suites de schémas correspondant aux inégalités $f \geq j$. Ce subterfuge syntaxique est sans rapport avec la structure schématique choisie dans le cas général qui est celle suggérée par le travail de M. Green et R. Lazarsfeld ([GL].). On complète leur contribution à ce problème en montrant que la structure schématique qu'ils introduisent, qui dépend a-priori d'un complexe représentant l'image directe totale de F, en est en fait indépendante (§ 2.3). Au § 3, on essaie de justifier cette structure schématique en démontrant à peu de choses près que c'est la plus fine (i.e. la moins nilpotente) qui soit compatible aux changements de base. Cette preuve est une application convaincante de la théorie des monômes standard.

La formule de Riemann-Roch, par exemple, met en jeu des différences de fonctions scs, c'est pourquoi on introduit ici le symétrisé du monoïde des (nouvelles) fonctions scs, qui est le groupe des (nouvelles) fonctions constructibles. La mauvaise surprise est alors que le monoïde ne s'injecte pas dans son symétrisé. Ce fâcheux phénomène ôte à ce symétrisé beaucoup de son intérêt et empêche qu'on puisse parler confortablement d'un ordre sur les (nouvelles) fonctions constructibles ou scs.

Je remercie Joël Briançon et Philippe Maisonobe pour des conversations utiles pendant la préparation de ce travail.

1. - NOUVELLES FONCTIONS CONSTRUCTIBLES

1.1 Anciennes fonctions sci. A toute fonction f semi-continue inférieurement d'un schéma X dans $\mathbf{Z}$ on associe la suite des sous-schémas (réduits) de X définis par les inégalités $f < i$. Cette suite croissante caractérise la fonction f, tout comme la suite décroissante f_i des idéaux correspondants.

1.2 Nouvelles fonctions sci. On appellera désormais (nouvelle) fonction sci de Y dans **Z** toute suite $(f_i)_{i\in \mathbf{Z}}$ décroissante d'idéaux cohérents de X vérifiant

i) $f_n = \mathcal{O}$ pour n suffisamment petit ;
ii) $f_n = 0$ pour n suffisamment grand ;
iii) pour $p, q \geq 0$, $f_n = \mathcal{O}$ implique $f_{n+p+q} \subset f_{n+p} \cdot f_{n+q}$.

Il convient sans doute de justifier la condition iii) : d'une part elle est vérifiée par l'exemple (1.4), qui est aussi le seul auquel on soit vraiment attaché et d'autre part elle intervient pour limiter les dégats en matière de régularité de l'addition (§ 1.13-1.15).

1.3 Fonction classique associée. Si f est une fonction sci, on définit, pour x dans X, f(x) comme le plus petit entier i tel que l'idéal défini par x ne contienne pas f_i.

1.4 Exemple. Soit $u : E \longrightarrow F$ un morphisme de faisceaux localement libres. Par définition le rang de u, noté rku, est la suite des idéaux définis par les puissances extérieures $\Lambda^0 u, \Lambda^1 u, \ldots$. On a $\mathrm{rku} = \mathrm{rku}^\vee$.

1.5 Notations. Si $f = (f_i)$ est une nouvelle fonction sci, on notera encore $\{f < i\}$ (ou bien sûr $\{f \leq i-1\}\{i-1 \geq f\}\{i > f\}$) le sous-schéma d'idéal f_i. Ses points sont les x vérifiant $f(x) < i$.

1.6 Changement de base. Comme les sous-schémas, les fonctions constructibles constituent un faisceau sur la topologie de Zariski. Et comme sur les sous-schémas, les changements de base quelconques opèrent sur les fonctions constructibles. Si $m : X \longrightarrow Y$ est un morphisme et f une fonction sci sur Y, on notera m^*f la fonction correspondante sur X. Par exemple si $u : E \longrightarrow F$ est un morphisme de faisceaux localement libres sur Y, on a évidemment $\mathrm{rk}(m^*u) = m^*(\mathrm{rku})$.

1.7 Addition. Si f et g sont deux fonctions sci sur X à valeurs dans **Z**, on définit leur somme f + g par les formules

$$(f+g)_i = \bigoplus_{p+q=i} f_p \cdot g_q .$$

1.8 Pourquoi ? La formule précédente a été choisie pour vérifier :

Addition des rangs. Si $u : E \longrightarrow F$ et $v : G \longrightarrow H$ sont deux morphismes de faisceaux localement libres sur X, alors on a $\mathrm{rk}(u \oplus v) = \mathrm{rku} + \mathrm{rkv}$, où $u \oplus v$ désigne le morphisme diagonal de $E \oplus G$ dans $F \oplus H$.

1.9 Fonctions localement constantes. On définit pour n dans $\mathbf{Z}$ la fonction sci notée abusivement n par $n_i = \mathcal{O}$ pour $i \leqslant n$ et $n_i = 0$ pour $i > n$. On vérifie que pour toute fonction sci f, on a $(f+n)_i = f_{i-n}$. La fonction 0 est évidemment élément neutre. Les fonctions localement constantes sont les seules à admettre un opposé.

1.10 Addition et changement de base. L'addition commute aux changements de base quelconques, ce que rappelle la formule : $m^*(f+g) = m^*f + m^*g$.

1.11 Associativité - Commutativité. Ces propriétés sont évidentes. On a donc

$$(f+g)+h = f+(g+h) \text{ et } f+g = g+f.$$

1.12 Warning. On peut donc définir nf pour n entier et f semi-continue inférieurement. Il faut cependant être prévenu qu'on n'a pas en général $\{f < i\} = \{nf < ni\}$, ni $\{f \leqslant i\} = \{nf \leqslant ni\}$. Par exemple si f est le rang de la matrice scalaire zI, où I est l'identité de rang n, l'idéal de $\{f \leqslant 1\}$ est (z^2) et celui de $\{2f \leqslant 2\}$ est (z^3).

1.13 Régularité en général. Le défaut majeur de l'addition qu'on a introduite est qu'elle n'est pas régulière en général, c'est-à-dire qu'on n'a pas : $f+g = f+h \Rightarrow g = h$.

Il est facile de construire des contre-exemples sur un schéma ponctuel non réduit. L'exemple suivant, proposé par Joël Briançon, est plus significatif puisque X y est non-singulier:

$$g = (\mathcal{O} \supset (x^2, xy, y^2) \supset 0)\,;$$
$$h = (\mathcal{O} \supset (x^2, y^2) \supset 0)\,;$$
$$f = (\mathcal{O} \supset (x, y) \supset 0)\,.$$

On a $f+g = f+h = (\mathcal{O} \supset (x,y) \supset (x^3, x^2y, xy^2, y^3) \supset 0)$.

Cependant, les fonctions localement constantes, admettant des opposées, sont évidemment régulières.

1.14 Régularité en dimension un. Si X est non singulier de dimension un, alors l'addition est régulière. En effet appelons élémentaire toute fonction sci f vérifiant $0 \leqslant f \leqslant 1$ (c'est-à-dire $\{f \leqslant 1\} = X = \{0 \leqslant f\}$). Alors toute fonction sci sur X à valeurs dans $\mathbf{N}$ se décompose localement de manière unique en somme de fonctions élémentaires : si on note e_i la fonction élémentaire en $z = 0$ définie par $\mathcal{O} \supset (z^i) \supset 0$ et si $f = (\mathcal{O}, (z^{n_1}), \ldots, (z^{n_i}), \ldots)$, alors f est la somme des $e_{n_{i+1}-n_i}$. C'est ici qu'on utilise la condition iii) de 1.2; le présent énoncé serait faux sans cette condition.

1.15 Semi-régularité. Il résulte de l'énoncé précédent et de (1.10) que si $f+g$ égale $f+h$, alors g et h ont mêmes traces sur les courbes lisses. Si X est non singulier cela signifie (cf. [T] Prop. 0.4) que g et h ont même clôture intégrale, la clôture intégrale de g étant la suite des clôtures intégrales des g_i.

1.16 Nouvelles fonctions constructibles. Le groupe des (nouvelles) fonctions constructibles sur X, à valeurs dans $\mathbf{Z}$ est le symétrisé $C(X)$ du monoïde $I(X)$ des fonctions sci à valeurs dans $\mathbf{Z}$. Un grave défaut de ce groupe est que $I(X)$ ne s'y injecte pas.

1.17 Nouvelles fonctions scs. On ne peut donc pas définir les nouvelles fonctions scs dans $C(X)$ sous peine de perdre de l'information. On appellera désormais fonction scs de X dans $\mathbf{Z}$ toute suite $(f_i)_{i\in\mathbf{Z}}$ croissante d'idéaux cohérents de X telle que la suite notée $-f$ et définie par $(-f)_i = f_{-i}$ soit une fonction sci. De même, si g est sci on notera $-g$ la fonction scs telle que $-(-g) = g$.
La fonction classique associée à une fonction scs f est définie par $f(x) = -((-f)(x))$. Les fonctions scs sur X constituent un monoïde $S(X)$ isomorphe à celui $I(X)$ des fonctions sci.
L'injection relevante de $\mathbf{Z}$ dans $S(X)$ est celle qui commute à la formation de la fonction classique associée.
Si f est une fonction scs (ou sci) on notera f' la fonction constructible associée.

1.18 Différences. Si f est scs et g est sci, on notera $f-g$ la somme $f+(-g)$ et de même $g-f$ la somme $g+(-f)$, et on étendra évidemment cette convention au cas de combinaisons de plusieurs termes.

Si maintenant f et g sont deux fonctions disons scs, la différence $f-g$ n'a de sens que comme fonction constructible. Cependant on convient d'écrire $f-g \equiv h-l$ si les fonctions scs $f+l$ et $g+h$ sont égales. Il convient évidemment de manipuler cette convention (qu'on étend bien sûr aux combinaisons de plusieurs termes) avec prudence.

1.19 Exemple. Soit X un schéma non singulier et $\ldots E^i \xrightarrow{d^i} E^{i+1} \longrightarrow \ldots$ un complexe de faisceaux localement libres sur X. On définit le rang de la cohomologie du complexe E^* par la formule

$$h^i(E^*) = r_i - rkd^i - rkd^{i-1}$$

où r_i désigne le rang de E^i. Si le complexe est borné, on a la formule évidente

$$\Sigma(-1)^i h^i(E^*) \equiv \Sigma(-1)^i r_i .$$

1.20 Changement de base. D'après 1.10, le changement de base s'étend aux fonctions constructibles et fait apparaître le groupe des fonctions constructibles comme un foncteur de la catégorie des schémas dans celle des groupes abéliens.

1.21 Dérivation. Si f est une fonction sci sur X, sa dérivée $d_x f$ au point x est une fonction sci sur l'espace tangent et définie de la façon suivante :
si $f(x) = n$, l'idéal de $\{d_x f < p\}$ est celui engendré par les formes initiales de degré p de celui de $\{f < n+p\}$. La dérivation ainsi définie est additive de sorte qu'elle s'étend aux fonctions constructibles. Elle a la propriété que la dérivée du rang d'un morphisme de faisceaux localement libres est égale au rang de la dérivée du morphisme : ici il s'agit plutôt de la dérivée intrinsèque du morphisme, au sens de Porteous (cf. [jmB]).

2. - RANG DES FAISCEAUX U-COHERENTS.

2.1 Rang des faisceaux cohérents. Si F est un faisceau cohérent, on définit son rang rkF à l'aide de ses idéaux de Fitting.
C'est une fonction scs dont la formation commute aux changements de base
(i.e. $rk(m^*F) = m^*(rkF)$).

2.2 Rang des faisceaux U-cohérents. Soit F un faisceau U-cohérent (cf. [H]). Alors F admet des résolutions locales de la forme

$$0 \longrightarrow F \longrightarrow C^0 \longrightarrow C^1 \longrightarrow C^2 \longrightarrow 0$$

où les C^i sont cohérents.

La fonction constructible $rkC^0 - rkC^1 + rkC^2$ ne dépend pas de la résolution locale choisie. En effet soit

$$0 \longrightarrow F \longrightarrow D^0 \longrightarrow D^1 \longrightarrow D^2 \longrightarrow 0$$

une seconde résolution de F sur le même ouvert.

Alors, du fait que les faisceaux cohérents sont injectifs dans la catégorie des faisceaux U-cohérents ([H]), l'identité de F provient d'un morphisme entre ces deux résolutions. Et le cylindre d'application de ce morphisme est donc exact

$$0 \longrightarrow C^0 \longrightarrow D^0 \oplus C^1 \longrightarrow D^1 \oplus C^2 \longrightarrow D^2 \longrightarrow 0 .$$

Comme les suites exactes de U-faisceaux cohérents sont localement scindées ([H] 7.3), les faisceaux $D^1 \oplus C^2 \oplus C^0$ et $D^0 \oplus C^1 \oplus D^2$ sont localement isomorphes. D'où l'égalité $rkC^0 - rkC^1 + rkC^2 \equiv rkD^0 - rkD^1 + rkD^2$. On a ainsi montré qu'il existe une unique fonction constructible qu'on note $rk'F$ vérifiant pour toute résolution comme plus haut

$$rk'F = rkC^0 - rkC^1 + rkC^2 .$$

2.3 Rang des U-cohomologies. Si F est une U-cohomologie (cf. [H]), alors F admet des résolutions locales de la forme

$$0 \longrightarrow F \longrightarrow C^0 \longrightarrow L^1 \longrightarrow C^2 \longrightarrow 0$$

avec C^0 et C^2 cohérents et L^1 localement libre (cf.[H] 2.2). Alors on montre comme en 2.2 que la fonction scs $rkC^0 - rkL^1 + rkC^2$ ne dépend pas de la résolution de ce type choisie. En effet si

$$0 \longrightarrow F \longrightarrow D^0 \longrightarrow M^1 \longrightarrow D^2 \longrightarrow 0$$

est une résolution de la même forme, on a comme en 2.2 :

$$rkC^0 + rkM^1 + rkC^2 = rkD^0 + rkL^1 + rkD^2 .$$

De la régularité des fonctions localement constantes on déduit l'égalité de fonctions scs :

$$rkC^0 - rkL^1 + rkC^2 = rkD^0 - rkM^1 + rkD^2 .$$

On a ainsi montré qu'il existe une unique fonction scs qu'on note rkF vérifiant, pour toute résolution locale comme plus haut, l'égalité entre fonctions scs

$$rkF = rkC^0 - rkL^1 + rkC^2 .$$

Cette notion est compatible avec la définition donnée en (2.2). En effet si F est la cohomologie de $L^0 \xrightarrow{u} L^1 \xrightarrow{v} L^2$, alors F admet la résolution $0 \longrightarrow F \longrightarrow \operatorname{coker} u \longrightarrow L^2 \longrightarrow \operatorname{coker} v \longrightarrow 0$ qui permet de constater que rkF et $rkh^1(L^*)$ sont égaux.

2.4 Rang des images directes. Les images directes $R^i \pi_* F$ des faisceaux cohérents propres et plats sont des U-cohomologies d'après un théorème de Grothendieck (cf. [M] §5). On a donc défini leur rang au passage.

2.5 Rang et suites exactes de faisceaux U-cohérents. Si F^* est une suite exacte bornée de faisceaux U-cohérents, on a l'égalité

$$\Sigma (-1)^i rk' F^i = 0 .$$

En effet, du fait de l'injectivité des cohérents, on peut construire un complexe double C^{**} borné tel que chaque complexe C^{i*} soit une résolution à trois termes de F^i, comme en 2.2. Le complexe simple associé D^* est exact ([K], Appendix, Prop.1). Comme toute suite exacte de U-faisceaux cohérents est localement scindée, on a $\Sigma (-1)^i rk D^i \equiv 0$, c'est-à-dire $\Sigma (-1)^{i+j} rk C^{ij} \equiv 0$ ou encore $\Sigma (-1)^i rk' F^i = 0$.

2.6 Rang et suites exactes de U-cohomologies. Si F^* est une suite exacte bornée de U-cohomologies, on a l'égalité

$$\Sigma (-1)^i rk F^i \equiv 0 .$$

En effet, en utilisant des résolutions

$$0 \longrightarrow F^i \longrightarrow C^{i_0} \longrightarrow L^{i_1} \longrightarrow C^{i_2} \longrightarrow 0$$

comme en 2.3, on trouve comme en 2.5,

$$\Sigma\,(-1)^i\,(\mathrm{rk}\,C^{i_0} - \mathrm{rk}\,L^{i_1} + \mathrm{rk}\,C^{i_2}) \equiv 0 \;.$$

2.7 Rang et U-dualité. Pour un U-faisceau U-cohérent, on a $\mathrm{rk}'F = \mathrm{rk}'F^U$. Si, de plus F est une U-cohomologie, on a $\mathrm{rk}\,F = \mathrm{rk}\,F^U$. En effet, on commence par s'en convaincre pour F cohérent : si $u : L_1 \longrightarrow L_0$ est une présentation locale de F, on a $\mathrm{rk}\,F = \mathrm{rk}\,L_0 - \mathrm{rk} u$. Pour F^U, on a la résolution $0 \longrightarrow F^U \longrightarrow \check{L}_0 \longrightarrow \check{L}_1 \longrightarrow C \longrightarrow 0$ où C est le faisceau cohérent conoyau de $u^\vee : \check{L}_0 \longrightarrow \check{L}_1$. On a donc $\mathrm{rk}\,F^U = \mathrm{rk}\,L_0 + \mathrm{rk}\,C - \mathrm{rk}\,L_1$,

avec $\mathrm{rk}\,C = \mathrm{rk}\,L_1 - \mathrm{rk}\,u^\vee = \mathrm{rk}\,L_1 - \mathrm{rk} u$ (cf. 1.4). D'où $\mathrm{rk}\,F^U = \mathrm{rk}\,F$.
Dans le cas général, on prend une résolution locale

$$0 \longrightarrow F \longrightarrow C^0 \longrightarrow C^1 \longrightarrow C^2 \longrightarrow 0 .$$

On a $\mathrm{rk}'\,F = \mathrm{rk}\,C^0 - \mathrm{rk}\,C^1 + \mathrm{rk}\,C^2$.

Par exactitude de la U-dualité, ou a une suite exacte :

$$0 \longrightarrow C^{2U} \longrightarrow C^{1U} \longrightarrow C^{0U} \longrightarrow F^U \longrightarrow 0$$

et d'après 2.5 :

$$\mathrm{rk}'\,F^U = \mathrm{rk}\,C^{0U} - \mathrm{rk}\,C^{1U} + \mathrm{rk}\,C^{2U} = \mathrm{rk}\,C^0 - \mathrm{rk}\,C^1 + \mathrm{rk}\,C^2 .$$

Ce raisonnement s'adapte sans difficulté au cas des U-cohomologies.

2.8 Rang et dualité relative. Si A et B sont S-propres et S-plats sur X, on sait que $\mathrm{Ext}^i_S(A, B)$ et $\mathrm{Ext}^{n-i}_S(B, A \otimes \omega_{X/S})$ sont U-duaux, X étant lisse de dimension relative n sur S (cf.[H]). Par suite, d'après (2.7), on a

$$\mathrm{rk}\,\mathrm{Ext}^i_S(A,B) = \mathrm{rk}\,\mathrm{Ext}^{n-i}_S(B, A \otimes \omega_{X/S}) .$$

3. - UN EXERCICE DE THEORIE DES MONOMES STANDARD

On a choisi de définir le rang d'une U-cohomologie en s'inspirant du calcul matriciel. Une autre approche aurait pu être la suivante : on veut d'une part que le rang soit invariant par changement de base. Par ailleurs, bien que le groupoïde des U-cohomologies ne soit pas un modèle (cf.[H']), il présente des objets plus universels que les autres à savoir les schémas de complexes. Et on aurait pu exiger que dans les schémas de complexes, le rang de la cohomologie définisse des sous-schémas réduits. Heureusement la notion qu'on a choisie a également cette vertu :

3.1 Théorème. Soit $E_{p,q,r}$ le schéma des couples d'une matrice (p, q) et d'une matrice (q, r) de produit nul (cf. [DC-S]). Si H désigne la U-cohomologie naturelle sur $E_{p,q,r}$ alors les sous-schémas $\{\mathrm{rk}\,H \geqslant x\}$ de $E_{p,q,r}$ sont réduits.

Preuve. Elle repose sur l'étude de $E_{p,q,r} =: E$ en termes de théorie des monômes standard, faite par De Concini – Strickland [DC – S] (cf. [DC – E – P] § 16). Soit A l'anneau du schéma affine E. Les mineurs des matrices du complexe forment une partie finie H de A. Les produits finis d'éléments de H sont par définition les monômes. Ils forment une partie de A indexée par $\mathbb{N}^H$. On définit une partie Σ de $\mathbb{N}^H$ telle que A soit une algèbre de Hodge gouvernée par Σ. Cela signifie en particulier que si on pose $S = \mathbb{N}^H - \Sigma$ (c'est l'ensemble des monômes standard), alors S est une base de A (comme espace vectoriel). Cela signifie aussi que si I(u, v) est la réunion dans H des mineurs de rang au moins u +1 dans la première matrice et des mineurs de rang au moins v + 1 dans la seconde, alors I(u, v) A, qui contient évidemment l'ensemble S(u, v) des monômes standard non identiquement nuls sur I(u, v), est en fait engendré par S(u, v).

Notons E(u,v) le sous-schéma de E défini par I(u,v). Il est réduit (loc.cit. 1.6.2). Et notons E(s) le sous-schéma des complexes dont la cohomologie a rang au moins q – s. Ensemblistement, E(s) est la réunion des E(u, v) pour u + v = s. Et par définition, l'idéal I(s) de E(s) est engendré par les monômes de la forme xy où x est un mineur de rang u de la première matrice et y un mineur de rang v de la seconde avec u + v = s + 1. Il nous faut montrer que cet idéal est l'intersection I'(s) des I(u,v) A pour u + v = s. Comme chacun des I(u,v) A est engendré par une partie de la base S, leur intersection est engendrée par l'intersection correspondante S(s) des S(u, v) pour u + v = s. Soit donc M un monôme de S(s) et montrons que M est dans I(s). Soit t la taille maximale d'un mineur de la première matrice x divisant M. Alors M, n'étant pas identiquement nul sur I(t, s-t), doit ne pas s'annuler sur un mineur y de rang au moins s – t + 1 de la deuxième matrice. Autrement dit M est divisible par xy.

3.2 **Remarque.** On aurait été encore plus convaincant si on avait montré qu'il existe un faisceau par exemple sur $E \times \mathbb{P}^2$, E-propre et E-plat, dont la première image directe sur E soit la cohomologie du complexe universel.

Bibliographie.

[B] W. Barth: Some properties of stable rank-2 vector bundles on $\mathbb{P}^n$. Math. Ann. 226, 125-150 (1977).

[jmB] J.M.Boardman: Singularities of differentiable maps. Publ. I.H.E.S. 33, 21-57 (1967).

[DC – E – S] C. De Concini – D.Eisenbud – C.Procesi: Hodge Algebras. Astérisque 91, SMF.

[DC – S] C.De Concini – E. Strickland: On the variety of complexes. Adv. in Math. 41, 57-77 (1981).

[GL] M. Green – R. Lazarsfeld: Deformation theory, generic vanishing theorems and some conjectures of Enriques, Catanese and Beauville. Invent. Math. 90, 389-407 (1987).

[H] A. Hirschowitz: Cohérence et dualité sur le gros site de Zariski. Dans ce volume.

[H']----------: Sections planes et multisécantes pour les courbes gauches génériques principales. Space curves, Proceedings Rocca di Pappa 1985, Lect. Notes in Math. 1266, 124-155 (1987).

[K] M. Kashiwara: Systems of microdifferential equations. Progress in Math. 34 (1983).

[M] D.Mumford: Abelian varieties. Oxford University Press (1970).

[T] B. Teissier: Cycles évanescents, sections planes et conditions de Whitney. Singularités à Cargèse, Astérisque 7/8,285-362 (1973).

Generating six skew lines in $\mathbb{P}^3$

Monica Idà
Dipartimento di Scienze Matematiche
Piazzale Europa 1 - 34127 Trieste -I

Introduction. We work over an algebraically closed field of characteristic zero.
Let C be a non special curve in $\mathbb{P}^3$, and assume that the postulation of C, $\{h^\circ(\mathcal{I}_C(k))\}_{k\geq 1}$, is known. For example, if C is of maximal rank, i.e. if the map $\rho_k: H^\circ(\mathcal{O}_{\mathbb{P}^3}(k)) \to H^\circ(\mathcal{O}_C(k))$ is of maximal rank for each k, then $h^\circ(\mathcal{I}_C(k)) = \dim\ker\rho_k = \max\{0, h^\circ(\mathcal{O}_{\mathbb{P}^3}(k)) - h^\circ(\mathcal{O}_C(k))\}$, and $h^\circ(\mathcal{O}_C(k))$ is given by Riemann-Roch. So we know the dimensions of the components of degree k of the graded ideal of C, $I_C = \oplus H^\circ(\mathcal{I}_C(k))$. To say something more about this ideal, we can study the maps:
$\sigma_k: H^\circ(\mathcal{I}_C(k)) \otimes H^\circ(\mathcal{O}_{\mathbb{P}^3}(1)) \to H^\circ(\mathcal{I}_C(k+1))$.
In particular,if we know dim ker σ_k for each k, we have the first step of the minimal free resolution of the homogeneous ideal of C; the simplest case, that is, the smallest number of generators, occurs when the σ_k's are all of maximal rank; in this case we say that C is minimally generated.
Let C be a union of d skew lines in $\mathbb{P}^3$. If C is generic, we know everything about the first step of the minimal free resolution of I_C; in fact, the generic C is of maximal rank for any d ([H-H]), and minimally generated for $d \neq 4$ ([I,1]). In this paper we study the non generic case.
If X is a union of d skew lines such that $h^1(\mathcal{I}_X(n)) = 0$, $n>0$, then by Castelnuovo - Mumford lemma, $h^1(\mathcal{I}_X(k)) = 0$, and the maps σ_k relative to X are all surjective, for $k \geq n+1$ (see [Mu] p. 99).
The last n for which this can fail is d - 1.In fact, for any union of d skew lines X, $\mathcal{I}_X$ is d-regular (hence $h^1(\mathcal{I}_X(k)) = 0$ for $k \geq d-1$ and σ_k is surjective for $k \geq d$); this follows from a very general theorem of Gruson-Lazarsfeld-Peskine (see[G-L-P], remark 1.1). For a fixed d, there is hence only a finite number of possibilities about the ρ_k's, as well as for the σ_k's, being or not of maximal rank.
What is clearly expected is a strong relation among secants, postulation and generation (see [GLP]).
In section 2 we find, for a union of d skew lines X, an inferior bound for $h^\circ(\mathcal{I}_C(k))$, resp. for dim ker σ_k, depending on a constant Λ measuring the surplus of condition imposed by the secant lines of X to $\mathcal{O}_{\mathbb{P}^3}(k)$, respectively to the cotangent bundle Ω_3 of $\mathbb{P}^3$ twisted by k+1 (recall that $\ker\sigma_k \cong$ $\cong H^\circ(\Omega_3(k+1))$.
What we would like would be an equality, at least for a generic C (generic in the subscheme of the unions of d skew lines with assigned secants). But what is immmediately clear from examples, is that we have to take care not only of secant lines, but also of surfaces of secants, and of secant curves of degree > 1 ([I,2]). So, in order to give a general conjecture, the first thing to do is to find a good way of computing the number of independent conditions imposed to the suitable twist of Ω_3 by all these secant objects.
In section 3 we study the maps ρ_{d-2} and σ_{d-1} for a union of d skew lines in rapport to the number of d-secants, and in fact we see that for each d-secant, we need a generator in degree d more for the homogeneous ideal, and $h^\circ(\mathcal{I}_C(d-2))$ increases of one (for C generic in a suitable sense). This means that the previous inequality has become an equality in these extremal cases.
In the remaining sections of this paper, we study the stratification induced by postulation on the open subset $U \subset Hilb_{6,-5}(\mathbb{P}^3)$ corresponding to disjoint unions of six lines, and we study the maps σ_k's on each stratum. In fact, d = 6 is the first interesting case (for $d \leq 5$, the behaviour of the maps σ_k, for each possible postulation of X, is clear).

It is easy to see that, for the extremal maps (i.e. ρ_4 and σ_5) relative to any union of six skew lines, the dimension of the kernel reaches the inferior bound depending on Λ we were discussing above (we don't need genericity here). Moreover, for C generic in a suitable sense, σ_4 is surjective if and only if there are no proper 5-secants (6-secants are allowed; notice that for any C, $h^1(\mathcal{I}_C(3)) \neq 0$) (section 5); this means (4.3, 4.4) that if σ_4 is surjective,again the kernel of σ_4 reaches the expected inferior bound.
A summery of the situation in $\mathrm{Hilb}_{6,-5}(\mathbb{P}^3)$ is given in th.4.3. The study of the behaviour of the maps σ_k for six skew lines allows us, in particular, to check some strange behaviour of these maps; for example, we see that the property "σ_k is surjective" is not preserved by generalization (while this is clearly true for ρ_k).
I wish to thank A.Hirshowitz for the useful conversations on this subject.

<u>**Notations and preliminaries.**</u>

0.1 We set $V := H^0(\mathcal{O}_{\mathbb{P}^3}(1))$.

0.2 We set, for a disjoint union C of d lines,

$$\omega(C) := \min \{k \mid 4\, h^0(\mathcal{I}_C(k)) \geq h^0(\mathcal{I}_C(k+1)) > 0\},$$

and if Y is the generic union of d lines, $\omega(d) := \omega(Y)$.

0.3 Let C be a disjoint union of d lines. With σ_k (or $\sigma_k(C)$) we always denote the map:

$$\sigma_k : H^0(\mathcal{I}_C(k)) \otimes V \rightarrow H^0(\mathcal{I}_C(k+1)).$$

We denote by Ω_3 the cotangent sheaf of $\mathbb{P}^3$. From the Euler sequence we have $\ker \sigma_k \cong$ $\cong H^0(\Omega_3(k+1)\otimes\mathcal{I}_C)$, and if $h^1(\mathcal{I}_C(k)) = 0$, $\mathrm{coker}\, \sigma_k \cong H^1(\Omega_3(k+1)\otimes\mathcal{I}_C)$ ([I,1] ,2.1).

0.4 We sometimes set $I_k := H^0(\mathcal{I}_C(k))$, the k-th component of the homogeneous ideal I of C, if no confusion arises about C.

0.5 If σ_k is surjective for $k \geq k_0$, I is generated in degrees $\leq k_0$.

0.6 We recall that Castelnuovo-Mumford lemma says in particular that, if C is a non special curve in $\mathbb{P}^3$, and $h^1(\mathcal{I}_C(m-1)) = 0$, then (see [Mu] p.99):

a) $H^1(\mathcal{I}_C(k)) = 0$, for $k \geq m-1$

b) σ_k is surjective, for $k \geq m$.

0.7 For the techniques related with the use of "la méthode d'Horace", namely,when we bound the postulation or the dimension of $\ker \sigma_k$, we send to [H,1],[H,2], and [I,1].
In particular, we recall that if C is a fixed curve in $\mathbb{P}^3$, and if T denotes the union of C with n points, then $h^0(\mathcal{I}_T(k)) = 0$ implies $h^0(\mathcal{I}_C(k)) \leq n$. Moreover, if C is a curve and C' a specialization of C, with $h^0(\mathcal{I}_{C'}(k)) = n$, then $h^0(\mathcal{I}_C(k)) \leq n$ by semicontinuity.
Since $\ker \sigma_k \cong H^0(\Omega_3(k+1) \otimes \mathcal{I}_C)$, when we apply la méthode d'Horace to bound dim ker σ_k, we are obliged to work in $\mathbb{P}(\Omega_3)$ ([H,3], [I,1], 1.1, 1.9). In the following we use the same notations as in [I,1], (1.2). In particular, with s-point, resp. d-point, t-point we mean a point of $\mathbb{P}(\Omega_3)$, resp. two points of $\mathbb{P}(\Omega_3)$ lying in the same fiber $\pi^{-1}(x)$, resp. three points of $\mathbb{P}(\Omega_3)$ lying in the same fiber $\pi^{-1}(x)$ but not on a line of $\pi^{-1}(x)$.
Moreover, we denote the invertible sheaf $\mathcal{O}_{\mathbb{P}(\Omega_3)}(1) \otimes \pi^*\mathcal{O}_{\mathbb{P}^3}(k+1)$ on $\mathbb{P}(\Omega_3)$ with L_k, and if Q is a smooth quadric in $\mathbb{P}^3$, the sheaf $\mathcal{O}_{\mathbb{P}(\Omega_3)}(1) \otimes \pi^*\mathcal{O}_Q(a,b)$ is denoted by $F_{a,b}$.
Finally, for the definition of residual scheme with the related exact sequence we send to [H,2].

0.8 Let C be a union of 6 skew lines in $\mathbb{P}^3$; we denote by c the corresponding point in $\mathrm{Hilb}_{6,-5}(\mathbb{P}^3)$. We denote by U the set of such points of $\mathrm{Hilb}_{6,-5}(\mathbb{P}^3)$.

0.9 With Y we denote the generic union of six lines in $\mathbb{P}^3$.

0.10 We recall that the generic union of six lines Y is of maximal rank, hence its postulations is given as follows:

k	$h^\circ(\mathcal{I}_Y(k))$	$h^\circ(\mathcal{O}_{\mathbb{P}^3}(k))$	$h^\circ(\mathcal{O}_Y(k))$	$h^1(\mathcal{I}_Y(k))$
3	0	20	24	4
4	5	35	30	0
5	20	56	36	0
6	42	84	42	0

Moreover, we recall that the maps σ_k are all of maximal rank for Y. Hence, σ_k is injective for $k \le 4$ and surjective for $k \ge 4$.

0.11 By [G-L-P] remark 1.1, if C is any union of 6 skew lines, then $H^1(\mathcal{I}_C(5)) = 0$.

Hence a) $H^1(\mathcal{I}_C(k)) = 0$ for $k \ge 5$,

b) σ_k is surjective for $k \ge 6$.

In particular, for any $c \in U$, $h^\circ(\mathcal{I}_C(k)) = h^\circ(\mathcal{I}_Y(k))$ for $k \ge 5$; moreover, $h^\circ(\mathcal{I}_C(k)) \ge h^\circ(\mathcal{I}_Y(k))$ for $2 \le k \le 4$, and $h^\circ(\mathcal{I}_C(1)) = 0$. Finally for any c in U $h^0(\mathcal{I}_C(2)) \le 1$, hence σ_1 and σ_2 are always injective.

0.12 We set $S_k{}^i := \{c \in U \mid h^\circ(\mathcal{I}_C(k)) \ge h^\circ(\mathcal{I}_Y(k)) + i\}$.

0.13 Let Z be a scheme with irreducible components $Z_1,\dots,Z_t$. When we say that x is generic in Z, we mean that x is the generic point of some Z_i.

Section 1

The aim of this section is to define the subschemes of $\mathrm{Hilb}_{d,1-d}\,\mathbb{P}^3$, whose generic points are unions of d skew lines with assigned number λ_j of proper and distinct j-secant lines.

1.1 *Definition.* Let C be a union of d skew lines $r_1,\dots,r_d$ in $\mathbb{P}^3$. A k-secant for C is a point of

$$X^k(C) := \mathrm{Al}^k\mathbb{P}^3 \cap (\mathrm{Hilb}^k C \smallsetminus (\mathrm{Hilb}^k r_1 \cup \dots \cup \mathrm{Hilb}^k r_d))$$

(the intersection is in $\mathrm{Hilb}^k\,\mathbb{P}^3$; $\mathrm{Al}^k\mathbb{P}^3$ is the subscheme of the curvilinear k-uples of $\mathbb{P}^3$ which are contained in a reduced line. For details, see [LB],I).

1.2 *Remark.* Consider the natural morphism $f\colon \mathrm{Al}^k\mathbb{P}^3 \to G(1,3)$ ([LB],I). If lenght $(X^k \cap f^{-1}(r)) \ge 2$, then the line r is the support for a m-secant, with $m \ge k+1$ (this is due to the transversality of intersection of two lines).

1.3 *Remark.* It is possible to relativize the construction in 1.1 ([LB]). By semicontinuity of the dimension of the fiber and of the cardinality of the fiber of a finite morphism, it is hence possible to define the subscheme of the subset of $\mathrm{Hilb}_{d,1-d}\,\mathbb{P}^3$ containing unions of d skew lines C, with $\dim X^k(C) \ge 1$ or $\dim X^k(C) = 0$ and lenght $X^k(C) \ge n$, where n is a fixed integer.

1.4 *Definition.* Assume a sequence of natural numbers is given: $\lambda_t,\dots,\lambda_d$. We define a new sequence, setting:

$$n_{d-k} := \sum_{i=0}^{k} \binom{d-k+i}{i} \cdot \lambda_{d-k+i}$$

Using remark 1.3, we can define the subscheme $T^{t,\dots,d}{}_{\lambda_t,\dots,\lambda_d}$ of the subset of $\mathrm{Hilb}_{d,1-d}\mathbb{P}^3$ containing unions of d skew lines, whose points are unions of d skew lines C, with $\dim X^j(C) \ge 1$ or with $\dim X^j(C) = 0$ and lenght $X^j(C) \ge n_j$, $j=t,\dots,d$.

Now let Y be an irreducible component of $T^{t,\dots,d}{}_{\lambda_t,\dots,\lambda_d}$, such that, denoting by C the generic point of Y, $X^j(C)$ is reduced and finite, of lenght n_j, for $j=t,\dots,d$. We shall denote by $W^{t,\dots,d}{}_{\lambda_t,\dots,\lambda_d}$ the

union of such irreducible components Y.
If $t = d-1$, we shall make use of the more concise notation: $W_{\lambda_{d-1},\lambda_d} := W^{d-1,d}_{\lambda_{d-1},\lambda_d}$.

1.5 *Definition.* Let $c \in W^{t,..,d}_{\lambda_t,...,\lambda_d}$. We say that the union of lines C corresponding to c has λ_t proper and distinct t-secants if $X^t(C)$ is reduced and finite of lenght n_t.
If f: $X^t(C) \to G(1,3)$ is the natural morphism (see 1.2), there are λ_t (reduced) points p_i in the Grassmann variety such that $f^{-1}(p_i)$ has lenght one. We shall always denote by X_t, or $X_t(C)$ if confusion may arise, the set of λ_t lines of $\mathbb{P}^3$ corresponding to $p_1,...,p_{\lambda_t}$.
Hence $X_t(C)$ is the set of t-secant lines for C, which are not (t+1)-secants.

Section 2.

In this section we give an inferior bound for the postulation and the number of generators of a union of d skew lines C; this bound depends on the multisecant lines of C.

2.1 *Definition.* Let F be a rank r uniform vector bundle over $\mathbb{P}^3$, with splitting type $a_1 \leq ... \leq a_r$. Let n be a positive integer, such that $r\cdot(n+1)+a_1+...+a_r \geq 0$, and let t and u be the unique integers such that $r\cdot(n+1)+a_1+...+a_r = r\cdot t+u$, $0 \leq u < r$.
Let C be a union of d skew lines in $\mathbb{P}^3$, and assume that C has λ_i proper and distinct i-secants, for i = =t+1,..., d. We shall denote by $\Lambda(F,C,n)$ the following integer:
$\Lambda(F,C,n) := \max\{0, r\cdot((t+1)\cdot\lambda_{t+1}+...+d\cdot\lambda_d) - (r\cdot t+u)\cdot(\lambda_{t+1}+...+\lambda_d)\}$.

2.2 *Remark.* The integer t is the rest class modulo r of the global sections of the restriction of F(n) to a line. The integer $\Lambda(F,C,n)$ gives the number of superfluos conditions imposed to F(n) by distinct and proper multisecants.

2.3 *Proposition.* Let F, n, C, t be as in 2.1, and assume that C has λ_i proper and distinct i-secants, for i = t+1,..., d. Moreover, assume lenght $((X_{t+1} \cup ... \cup X_d) \cap L) \leq n+a_1+1$ for each line $L \subset C$.
Then,
$$h^\circ(\mathcal{I}_C \otimes F(n)) \geq h^\circ(F(n)) - h^\circ(F(n)|_C) + \Lambda(F,C,n).$$

2.4 *Remark.* F being uniform, $h^\circ(F(n)) - h^\circ(F(n)|_C)$ depends only on F, n and d.

***Proof of* 2.3**. We look at the exact sequence:
$$0 \to H^\circ(\mathcal{I}_C \otimes F(n)) \to H^\circ(F(n)) \xrightarrow{f} H^\circ(F(n)|_C) \to \operatorname{coker} f \to 0.$$
It is enough to prove dim coker f $\geq \Lambda(F,C,n)$ when $\Lambda(F,C,n) > 0$. We now denote by Z the subscheme of $\mathbb{P}^3$ union of $X_{t+1},...,X_d$. Let L be a line contained in C. The assumption lenght $(Z \cap L) \leq n+a_1+1$ assures that the map $H^\circ(F(n)|_L) \to H^\circ(\mathcal{O}_{Z\cap L}^{\oplus r})$ is surjective.
Equivalently, the map g: $H^\circ(F(n)|_C) \to H^\circ(F(n)|_{Z\cap C})$ is surjective. We have a commutative diagram:
$$\begin{array}{ccc} H^\circ(F(n)) & \xrightarrow{f} & H^\circ(F(n)|_C) \\ & h \searrow & \downarrow g \\ & & H^\circ(F(n)|_{Z\cap C}) \end{array}$$
and g being surjective, we have dim coker f $\geq$ dim coker h.
Let us denote by K a set of $(\lambda_{t+1}+...+\lambda_d)\cdot(r\cdot t+u)$ points of $\mathbb{P}(F(n))$, such that, denoting by π the canonical projection $\mathbb{P}(F(n)) \to \mathbb{P}^3$, and by E the invertible sheaf $\mathcal{O}_{\mathbb{P}(F(n))}(1)$:

i) $K \subset \pi^{-1}(Z\cap C)$ (this is possible since $\Lambda(F,C,n) > 0$);

ii) $h^{\cdot}(I_K \otimes E|_{\pi^{-1}(R)}) = 0$ for each line $R \subset Z$.

Hence, $H^{\cdot}(E \otimes I_K) \cong H^{\cdot}(F(n) \otimes I_{Z \cap C})$. On the other hand, $h^{\cdot}(E \otimes I_K) \geq h^{\cdot}(F(n))$ - lenght K. Hence, coker $h = h^{\cdot}(F(n)|_{Z \cap C}) - h^{\cdot}(F(n)) + h^{\cdot}(I_K \otimes E) \geq \Lambda(F,C,n)$.

2.5 *Remark.* Proposition 2.3 with $F = \mathcal{O}_{P^3}$, respectively $F = \Omega_3$, gives a bound on the postulation of C at level n, respectively on the number of generators of degree n of C.

Section 3.

In this section we study the behaviour of the map σ_{d-1} for a union of d skew lines C with d-secants. If dim $X^d(C) > 0$, C lies on a quadric, and the first step of the minimal free resolution is described in lemma 3.1. If $X^d(C)$ is reduced and finite, the generic behaviour of σ_{d-1} is described in prop.3.2.

3.1 *Lemma.* Let T be a union of $d \geq 4$ skew lines on a smooth quadric Q. Then, the maps σ_k relative to T are as follows: σ_1, σ_2 are injective, σ_k is surjective for $k \geq 3$, and $k \neq d-1$; σ_{d-1} is not of maximal rank. Moreover, dim coker $\sigma_{d-1} = d+1$.

Proof. Since $h^{\circ}(I_T(1)) = 0$, $h^{\circ}(I_T(2)) = 1$, σ_1 and σ_2 are injective. On the other hand, from the exact sequence $0 \to I_Q(d-1) \to I_T(d-1) \to I_{T,Q}(d-1) \to 0$ it follows $h^1(I_T(d-1)) = 0$, hence σ_k is surjective for $k \geq d$.

Hence σ_{d-1} cannot be surjective; in fact, if it were so, the homogeneous ideal I_T would be generated in degree $\leq d-1$. But a surface S_t of degree $t \leq d-1$ containing T contains also the quadric (otherwise the intersection S_t would be a curve on Q of type (t,t)),but for degree reasons, such a curve can not contain C). So σ_{d-1} is not surjective. Moreover, we have seen that a surface S_t, $d-1 \geq t \geq 3$, containing T is of the form $S_t = Q \cup S_{t-2}$ with S_{t-2} any surface of degree t-2. Hence, setting $I_j := H^{\circ}(I_T(j))$, we have $I_j = q(k[x_0, ..., x_3])_{j-2}$ for $3 \leq j \leq d-1$, where q is an equation for the quadric. It follows that studying the maps σ_j: $I_j \otimes (k[x_0, ..., x_3])_1 \to I_{j+1}$ $3 \leq j \leq d-2$ is the same as studying the natural maps: $(k[x_0, ..., x_3])_{j-2} \otimes (k[x_0, ..., x_3])_1 \to (k[x_0, ..., x_3])_{j-1}$, which are all surjective.

Notice that, for $d \geq 4$, if σ_{d-1} were of maximal rank, σ_{d-1} should be surjective, which is not ; in fact, since $h^1(I_T(k)) = 0$ for $k \geq d-1$, $4h^{\circ}(I_T(d-1)) - h^{\cdot}(I_T(d)) =$

$$= 4\left[\binom{d+2}{3} - d^2\right] - \left[\binom{d+3}{3} - d(d+1)\right] = 3(d^3 - 4d^2 + d - 2)$$ which is positive for $d \geq 4$.

We conclude by computing dim coker σ_{d-1}. Let f: $I_2 \otimes k[x_0,...,x_3]_{d-2} \to I_d$ be the natural map; f is injective, coker f = coker σ_{d-1}, and dim coker f = dim I_d - dim $k[x_0,...,x_3]_{d-2}$ = d+1.

3.2. *Proposition.* Let C be a union of d skew lines and assume C generic in some irreducible component of $W_{0,s}$. Then $s \leq 2$, and

a) $h^1(I_C(d-2)) = s$;

b) dim coker $\sigma_{d-1} = s$ (i.e. the generators of I_C of degree d are s).

Proof. Observe that $d \geq 4$, and $0 \geq s \geq 2$ (otherwise dim $X^d(C) = 1$, that is, C lies on a quadric). The inequalities "$\geq$" in both a) and b) are consequence of 2.4 and 2.5. Now we prove the reverse inequalities; we assume for the moment $d \geq 5$.

We choose a specialization of C, X, union of d-2 skew lines on a quadric Q (as lines, say, of the first family) and two, say r and t, outside.

a) Let Z denote the 4 points $(r \cup t) \cap Q$; since the quadric contains the d-secants to X, Z impose 4-s independent conditions to $\mathcal{O}_Q(0,d-2)$, hence $h^1(I_{Z,Q}(0,d-2)) = -h^0(\mathcal{O}_Q(0,d-2)) + h^0(\mathcal{O}_Z) + h^0(I_{Z,Q}(0,d-2)) =$

=s. By the residue sequence ($res_Q X = r\cup t$) :
$0 \to I_{r\cup t}(d\text{-}4) \to I_X(d\text{-}2) \to I_{X\cap Q,Q}(d\text{-}2) \to 0$ we get for $d \geq 5$ (since $h^1(I_{r\cup t}(d\text{-}4)=0)$: $h^1(I_X(d\text{-}2)) =$
$= h^1(I_{X\cap Q,Q}(d\text{-}2)) = h^1(I_{Z,Q}(0,d\text{-}2)) = s$.
b) Since $h^1(I_T(d\text{-}1)) = 0$ for any union T of d skew lines ([GLP] remark 1.1), coker $\sigma_{d-1}=$
$=H^1(\Omega_3(d)\otimes I_C)$. We work in $\mathbb{P}(\Omega_3)$. We look at the residual sequence :
$0 \to L_{d-3}\otimes I_{r\cup t} \to L_{d-1}\otimes I_X \to L_{d-1}\otimes I_{X\cap Q,Q} \to 0$ (*).
Two lines being of maximal rank and minimally generated, $h^1(\Omega_3(d\text{-}2)\otimes I_{r\cup t}) = 0$ for $d \geq 5$. Moreover, since $h^1(\Omega_3(d\text{-}2)|_{r\cup t}) = 0$ for $d \geq 3$, and $h^2(\Omega_3(d\text{-}2)) = 0$, we have $h^2(\Omega_3(d\text{-}2)\otimes I_{r\cup t}) = 0$. Hence (*) gives (for $d \geq 5$): $h^1(L_{d-1}\otimes I_X) = h^1(L_{d-1}\otimes I_{X\cap Q,Q}) = h^1(F_{2,d}\otimes I_Z) = s$, since Z imposes 3·4 - s independent conditions to the sections of $F_{2,d}$ (recall that, if L is a line of type (0,1) on Q, $\Omega_3|_Q(2,d)|_L \equiv$ $\equiv \mathcal{O}_L \oplus \mathcal{O}_L(1)^{\oplus 2}$, hence if L is a d-secant for C, there are two t-points of Z on L, which imposes one superfluos condition to $\Omega_3|_Q(2,d)$).

If d = 4, since any union C of 4 skew lines not on a quadric has exactly two 4-secants, and it is of maximal rank, $h^1(I_C(2)) = 2$ so a) is true. To prove "≤" in b) we proceed as above, except that a unique line r is out of the quadric; notice that $h^1(F_{1,4}\otimes I_{r\cap Q}) = 2$.

Section 4

In lemma 4.1 we prove the equalities in 3.2 for d = 6 and dropping genericity. We then state two theorems summarizing the results obtained for six lines in $\mathbb{P}^3$.

4.1 *Lemma.* Let C be a union of six lines in $W_{0,s}$. If s = 0,1, assume that C has exactly s 6-secants; if s = 2, assume that C has 2 distinct 6-secants or that there exists a a double structure η on a line such that lenght $(\eta\cap C) = 12$. Then, a) $h^\circ(I_C(4)) = 5 + s$,

b) $h^\circ(\Omega_3(6)\otimes I_C) = 38 + s$.

Proof. Let C be the union of $r_1,\dots,r_6$, let Q be the quadric generated by r_1,r_2,r_3, and set $Z := res_Q C$ (Z can consist of one, or two, or three lines). Notice that if there exists a double structure η on a line with lenght $(\eta\cap C) = 12$, then $\eta \subset Q$. We consider the residue sequence:

$$0 \to I_Z(2) \to I_C(4) \to I_{C\cap Q,Q}(4) \to 0.$$

Since $h^1(I_Z(2)) = h^2(I_Z(2)) = 0$, we have $h^1(I_C(4)) = h^1(I_{C\cap Q,Q}(4,4))$. If Q contains five of the lines r_i, $h^1(I_{C\cap Q,Q}(4,4) = h^1(I_{Z\cap Q,Q}(-1,4)) = 2$. If Q contains four of the lines r_i, then exactly the same argument used in the proof of 3.2 gives $h^1(I_C(4)) = s$. If Q contains only r_1,r_2,r_3, it is enough to observe that the six points (counted with multiplicity) $Z\cap Q$ imposes 6-s independent conditions to $\mathcal{O}_Q(1,4)$, hence $h^1(I_{C\cap Q,Q}(4,4)) = h^1(I_{Z\cap Q,Q}(1,4)) = s$.
The b) part of the proof is a replay of the a) part, with the caution of working in $\mathbb{P}(\Omega_3)$ (see proof of 3.2 part b)).

4.2 *Theorem.* Let $k \geq 4$, and let C be a union of six lines with λ_i proper and distinct i-secants for $i \geq k+1$ Then, if σ_k is surjective,

$$4\,h^\circ(I_C(k)) - h^\circ(I_C(k+1)) \geq h^\circ(\Omega_3(k+1)) - h^\circ(\Omega_3(k+1)|_C) + \Lambda(\Omega_3,C,k+1). \qquad (*)$$

Conversely, if (*) holds and, when k =4, if C is generic in W_{λ_5,λ_6} , then σ_k is surjective.

4.3 *Remark.* For any C as in 4.2, the inequality:

$$4\,h^\circ(I_C(k)) - h^\circ(I_C(k+1)) \leq h^\circ(\Omega_3(k+1)) - h^\circ(\Omega_3(k+1)|_C) + \Lambda(\Omega_3,C,k+1) \qquad (**)$$

always holds. In fact, if Y denotes the generic union of six lines, the following holds:

$$h^\circ(\Omega_3(k+1)) - h^\circ(\Omega_3(k+1)|_C) = 4\,h^\circ(I_Y(k)) - h^\circ(I_Y(k+1)) .$$

Hence by 4.1 , 0.10 and 0.11, (**) becomes :
for k =4 : $4\cdot\lambda_6 \le \Lambda(\Omega_3,C,5) = \lambda_5 + 4\cdot\lambda_6$
for k =5 : $0 \le \Lambda(\Omega_3,C,6) = \lambda_6$.
So th.4.2 says that for any C σ_5 is surjective if and only if there are no 6-secants, and for C generic σ_4 is surjective if and only if there are no proper 5-secants.

***Proof of* 4.2.** For $k \ge 6$, there is nothing to prove, and for k = 5, this is a particular case of 4.1 b). So now assume σ_4 surjective for C; if it is possible to apply 1.4, we get

$$4h^\circ(I_C(4)) - h^\circ(I_C(5)) = h^\circ(\Omega_3(5)\otimes I_C) \ge h^\circ(\Omega_3(5)) - h^\circ(\Omega_3(5)|_C) + \Lambda(\Omega_3,C,5).$$

The assumptions in 1.4 are (L denotes any line contained in C): lenght $((X_5 \cup X_6) \cap L) \le 4$, which is satisfied unless $\lambda_6 = 0$ and $\lambda_5 \ge 5$ (which is possible; for example, six lines on a smooth cubic). But by 4.1 $\lambda_6 = 0$ implies I_C generated in degrees ≤ 5. So, if there are 5-secants, σ_4 cannot be surjective, against assumption.
Conversely, (*) for k =4 gives (see 4.3) $\lambda_5 = 0$, so we conclude, C being generic in W_{λ_5,λ_6}, by 5.1, 5.2 and [I,1], 5.6, that σ_4 is surjective.

4.4 *Remark.* Let C be as in 4.2; then, 4.1, 4.2 and 4.3 say that the inferior bound for $h^\circ(\Omega_3(k+1)\otimes I_C)$ given by 1.4 is effetively atteint for k = 4 if σ_4 surjective , and for k = 5 in any case.

4.5 *Notation.* Let C be a union of six skew lines $r_1, ..., r_6$. There are 4 possibilities:

a) any quadric containing 3 among the lines r_i does not contain one of the others;

b) there exists some quadric containing 4 of the lines, and no quadric containing 5 of them;

c) there exists a quadric Q containing exactly 5 of the lines;

d) $r_1, ..., r_6$ lie on a quadric Q.

We say that C is of type a, resp. b, c, d.

We now describe graphically the various possibilities of cases a and b, with respect to the intersection with the quadric Q generated by, say, r_1,r_2,r_3 (if C is of type b, we assume that r_4 lies in Q). Namely, if r_1,r_2,r_3 are lines of type (1,0) on Q, we distinguish the cases accordingly to how many points of $C \cap Q$ lie on the same line t_i (of type (0,1)). Also if not graphically specified, two distinct lines t_1 , t_2 can become a double line, and so on. For example, in b_3: if instead of $t_1 \cup t_2$ we have a double line on Q, this means r_5, r_6 are tangent to Q. In the following we utilize the formula " the intersection of C with Q is of type a_i (i =1,...,8) or b_j (j =1,2,3)" with the obvious meaning.

Case b):

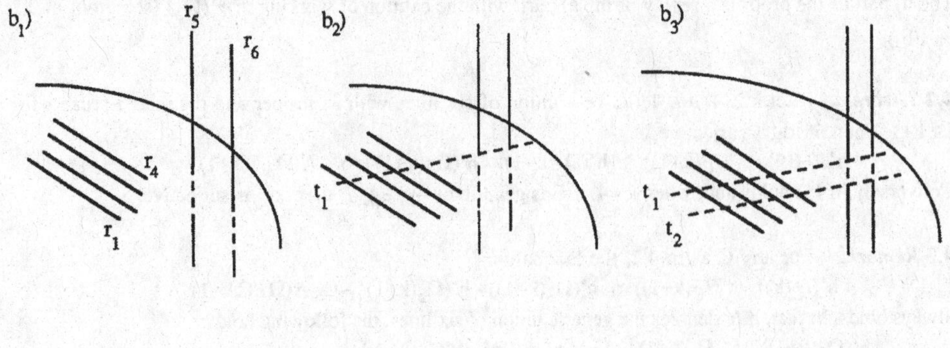

Case a):

a_1)

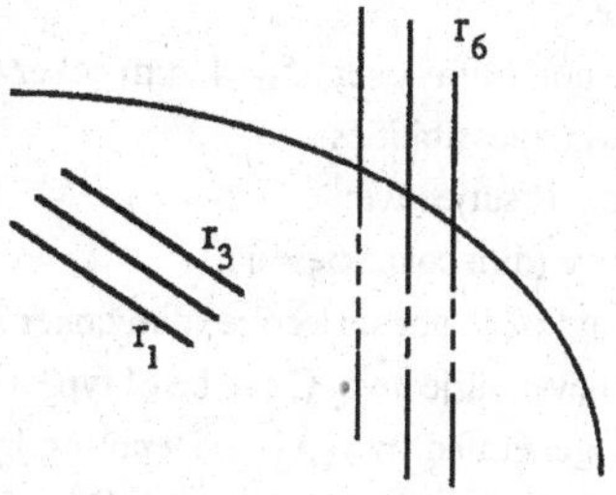

a_2)

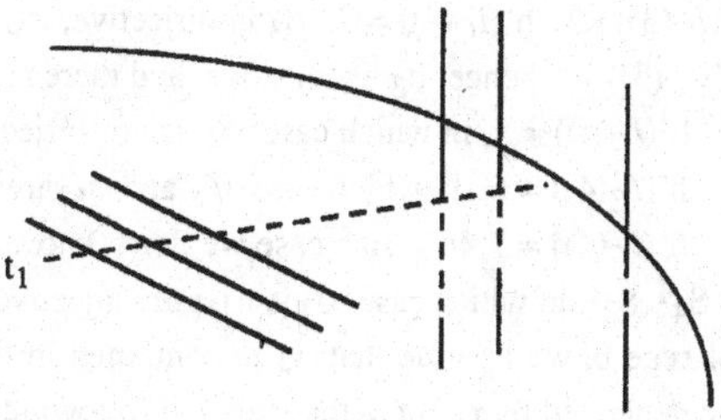

a_3)

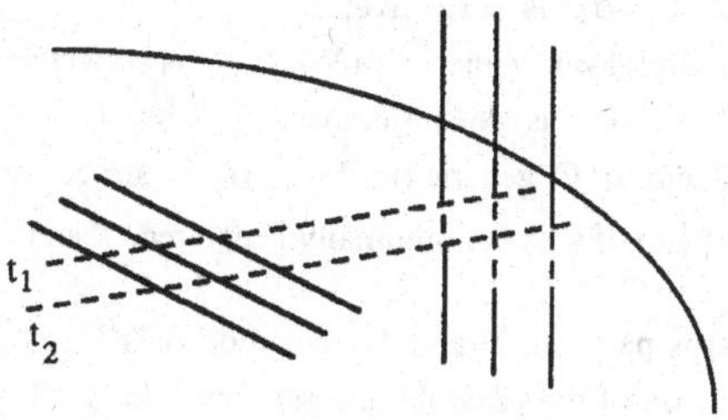

a_4)

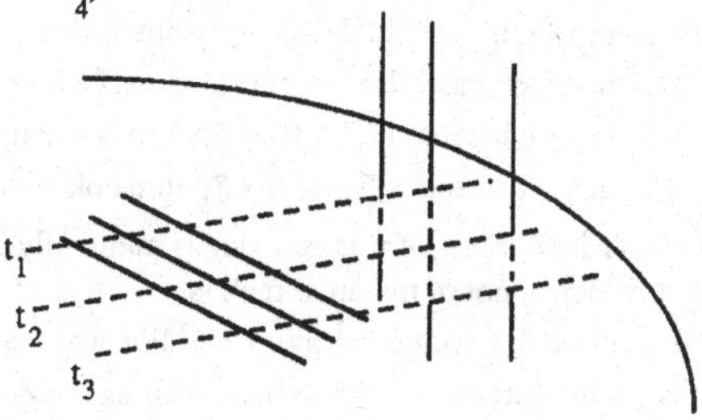

a_5)

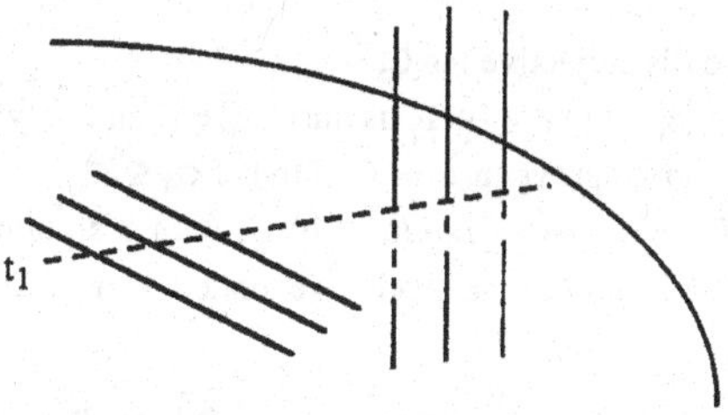

a_6)

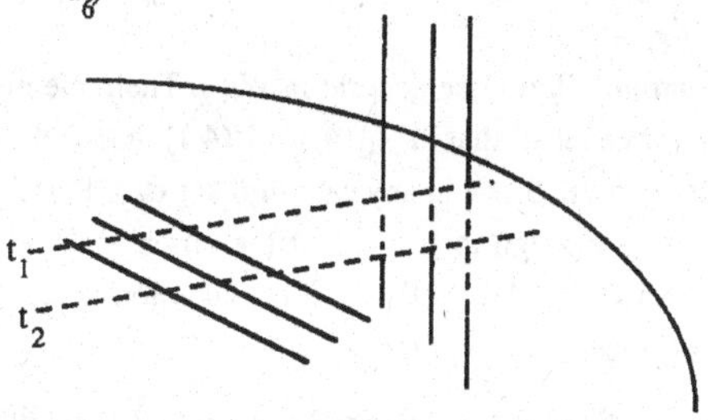

a_7)

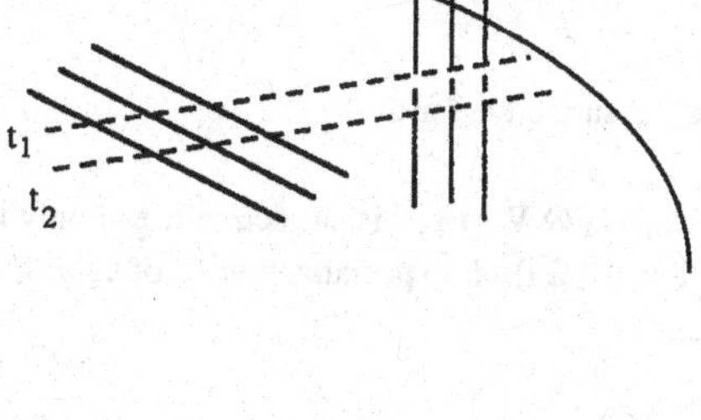

a_8)

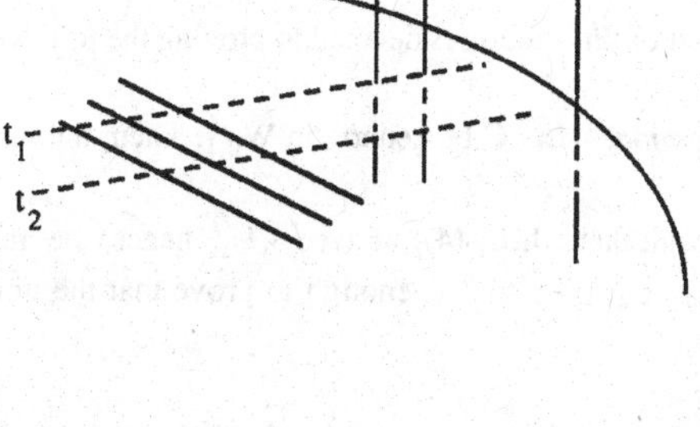

4.6 *Theorem.* Let C be a union of six skew lines $r_1,...,r_6$. Then there are the following possibilities:
1) $c \in S_2{}^1$, in which case: $h^\circ(I_C(2)) = 1$, $h^\circ(I_C(3)) = 4$, $h^\circ(I_C(4)) = 10$, σ_3, σ_4 are surjective, and σ_5 is not surjective (dim coker $\sigma_5 = 7$).
2) $c \in S_3{}^1 \setminus S_2{}^1$, in which case there are the following possibilities:
 i) $h^\circ(I_C(3)) = 2$, $h^\circ(I_C(4)) = 7$, σ_3 is surjective, σ_4 and σ_5 are not (dim coker $\sigma_4 = 4$, dim coker $\sigma_5 = 2$)
 ii) $h^\circ(I_C(3)) = 1$, hence σ_3 is injective, and there are the following possibilities:
  $h^\circ(I_C(4)) = 5$, in which case σ_4 is not surjective, and σ_5 is surjective;
  $h^\circ(I_C(4)) = 6$, in which case σ_4 and σ_5 are not surjective (dim coker $\sigma_5 = 1$);
  $h^\circ(I_C(4)) = 7$, in which case we don't know about σ_4, and σ_5 is not surjective (dim coker $\sigma_5 = 2$).
3) $c \in S_4{}^1 \setminus S_3{}^1$, in which case σ_3 is trivially injective and σ_5 is never surjective. C can be of type a or b; if C is of type b, we assume that r_4 is contained in the quadric generated by r_1,r_2,r_3. Denoting by Q the quadric generated by r_1,r_2,r_3, there are the following possibilities: the intersection of C with Q is of type
  b_2, in which case $h^\circ(I_C(4)) = 6$, dim coker $\sigma_5 = 1$ and σ_4 is not surjective;
  b_3, in which case $h^\circ(I_C(4)) = 7$, dim coker $\sigma_5 = 2$ and σ_4 is surjective;
  a_5, in which case $h^\circ(I_C(4)) = 6$, dim coker $\sigma_5 = 1$ and for C generic (in $W_{0,1}$) σ_4 is surjective;
  a_6, in which case $h^\circ(I_C(4)) = 6$, dim coker $\sigma_5 = 1$ and σ_4 is not surjective;
  a_7, in which case $h^\circ(I_C(4)) = 7$, dim coker $\sigma_5 = 2$ and for C generic (in $W_{0,2}$) σ_4 is surjective;
4) C is of maximal rank; if it is generic, is minimally generated. If C is not minimally generated, then the only map which is not of maximal rank is σ_4.

Proof. Lemma 3.1 with d =6 gives 1). For what concerns parts 2), 3) and 4), the study of $h^\circ(I_C(3))$, σ_3 and σ_4 is in sections from 5 to 8, as well as the description of the possible intersection with Q. This, together with 4.1, give $h^\circ(I_C(4))$ and dim coker σ_5.

Section 5

Prop.3.2 gives the generic behaviour in $W_{0,1}$ and $W_{0,2}$ for the map σ_5. In this section, we study σ_4.

5.1 *Lemma.* Let C be generic in $W_{0,2}$. Then, the map σ_4 is surjective for C.
Proof. We recall that $h^\circ(I_C(4)) = 7$ (4.1), hence the map $\sigma_4 : I_4 \otimes V \to I_5$ is surjective if and only if dim ker $\sigma_4 = 8$. Since in any case dim ker $\sigma_4 \geq 8$, it is enough to prove that for C dim ker $\sigma_4 \leq 8$.
It is hence enough to prove that the union in $\mathbb{P}(\Omega_3)$, R, of a specialization C' of C, with 8 s-points, satisfies $h^\circ(L_4 \otimes I_R) = 0$. Let C be the union of $r_1,...,r_6$, and let Q be the quadric generated by $r_1,...,r_3$; we specialize r_4 and 4 of the s-points on Q.

Since $res_Q R$ that is, the union of 2 skew lines and 4 s-points satisfies $h^\circ(I_{res_Q R} \otimes L_2) = 0$

(see [I,1], 5.2), it is enough to prove $H^\circ(F_{5,5} \otimes I_{R \cap Q,Q}) = 0$, that is, $H^\circ(F_{1,5} \otimes I_A) = 0$, where A is the union of 4 s-points and of $(r_5 \cup r_6) \cap Q$. Let s be a 6-secant for C; then, s is contained in Q, and since $F_{1,5}|_s \cong \mathcal{O}^{\oplus 2} \oplus \mathcal{O}(-1)$, a section in $H^\circ(F_{1,5} \otimes I_A)$ vanishes identically on s. Since $h^\circ(F_{1,3}) = 4$, a section $\sigma \in H^\circ(F_{1,3})$ vanishing on the 4 residual s-points vanishes.

The rest of this section is devoted to proving the following:

5.2 *Lemma.* Let C be generic in $W_{0,1}$. Then, the map σ_4 is surjective for C.

We recall that $h^\circ(I_C(4)) = 6$ (see 4.1), hence the map $\sigma_4 : I_4 \otimes V \to I_5$ is surjective if and only if dim ker $\sigma_4 = 4$; so it is enough to prove that the union T in $\mathbb{P}(\Omega_3)$ of a specialization C' of C with 4 general points satisfies: $h^\circ(\mathcal{I}_T \otimes L_4) = 0$. This is proved in 5.5. We shall use"la méthode d'Horace" in a blow up of $\mathbb{P}(\Omega_3)$. For the related tecniques, we send to [I,1], 1.9.3.

We shall need some preliminary lemmas and notations.

5.3 ***Notations.*** We fix a plane H in $\mathbb{P}^3$, a point $x \in H$ and a line s such that $s \not\subset H$, $x \in s$. With $\eta_s(x)$ we denote the first infinitesimal neighbourhood of x in s.

If $q: F \to \mathbb{P}^3$ is a rank 3 vector bundle, $\mathbb{P}(F)$ is locally $\mathbb{A}^3 \times \mathbb{A}^2$. If $b \in \mathbb{P}(F)_x$, with $\tau(b)$, resp. $\chi_s(b)$ we denote the first infinitesimal neighbourhood in $H \times \{b\}$, resp. $s \times \{b\}$.

5.4 ***Lemma.*** Let H be a plane in $\mathbb{P}^3$, $r_1,\dots,r_4$ skew lines with $r_4 \subset H$, such that $x \in r_4$ is the intersection of r_4 with a quadrisecant s to the four lines.

We denote by Z the general union in $\mathbb{P}(\Omega_3)$ of r_1,r_2,r_3, of $r_4 \cap \mathbb{P}(\Omega_H)$, of $\pi^{-1}(\eta_s(x))$, of $\tau(b)$, where b denotes an s-point in $\pi^{-1}(x)$, not in $\mathbb{P}(\Omega_H)_x$, and of 4 s-points, two among which on $\mathbb{P}(\Omega_H)$.Then, $H^\circ(\mathcal{I}_Z \otimes L_3) = 0$.

Proof. We work with the following elementary transformations:

$$0 \to \Omega_3(3) \to \mathcal{O}_{\mathbb{P}^3}(2)^{\oplus 3} \to \mathcal{O}_H(3) \to 0$$

$$0 \to \mathcal{O}_{\mathbb{P}^3}(2)^{\oplus 3} \to \Omega_3(4) \to \Omega_H(4) \to 0$$

and we set $\overline{V} = \mathbb{P}(\Omega_3(4))$, $V' = \mathbb{P}(\mathcal{O}_{\mathbb{P}^3}(2)^{\oplus 3})$, $Y = \mathbb{P}(\Omega_H(4))$, $Y' = \mathbb{P}(\mathcal{O}_H(3))$, and:

$$\begin{array}{ccccc} & & \widetilde{V} & & \\ & g\swarrow & & \searrow f & \\ V' & & & & \overline{V} \\ & p\searrow & & \swarrow \pi & \\ & & \mathbb{P}^3 & & \end{array}$$

blow up along Y' with exc. div.G' (for g); blow up along Y with exc. div. G (for f).

Denoting by Q the (unique) quadric containing r_1, r_2, r_3, we see that Q has to contain s, hence, $p^{-1}Q$ contains $p^{-1}(x)$ and $\chi_s(c)$, with $c = p^{-1}(x) \cap Y'$. Moreover, $p^{-1}(Q)$ contains a double point contained in $\tau(c)$ (i.e. the tangent direction to $Q \cap H$). With ϕ we denote an equation for Q.

We denote by K' the following subscheme of V': the union of $p^{-1}(r_1 \cup r_2 \cup r_3)$ with $p^{-1}(x)$, $\chi_s(c)$ and $\tau(c)$, plus 2 s-points y,z in Y' not belonging to $p^{-1}Q$. We set $K := g^{-1}K'$. We wish to prove $H^\circ(\mathcal{I}_K \otimes g^*\mathcal{O}_{V'}(1)) = 0$. From the exact sequence:

$$0 \to g^*\mathcal{O}_{V'}(1) \otimes \mathcal{I}_{G'} \otimes \mathcal{I}_{\mathrm{res}_{G'}K} \to g^*\mathcal{O}_{V'}(1) \otimes \mathcal{I}_K \to g^*\mathcal{O}_{V'}(1) \otimes \mathcal{I}_{K\cap G', G'} \to 0$$

we see that it is enough to prove the following i) and ii).

i) $H^\circ(g^*\mathcal{O}_{V'}(1) \otimes \mathcal{I}_{G'} \otimes \mathcal{I}_{\mathrm{res}_{G'}K}) = 0$

Since the map σ_2 is injective for 3 skew lines, $\ker \sigma_2 \cong H^\circ(\Omega_3(3) \otimes \mathcal{I}_{r_1 \cup r_2 \cup r_3}) = 0$.

On the other hand, $g^{-1}p^{-1}(r_1 \cup r_2 \cup r_3) \subset \mathrm{res}_{G'}K$.

Hence: $H^\circ(g^*\mathcal{O}_{V'}(1) \otimes \mathcal{I}_{G'} \otimes \mathcal{I}_{\mathrm{res}_{G'}K}) \subset H^\circ(g^*\mathcal{O}_{V'}(1) \otimes \mathcal{I}_{G'} \otimes \mathcal{I}_{g^{-1}p^{-1}(r_1 \cup r_2 \cup r_3)}) \cong$

$\cong H^\circ(\Omega_3(3) \otimes \mathcal{I}_{r_1 \cup r_2 \cup r_3}) = 0$.

ii) $\mathrm{Im}(H^\circ \gamma) = 0$.

As in [I,1], 5.5 p.118, it is easy to see that the last condition is equivalent to $\mathrm{Im}(i \circ \alpha) = 0$, where

$$H^\circ(\mathcal{O}_{V'}(1)\otimes \mathcal{I}_{K'}) \xrightarrow{\ i\ } H^\circ(\mathcal{O}_{V'}(1)\otimes \mathcal{I}_{R}) \xrightarrow{\ \alpha\ } H^\circ(\mathcal{O}_{V'}(1)\otimes \mathcal{I}_{R\cap Y',Y'}).$$

$$\sigma \mapsto (\lambda_0\phi, \lambda_1\phi, \lambda_2\phi) \mapsto \phi_{|Y'}\sum_{i=0}^{2}\lambda_i x_i$$

where λ_i are constants, and x_i are coordinates on $Y' \equiv H$ (α can be regarded as the map

$H^\circ(\mathcal{O}_{\mathbb{P}^3}(2)^{\oplus 3}\otimes \mathcal{I}_R) \to H^\circ(\mathcal{O}_H(3)\otimes \mathcal{I}_{R\cap H,H})$ coming from the first elementary transformation: see [I,1] 5.1.3). We now observe that ϕ vanishes neither on the s-points y,z nor on a tangent direction contained in $\tau(c)$ (and not in Q). Hence the linear form $\Sigma\lambda_i x_i$ vanishes identically, and we conclude that $\mathrm{Im}(i\circ\alpha) = 0$.

We denote by $\tilde{Z}$ the pull back through f of the scheme Z described in the statement of the lemma, thought as a subscheme of $\bar{V}$. We now prove $H^\circ(f^*\mathcal{O}_{\bar{V}}(1)\otimes \mathcal{I}_{\tilde{Z}}) = 0$ (this is equivalent to $H^\circ(\mathcal{I}_Z\otimes L_3) = 0$), exploiting the divisor G in $\tilde{V}$. We have:

i) $H^\circ(f^*\mathcal{O}_{\bar{V}}(1)\otimes \mathcal{I}_{G\cap\tilde{Z},G}) = H^\circ(\mathcal{O}_{\bar{V}}(1)\otimes \mathcal{I}_{Y\cap Z,Y}) = 0$

since a section of $\Omega_H(4)$ vanishing on a line, 3 d-points and 2 s-points vanishes.

ii) $H^\circ(f^*\mathcal{O}_{\bar{V}}(1)\otimes \mathcal{I}_G\otimes \mathcal{I}_{\mathrm{res}_G\tilde{Z}}) \cong H^\circ(\mathcal{O}_{V'}(1)\otimes \mathcal{I}_{\tilde{K}}) = 0,$

since $\tilde{K}$ is a generalization of the scheme K' described above. Hence we have the thesis.

5.5 ***Lemma.*** Let T be the general union in $\mathbb{P}(\Omega_3)$ of 4 skew lines, of a singular reduced conic with the singular point on a 4-secant, s, to the 4 lines, of the first infinitesimal neighbourhood of the singular point of the conic, and of 4 s-points. Then, $H^\circ(\mathcal{I}_T\otimes L_4) = 0$.

Proof. We fix a plane H in $\mathbb{P}^3$. We consider the following elementary transformations:

$$0 \to \Omega_3(4) \to \mathcal{O}_{\mathbb{P}^3}(3)^{\oplus 3} \to \mathcal{O}_H(4) \to 0$$

$$0 \to \mathcal{O}_{\mathbb{P}^3}(3)^{\oplus 3} \to \Omega_3(5) \to \Omega_H(5) \to 0$$

and we set $\bar{V} = \mathbb{P}(\Omega_3(5))$, $V' = \mathbb{P}(\mathcal{O}_{\mathbb{P}^3}(3)^{\oplus 3})$, $Y = \mathbb{P}(\Omega_H(5))$, $Y' = \mathbb{P}(\mathcal{O}_H(4))$, and:

$$\begin{array}{ccccc} & & \tilde{V} & & \\ & \swarrow g & & f \searrow & \\ V' & & & & \bar{V} \\ & p \searrow & & \swarrow \pi & \\ & & \mathbb{P}^3 & & \end{array}$$

blow up of V' along Y', with exceptional divisor G'.

blow up of $\bar{V}$ along Y, with exceptional divisor G.

Consider the following subscheme of $\mathbb{P}^3$: four skew lines r_1,r_2,r_3,r_4 outside H, and 2 lines r_5,r_6 in H meeting at the point x, with x lying on a 4-secant s for r_1,r_2,r_3,r_4. We denote by F the following subscheme of V': the union of $p^{-1}(r_1\cup\ldots\cup r_4)$, of $p^{-1}(r_5\cup r_6)\cap Y'$, $\chi_s(p^{-1}(x)\cap Y')$, and of $p^{-1}(x)$, plus 4 general s-points. We denote by K' the subscheme of V' obtained from F specializing r_4 in H (hence, it has to pass through x). We set $K := g^{-1}(K')$.

We exploit G' in $\tilde{V}$ to show $0 = H^\circ(g^*\mathcal{O}_{V'}(1)\otimes \mathcal{I}_K)$ $(\cong H^\circ(\mathcal{O}_{V'}(1)\otimes \mathcal{I}_{K'}))$. We have:

i) $H^\circ(g^*\mathcal{O}_{V'}(1)\otimes \mathcal{I}_{G'\cap K,G'}) \cong H^\circ(\mathcal{O}_{V'}(1)\otimes \mathcal{I}_{Y'\cap K',Y'}) = 0$, since a section of $\mathcal{O}_H(4)$ vanishing on 3 lines and 3 points vanishes.

ii) $H^\circ(g^*\mathcal{O}_{V'}(1)\otimes \mathcal{I}_{G'}\otimes \mathcal{I}_{\mathrm{res}_{G'}K}) \cong H^\circ(\mathcal{O}_{\bar{V}}(1)\otimes \pi^*\mathcal{O}_{\mathbb{P}^3}(-1)\otimes f_*\mathcal{I}_{\mathrm{res}_{G'}K}) \cong H^\circ(L_3\otimes \mathcal{I}_R)$

where R is a generalization of the subscheme Z of $\mathbb{P}(\Omega_3)$ described in lemma 5.4, hence the last cohomology group is zero. In fact, $\mathrm{res}_{G'}K$ contains the pull-back of the following subscheme of $\bar{V}$: the union of $r_1, r_2, r_3, r_4\cap\mathbb{P}(\Omega_H)$, $\pi^{-1}(\eta_s(x))$, $\tau(b)$, where b denotes an s-point in $\pi^{-1}(x)$, not in $\mathbb{P}(\Omega_H)_x$, and the 4 s-points outside H (see lemma 5.6).

We now denote by N the scheme described in the statement of the lemma, thought as a subscheme of $\overline{V}$. We set $M := f^{-1}N$, and we prove $H^{\cdot}(f^{*}\mathcal{O}_{\overline{V}}(1) \otimes \mathcal{I}_M) \cong H^{\cdot}(\mathcal{O}_{\overline{V}}(1) \otimes \mathcal{I}_N) \cong H^{\cdot}(L_4 \otimes \mathcal{I}_T) = 0$, exploiting the divisor G in $\widetilde{V}$. We have:

i) $H^{\cdot}(f^{*}\mathcal{O}_{\overline{V}}(1) \otimes \mathcal{I}_{G\cap M,G}) \cong H^{\cdot}(\mathcal{O}_Y(1) \otimes \mathcal{I}_{N\cap Y,Y}) = 0$, since a section of $\Omega_H(5)$ vanishing on a conic and 4 d-points in $\mathbb{P}(\Omega_H)$ vanishes.

ii) $H^{\cdot}(f^{*}\mathcal{O}_{\overline{V}}(1) \otimes \mathcal{I}_G \otimes \mathcal{I}_{res_G M}) \cong H^{\cdot}(\mathcal{O}_{V'}(1) \otimes g_{*}\mathcal{I}_{res_G M}) = 0$, since $res_G M = g^{-1}(F)$ (see [I,1] lemma 5.6.5).

5.6 *Lemma.* We keep the notations of 5.5. Let H be a plane in $\mathbb{P}^3$, s_1, s_2 two lines in H meeting at $\bar{x}$, and r, s lines outside H, with $\bar{x} \in s$ and $r \cap s \neq \phi$. We consider the union in V' of $p^{-1}(r \cup \{\bar{x}\})$, of $p^{-1}(s_1 \cup s_2) \cap Y'$ and of $\chi_s(p^{-1}(\bar{x}) \cap Y')$, and we denote by T_0 the scheme obtained specializing r in the plane so that $\bar{x} \in r$, with collision direction s.

Then, $res_{G'}\, g^{-1} T_0$ contains $g^{-1}p^{-1}(r) \cap G$, union with the pull back through g of $\tau(\{\bar{x}\} \cap Y')$, and of $p^{-1}(\eta_s(\bar{x}))$.

Proof. We work in local coordinates. Let A, resp. B, be the following open subsets of $\mathbb{P}^3$, resp. of $V'_{|A} \cong A \times \mathbb{P}^2$: $B = \mathrm{Spec}\, k\,[x,y,z]$, $A = \mathrm{Spec}\, k\,[x,y,z,u,v]$.

We may assume $I_H = (z)$, $I_{\bar{x}} = (x,y,z)$, $I_{Y'} = (u,v,z)$, $I_{s_1 \cup s_2} = (xy,z)$, $I_{\chi_s(p^{-1}(\bar{x}) \cap Y')} = (x,y,z^2,u,v)$

Let $I_{r_\lambda} = (z-\lambda, x-y)$. We first build a flat family with general fiber T_λ and special fiber T_0:

$$I_{T_\lambda} = I_{p^{-1}(r_\lambda)} \cap I_{p^{-1}(\bar{x})} \cap I_{p^{-1}(s_1 \cup s_2) \cap Y'} \cap I_{\chi_s(p^{-1}(\bar{x}) \cap Y')} =$$

$= (xy(z-\lambda), xz(z-\lambda), yz(z-\lambda), z^2(z-\lambda), ux(z-\lambda), uy(z-\lambda), uz(z-\lambda), vx(z-\lambda), vy(z-\lambda), vz(z-\lambda),$ $xy(x-y), xz(x-y), yz(x-y), z^2(x-y), ux(x-y), uy(x-y), uz(x-y), vx(x-y), vy(x-y), vz(x-y))$.

We get that $res_{p^{-1}(H)} T_0$ has associated ideal $(xy, xz, yz, z^2, ux, uy, uz, vx, vy, vz, x^2, y^2)$.

It is clear that the d-line $f^{-1}(\pi^{-1}(r) \cap Y)$ is contained in $res_{G'}\, g^{-1} T_0$. The ideal $\mathfrak{a}$ of the first infinitesimal neighbourhood of $\{\bar{x}\} \cap Y'$ in Y' is (x^2,y^2,xy,u,v,z). The ideal $\mathfrak{b}$ of $p^{-1}(\eta_s(\bar{x}))$ is (x,y,z^2) (notice that the collision direction is here the z-axis).

Since $\mathfrak{a} \cap \mathfrak{b} \supset I_{res_{p^{-1}H} T_0}$, and since $res_{G'}\, g^{-1}(T_0) \supset res_{g^{-1}p^{-1}H}\, g^{-1}(T_0) = g^{-1}\, res_{p^{-1}H}\, T_0$, we have the thesis.

5.7 *Remark.* There is no irreducible component of W_{02} lying in $S_3{}^1$; in fact given any union C of six lines with two 6-secants, no 5-secants, and $h^{\cdot}(\mathcal{I}_C(3)) = 1$, it is possible to build a family having C as a special fiber, and an element of W_{02} with $h^{\cdot}(\mathcal{I}_C(3)) = 0$ as generic fiber (C is of type a; now move a line, say r_4, until it lies in the quadric generated by r_1, r_2, r_3). The same holds for W_{01}.

Section 6: In $S_3{}^1 \setminus S_2{}^1$

In the following sections 6,7 and 8 we study the maps σ_k on the postulation strata for six skew lines. Lemma 3.1 takes care of $S_2{}^1$, so here we start by studying unions of six lines lying on a cubic surface but not on a quadric.

6.1 The case $h°(\mathcal{I}_C(3)) = 2$.

In this paragraph 6.1, C denotes a union of six lines $r_1,...,r_6$ contained in no quadrics and in at least one reducible cubic. We shall see (6.1.1, 6.2.2) that this condition is equivalent to the condition $h°(\mathcal{I}_C(3)) = 2$.

6.1.1 ***Lemma.*** Assume there is a reducible cubic containing C. Then, $h°(\mathcal{I}_C(3)) = 2$,and $h°(\mathcal{I}_C(4)) = 7$.

Proof. If there is a reducible cubic S_3 containing C, the only possibility is that $S_3 = Q \cup H$, where Q is a quadric containing 5 of the lines. Hence we have $h°(\mathcal{I}_C(3)) = 2$; in fact, a cubic S containing C has to contain Q, hence $S = Q \cup H$ with H plane through the line $r_6 \not\subset Q$. If S_4 is a quartic containing C, $S_4 = Q \cup Q'$ with Q' any quadric through r_6; and $h°(\mathcal{I}_{r_6}(2)) = 7$ (or see 4.1).

6.1.2 ***Lemma.*** Let C be as in 6.1.1. Then, all maps σ_k' s are of maximal rank, execpt σ_4 and σ_5; moreover, dim coker $\sigma_5 = 2$.

Proof. The map σ_2 is trivially injective, because $h°(\mathcal{I}_C(2)) = 0$. Now we consider the map $\sigma_3 : I_3 \otimes V \rightarrow$ $\rightarrow I_4$. If q is an equation for Q, we have (6.1.1) $I_3 \cong q\cdot W$, where $W \subset V$ is the dimension 2 vector space of the planes through r_6, and $I_4 \cong q\cdot \Lambda$, where $\Lambda \subset S^2 V$ is the dim 7 vector space of the quadrics through r_6. Hence the map σ_3 can be seen as $\bar{\sigma}_3 : W \otimes V \rightarrow S^2 V$, and by the following 6.1.3, dim Im $\bar{\sigma}_3 = 7$. Since Im $\bar{\sigma}_3 \subset \Lambda$, and dim Λ = dim Im $\bar{\sigma}_3$, we have that σ_3 is surjective. In particular, we have proved

$$I_4 \cong q\cdot\left(\frac{W\otimes V}{(x\otimes y - y\otimes x)}\right), \text{ where } \{x,y\} \text{ is a basis for } W.$$

Now we consider the map $\sigma_4 : I_4 \otimes V \rightarrow I_5$; since dim $(I_4 \otimes V) = 28$, dim $I_5 = 20$, σ_4 is of maximal rank if and only if dim ker $\sigma_4 = 8$. Since Im $\sigma_4 \subset q\cdot S^3V$, we can look at σ_4 as the natural map:

$$\bar{\sigma}_4 : \left(\frac{W\otimes V}{(x\otimes y - y\otimes x)}\right) \otimes V \rightarrow S^3V;$$

denoting by $\{x,y,z,t\}$ a basis for V, we see that there are at least twelve independent relations, namely:

$y\otimes y\otimes\beta - y\otimes\beta\otimes y$ $(\beta = x,z,t)$; $x\otimes x\otimes\beta - x\otimes\beta\otimes x$ $(\beta = y,z,t)$;

$x\otimes y\otimes\beta - x\otimes\beta\otimes y$ $(\beta = z,t)$; $y\otimes x\otimes\beta - y\otimes\beta\otimes x$ $(\beta = z,t)$;

$\beta\otimes z\otimes t - \beta\otimes t\otimes z$ $(\beta = x,y)$.

Finally, since σ_k is surjective for $k \geq 6$, the map σ_5 cannot be surjective, otherwise I_C would be generated by quintics, against the fact that the two 6-secants to C lie in each quintic surface containing C (lemma 4.1 gives in fact dim ker $\sigma_5 = 40$).

6.1.3 ***Lemma.*** Let V be a dimension 4 vector space, $W \subset V$ a dimension 2 subspace.Then, denoting by $\tau : W\otimes V \rightarrow S^2V$ the natural map, we have dim Im $\tau = 7$.

Proof. Let $\{e_i\}_{i=1,...,4}$, $\{e_i\}_{i=1,2}$ be a basis for V, resp.W. Then, ker $\tau = <e_1\otimes e_2 - e_2\otimes e_1>$. Since dim $W\otimes V = 8$ and dim ker $\tau = 1$, we have the thesis.

6.2 The case $h°(\mathcal{I}_C(3)) = 1$

All along this paragraph 6.2, C denotes a union of six skew lines contained in no quadric and in at least one irreducible cubic.

6.2.1 ***Lemma.*** If there is an irreducible cubic containing C, then $h°(\mathcal{I}_C(3)) = 1$.

Proof. Assume S is an irreducible cubic surface containing C; then, C of type a. Let Q be the quadric containing r_1, r_2, r_3 (as lines of type (1,0)); assume $h°(\mathcal{I}_C(3)) \geq 2$, and let S, S' be two irreducible cubic surfaces containing C (by 6.1.1, also S' is irreducible). Let t be a line of type (0,1) on Q through a point of $r_j \cap Q$, j = 4,5,6. Then, $t \subset S \cap S'$; in fact t intersects the cubic surface S in 4 points, so $t \subset S$ and the same for S'. The points $r_j \cap Q$, j = 4, ...,6, are contained in at least two (counted with multiplicity) lines of type (0,1), say t_1, t_2. We set $Y := r_1 \cup ... \cup r_6 \cup t_1 \cup t_2$, where $t_1 \cup t_2$ is eventually a double line on Q; we have $S \cap S' \supset Y$. Since S and S' are irreducible, they meet properly, and one has:

$0 \to \mathcal{O}_{\mathbb{P}^3}(-6) \to \mathcal{O}_{\mathbb{P}^3}(-3)^{\oplus 2} \to \mathcal{I}_{S \cap S'} \to 0.$

So we get $h^\circ(\mathcal{I}_{S \cap S'}(4)) = 2\, h^\circ(\mathcal{O}_{\mathbb{P}^3}(1)) = 8$. If we prove that $h^\circ(\mathcal{I}_Y(4)) \le 7$ we have a contradiction, since $S \cap S' \supset Y$. It is enough to prove $h^\circ(\mathcal{I}_T(4)) = 0$ for T generic union of Y and seven points in $\mathbb{P}^3$. Let S_4 be a quartic containing T, and specialize six of the points on Q. The curve $S_4 \cap Q$ if of type (4,4) on Q, and contains Y and the six points. Since $h^\circ(\mathcal{O}_Q(1,2)) = 6$, we conclude that S_4 contains Q, that is, $S_4 = Q \cup Q'$, with Q' a quadric containing the three remaining lines and the point; but there is no suoh Q'.

6.2.2 ***Remark.*** For such a C, all the maps σ_k are of maximal rank if $k \ne 4, 5$ by 0.11 (σ_k is trivially injective for $k \le 3$). Hence in the following we study σ_4 and σ_5.

6.2.3 ***Notation.*** Let C be the union of $r_1,...,r_6$; we denote in the following with Q the quadric generated by r_1, r_2, r_3. The number of lines t_i of type (0,1) on Q and passing through one of the points $r_j \cap Q$, j = 4,5,6, is at least 2 and at most 3 (counted with multiplicity), because $S \cap Q$ is of type (3,3), and contains r_1, r_2, r_3 and those t_i's. Hence, the intersection of C with Q is of type a_4, or a_6, or a_7.

6.2.4 ***Lemma.*** Let the intersecti on of C with Q be of type a_4; then, $h^\circ(\mathcal{I}_C(4)) = 5$, σ_5 is surjective, and σ_4 is not of maximal rank.

Proof. The first two assertions follow by 4.1. If $\sigma_4 : I_4 \otimes V \to I_5$ were of maximal rank, it should be bijective, hence I_C would be generated in degree ≤ 4, and this is impossible (there are 5-secants).

6.2.5 ***Lemma.*** Let the intersection of C with Q be of type a_6. Then, $h^\circ(\mathcal{I}_C(4)) = 6$, and the maps σ_4 and σ_5 are not of maximal rank; moreover, dim coker $\sigma_5 = 1$.

***Proof*.** Two of the assertions follow by 4.1. The claim on σ_4 follows directly from 4.2 and 4.3 for any C as in the assumptions, provided that two of the lines outside Q are not tangent to Q (i.e. as soon as the 5-secant and the 6-secant are supported by distinct lines). Now let C' be a limit case; it is enough to observe that the dimensions of domain and codomain of $\sigma_4(C')$ are equal to those of $\sigma_4(C)$ (see 4.1), while by semicontinuity dim $h^\circ(\Omega_3(5) \otimes \mathcal{I}_{C'}) \ge$ dim $h^\circ(\Omega_3(5) \otimes \mathcal{I}_C)$.

6.2.6 ***Lemma.*** Let the intersection of C with Q be of type a_7. Then, $h^\circ(\mathcal{I}_C(4)) = 7$, and σ_5 is not of maximal rank (dim coker $\sigma_5 = 2$).

Proof. See 4.1.

Section 7: In $S_4^1 \setminus S_3^1$

7.1 ***Remark.*** In the following, C always denotes a union of six lines $r_1,...,r_6$ such that $c \in S_4^1 \setminus S_3^1$, that is, $h^\circ(\mathcal{I}_C(3)) = 0$ and $h^\circ(\mathcal{I}_C(4)) > 5$. Since we assume $h^\circ(\mathcal{I}_C(3)) = 0$, C is of type a or b. If C is of type b, we may assume that $r_1,...,r_4$ lie in a quadric.

We shall denote all along this section by Q the quadric generated by r_1, r_2, r_3. Since $h^\circ(\mathcal{I}_C(4)) > 5$, the intersection of C with Q is of type b_2 or b_3, a_5, a_6 or a_7; we have respectively : $h^\circ(\mathcal{I}_C(4)) = 6$,

$h^\circ(\mathcal{I}_C(4)) = 7$, $h^\circ(\mathcal{I}_C(4)) = 6$, $h^\circ(\mathcal{I}_C(4)) = 6$, $h^\circ(\mathcal{I}_C(4)) = 7$ (see 4.1).
Moreover, for any such C, all the maps σ_k are of maximal rank, for $k \neq 4,5$, by 0.11 (σ_k is trivially injective for $k \leq 3$). On the other hand, σ_5 is never surjective, and $\dim \ker \sigma_5$ is given by 4.1. Hence in the following it is enough to study σ_4.

7.2 ***Lemma.*** Let the intersection of C with Q be of type b_3. Then, σ_4 is onto.
Proof. This is essentially the same proof of 5.1.

7.3 ***Lemma.*** Let the intersection of C with Q be of type b_2. Then, σ_4 is not onto.
Proof. It follows by 4.2 and 4.3 (there is a 5-secant); for the limit case, see proof of 6.2.5.

7.4 ***Remark.*** It is possible to build a flat family having a disjoint union of 6 lines with intersection with Q of type b_2 as general fiber, and b_3 as special fiber. This gives an explicit example of the following fact: the property "σ_k is surjective" is not preserved by generalization. On the other hand, it is clear that on a subvariety of V where the postulation is constant, this property is preserved by generalization. In fact, it is then equivalent to: "$h^\circ(L_k \otimes \mathcal{I}_T) = 0$, with T union of 6 lines and τ points, $\tau = 4 \dim I_k - \dim I_{k+1}$", and $h^\circ(L_k \otimes \mathcal{I}_T)$ is a semi- continuous function on V.

7.5 ***Lemma.*** Let the intersection of C with Q be of type a_6. Then, σ_4 is not surjective.
Proof. It follows by 4.2 and 4.3 (there is a 5-secant); for the limit case, see proof of 6.2.5.

7.6 ***Lemma.*** Let C be generic in W_{01} (respectively in W_{02}). Then, C is in $S_4{}^1 \setminus S_3{}^1$, the intersection with Q is of type a_5 (resp.a_7), and σ_4 is surjective.
Proof. See 5.2 (resp. 5.1) and 5.7.

Section 8: In the maximal rank locus

In the following, we denote by C a union of 6 skew lines $r_1,\dots,r_6$ of maximal rank. The postulation of C is completely described in 0.10.

8.1 ***Remark.*** i) We recall that if C is generic, all σ_k's are of maximal rank.
ii) If C is of maximal rank, C can be of type a or b. Denoting by Q the quadric generated by r_1,r_2,r_3, and assuming, if C is of type b, that r_4 lies in Q, the intersection of C with Q is of type b_1, resp. of type a_1,a_2,a_3,a_4 or a_8 (the other cases are excluded by 4.1). All these cases are effectively possible.
In fact, 4.1 gives $h^\circ(\mathcal{I}_C(4)) = 5$. Moreover, if C is of type b, we have $h^\circ(\mathcal{I}_C(3)) = 0$ (a cubic containing C should contain Q); if C is of type a, and if the intersection of r_4,r_5,r_6 with Q is not contained in three lines of type (0,1), then $h^\circ(\mathcal{I}_C(3)) = 0$ (a cubic containing C has to meet Q along a curve (3,3)), hence if the intersection of C with Q is of type a_2, a_3, or a_8, $h^\circ(\mathcal{I}_C(3)) = 0$. Finally, it is easy to build a family of unions of six lines whose special fiber is of type b, and whose generic fiber C has an intersection with the quadric of type a_4. By semicontinuity, we have $h^\circ(\mathcal{I}_C(3)) = 0$.

8.2 ***Lemma.*** If the intersection of C with Q is of type b_1, or of type a_2, a_3, a_4, or a_8 the maps σ_k are all of maximal rank, except σ_4.
Proof. Since C is of maximal rank, $h^\circ(\mathcal{I}_C(3)) = 0$ and $h^1(\mathcal{I}_C(4)) = 0$; hence σ_k is injective for $k \leq 3$, and surjective for $k \geq 5$. In all these cases, there exists at least a 5-secant. Hence σ_4 cannot be surjective.

Bibliography

[G-L-P] Gruson, L-Lazarsfeld, R-Peskine,C: *"On a theorem of Castelnuovo and the equations defining space curves"*, Invent. Math. **72**, 491-506 (1983).

[H-H] Hartshorne, R-Hirschowitz, A.: *"Droites en position générale dans l'espace projectif"* in Algebraic Geometry, Proceedings LaRabida, 1981. Lecture Notes in Math. **961**, Springer Verlag (1982).

[H,1] Hirschowitz, A.: *"Sur la postulation générique des courbes rationelles"*, Acta Math. **146**, 209-230 (1981).

[H,2] Hirschowitz,A: *"La méthode d'Horace pour l'interpolation a plusieurs variables"*, Manuscripta Math. **50**, 337-388 (1985).

[H,3] Hirschowitz,A.: Letter from Hirschowitz to Hartshorne of 12 August 1983.

[I,1] Idà, M: *"On the homogeneous ideal of the generic union of lines in* $\mathbb{P}^3$ ", Thesis. Nice University (1986). To appear on J.Reine Angew. Math.

[I,2] Idà, M: *"Maximal rank and minimal generation"*, Arch. Math. **52**, 186-190 (1989).

[LB] Le Barz, P: *"Formules multisecantes pour les courbes gauches quelconques"* , Enumerative Geometry. Progress in Mathematics,vol. 24 (1982).

[Mu] Mumford, D: *"Lectures on curves on an algebraic surface"*, Princeton University Press (1966).

(*) This paper was written in the ambit of GNSAGA of CNR, with support from MPI.

LIAISON OF FAMILIES OF SUBSCHEMES IN P^n.

Jan O. Kleppe
Oslo College of Engineering
Cort Adelersgt 30
OSLO - NORWAY

In this paper we study families of subschemes in a projective n-space $P = P^n_k$. One such family (of codimension r) is nothing more than a closed embedding $X \subset P \times S$ of flat schemes over S of relative codimension r. If $Y \subset P \times S$ is a family of complete intersections of multidegree (or type) $f_1, .., f_r$ and $Y \supset X$, then, under some weak conditions on X, it is possible to define linkage such that the linked subscheme $X' \subset P \times S$ is flat over S and such that for any $s \in S$, the fiber at s is linked to the fiber of $X' \subset P \times S$ in the usual sense, cf.(2.4). This natural, but fundamental result is the starting point of making more intuitive liaison considerations on the Hilbert scheme $H(p) := \mathrm{Hilb}^p(P^n_k)$ precise. Now recall that an S-point of the Hilbert flag functor is just a sequence $X \hookrightarrow Y \hookrightarrow P \times S$ of flat subschemes over S. For its representing object $D(p;\underline{f})$ the fundamental result above induces an isomorphism

$$(2.6) \qquad \tau : D(p;\underline{f})_{CM} \approx D(p';\underline{f})_{CM}$$

where the subscript CM denotes the restriction of $D(p;\underline{f})$ to equidimensionnal Cohen Macaulay subschemes of P, and where p, resp. p', is the Hilbert polynomial of $X \subset P$, resp. of the linked subscheme $X' \subset P$. Forgetting Y we get natural forgetful maps $pr_1 : D(p;\underline{f})_{CM} \longrightarrow H(p)_{CM}$ and $pr_1' : D(p';\underline{f})_{CM} \longrightarrow H(p')_{CM}$. If $U \subset H(p)_{CM}$, we define the $\underline{f}$-linked family U' to be $U' = pr_1'(\tau(pr_1^{-1}(U)))$. Of course U' consists of any linked subscheme $X' \subset P$ obtained from some $X \subset P$ of U using some complete intersection $Y \subset P$ of type $\underline{f} = (f_1, .., f_r)$ containing X. Studying pr_1 in detail we give in this paper explicite conditions under which the linked family U' is open (dense), irreducible and smooth, cf.(3.4), (3.8) and (3.9). For instance if U is an irreducible subset of $H(p)_{CM}$ consisting of closed points $(X \subset P)$ for which the cohomology of the sheaf ideal I_X satisfies $\dim H^0(I_X(f_i)) = \dim H^0(I_C(f_i))$ for any i, $(C \subset P)$ the "generic" point of U, then

U' is irreducible. Moreover suppose U is open in $H(p)_{CM}$ and $H^{n-r}(I_X(\Sigma_j f_j - n - 1 - f_i)) = 0$ for any i and any $X \subset P$ of U. Then U' is open in $H(p')_{CM}$, and a point $(X' \subset P)$ of U' is non-obstructed provided a corresponding point $(X \subset P)$ of U is non-obstructed.

In the same section we generalize the obstructedness criterion of Ellia and Fiorentini [EF] and obtain a rather general result for a doubly linked curve to be obstructed. For instance it follows from (3.19) that the even liaison class of any smooth non-special curve of maximal rank and of diameter (of the Hartshorne-Rao module) ≤ 6 contains smooth connected obstructed curves. Moreover in (3.22) we use other results of this paper to give an example of an obstructed (non-reduced) curve of maximal rank of degree $d = 5$ and arithmetic genus $g = 0$.

Another part of the paper is concerned with liaison-invariant cohomology groups. As a natural application of (2.6), we have an isomorphism of the corresponding tangent and obstruction spaces

$$A^i(X \subset Y) \approx A^i(X' \subset Y), \quad i = 1,2 \tag{2.14}$$

The tangent map $p_1 : A^1(X \subset Y) \longrightarrow H^0(N_X)$, N_X the normal sheaf of $X \subset P$, and the corresponding map of obstruction spaces $o_1 : A^2(X \subset Y) \longrightarrow H^1(N_X)$ of $pr_1 : D(p;\underline{f})_{CM} \longrightarrow H(p)_{CM}$ fit into an exact sequence (2.19.1), well suited for computing $\dim H^i(N_{X'}(v))$, v an integer, provided we know $\dim H^i(N_X(v))$. Moreover for curves in P^n we prove that the Euler-Poincaré characteristic $\chi(N_X)$ is liaison-invariant.

Even for arithmetically Cohen Macaulay curves, $A^2(X \subset Y)$ depends on Y and in Section 4 we introduce for curves in P^n a liaison-invariant subgroup $C(X \subset Y)$ of $A^2(X \subset Y)$ which is Y-independent in the arithmetically Cohen Macaulay case. The corresponding result for local rings of any dimension states that if A and A' are linked by a complete intersection $B = R/(F_1,..,F_r)$, R a regular local ring, then there is an isomorphism

$$H^2(k,A,A) \approx H^2(k,A',A') \tag{4.2}$$

of algebra cohomology (cotangent groups) provided A and A' satisfy the Serre condition S_3. Applying this, slightly modified, to the graded cones of $X \subset Y \subset P$ and $X' \subset Y \subset P$, we get precisely $C(X \subset Y) \approx C(X' \subset Y)$.

As one will observe the results of this paper apply particularly nicely if $A^2(X \subset Y)$ or $C(X \subset Y)$ vanish. We therefore include a vanishing criterion of $C(X \subset Y)$ for curves with small Rao module which implies the vanishing of $H^1(N_X)$ provided the index of speciality $e(X)$ is strictly smaller than the minimal degree $s(X)$ of a surface containing X.

This paper consists of four main sections;

1. Preliminaries. An infinitesimal study of the Hilbert scheme of nests
2. Deformations and liaison of families of CM - schemes
3. Irreducibility and smoothness of the linked family.
4. Further liaison invariants.

Some sections are rather long and they are therefore divided into subsections with a title indicating its main subject. Moreover we have tried to write the paper in such a way that it should not require profound knowledge of higher algebra cohomology before in Section 4.

I would like to thank A. Laudal, Chr. Peskine and G. Ellingsrud for suggestions, encouragements and help.

Terminology and assumptions.

Throughout this paper k is an algebraically closed field, and $I_{X/Y}$, resp. $N_{X/Y}$, is the sheaf ideal, resp. the normal sheaf, of a closed embedding $X \hookrightarrow Y$ of k-schemes. In case Y is the ambient space P (usually $P = P_k^n$), we omit Y in $I_{X/Y}$ and $N_{X/Y}$. A curve C in P^n is a closed 1-dimensional subscheme of P_k^n, locally Cohen Macaulay and equidimensional, i.e. a 1-dimensional CM-scheme. The curve C in P^n corresponds to a point, usually denoted by $(C \subset P)$, in the Hilbert scheme $H(d,g;n)_{CM}$. In this connection a point means a closed point. A l.c.i., resp. c.i., is a local complete intersection, resp. a (global) complete intersection. Iff means if and only if.

PRELIMINARIES. AN INFINITESIMAL STUDY OF THE HILBERT SCHEME OF NESTS.

The main results on liaison of this paper has its natural presentation in the Hilbert scheme of nests. In this section we study this scheme and we give a cohomologial description of its tangent and "obstruction" spaces. After having studied its connection with related Hilbert schemes, we finish with a theorem (1.27) which applies in a large number of situations because of its "liaison - invariant" nature.

The Hilbert scheme of nests (flags).

(1.1) Let K be any field and let P be a projective K-scheme endowed with a very ample invertible sheaf O(1). If F is a coherent O_P - Module, then let

$$X(F) = \Sigma(-1)^i h^i(F) \quad , \text{ where } \quad h^i(F) = \dim H^i(P,F),$$

be the Euler-Poincaré characteristic. Recall that $X(F(v)), v \in Z$, is a polynomial, called the Hilbert polynomial of F, and if $X \subset P$ is a closed subscheme, then $X(O_X(v))$ is the Hilbert polynomial of $X \subset P$, cf. [EGA,III,(2.5)].

(1.2) Let Sch be the category of locally noetherian k-schemes, let q_i for $i=1,2$ be polynomials in one variable with rational coefficients and let $S \in$ ob Sch. If $\hookrightarrow$ below means a closed embedding of flat S-schemes, we define the functor $\underline{D}(q_1,q_2)$ on Sch by letting $\underline{D}(q_1,q_2)(S)$ be the set

$$\{(X_1 \hookrightarrow X_2 \hookrightarrow P \times S) \mid \text{the fiber } (X_i)_s \text{ has Hilbert polynomial } q_i \text{ for any } s \in S\}$$

(1.3) Theorem. $\underline{D}(q_1,q_2)$ is representable and its representing object $D(q_1,q_2)$ is a projective k-scheme.

Proof. See [Kl2], remark 6.

(1.4) We will call $D(q_1,q_2)$ the Hilbert scheme of nests in P. Observe that if $\underline{H}(q_i)(-)=\mathrm{Mor}(-,H(q_i))$ is the usual Hilbert functor of P with Hilbert polynomial q_i, then there are natural forgetful maps $\underline{pr}_i$: $\underline{D}(q_1,q_2) \longrightarrow \underline{H}(q_i)$ inducing morphisms pr_i appearing in the diagram

$$\begin{array}{ccc} D(q_1,q_2) & \xrightarrow{\ pr_2\ } & H(q_2) \\ \downarrow{\scriptstyle pr_1} & & \\ H(q_1) & & \end{array}$$

For a k-point $(X_1 \subset X_2 \subset P)$ of $D(q_1,q_2)$ we have $pr_i(X_1 \subset X_2 \subset P) = (X_i \subset P)$ for $i=1,2$.

Tangent and obstruction space of $D(q_1,q_2)$ at $t = (X \subset Y \subset P)$.

(1.5) To study $D(q_1,q_2)$ locally, pick a closed point $t=(X \subset Y \subset P)$ of $D=D(q_1,q_2)$. Of course D has a well defined tangent space $A^1(X \subset Y)$ at t, and by [Kl2,A5] there is also an "obstruction group" $A^2(X \subset Y)$, i.e. a group containing all obstructions of infinitesimally deforming t. To simplify their description we will suppose that

i) $Y \subset P$ is a local complete intersection (l.c.i.) and $H(q_2)$ is smooth at $(Y \subset P)$.

ii) the 2. algebra cohomology sheaf $\underline{A}_X^2$ of $X \longrightarrow \mathrm{Spec}\,(k)$ with values in O_X, i.e. the cotangent sheaf $T^2(k,O_X,O_X)$ of Lichtenbaum-Schlessinger, cf.[LS], vanishes.

In the following we mostly refer to [Kl2] instead of the unpublished thesis [Kl1]. To do this we need to observe that, by the general theorems of [Kl1,ch.I], any result and proof of the appendix of [Kl2] (except the spectral sequence (A6) when $p + q \geq 3$) applies if we replace the assumption " X a l.c.i. in P" of [Kl2] by the weaker condition "$\underline{A}_X^2 = 0$".

(1.6) If $l^1 : H^o(N_X) = \mathrm{Hom}\,(I_X,O_X) \longrightarrow \mathrm{Hom}\,(I_Y,O_X) \approx H^o(N_Y \otimes O_X)$ is induced by the inclusion $I_Y \hookrightarrow I_X$ of sheaf ideals and $m^1 : H^o(N_Y) \longrightarrow H^o(N_Y \otimes O_X)$ by the restriction map, then there is a cartesian diagram

$$\begin{array}{ccc} A^1(X \subset Y) & \xrightarrow{\;p_2\;} & H^o(N_Y) \\ \downarrow{\scriptstyle p_1} & \square & \downarrow{\scriptstyle m^1} \\ H^o(N_X) & \xrightarrow{\;l^1\;} & H^o(N_Y \otimes O_X) \end{array}$$

cf. [Kl2,A3] . We may consider p_i as the tangent map of pr_i : $D(q_1,q_2) \longrightarrow H(q_i)$ of (1.4) at $t = (X \subset Y \subset P)$.

<u>(1.7) Remark.</u> a) We do not need the condition ii) of (1,5), nor the non - singularity condition on $H(q_2)$ of i), to prove (1.6), cf. [Kl1].
b) Moreover P need not to be a projective k - scheme. In this case, however, (1.6) is the tangent level description of a diagram as in (1.4) where we replace the schemes $D(q_1,q_2)$ and $H(q_i)$ by its corresponding local deformation functors, cf.(2.7) and (2.11) for details.

(1.8) Next by the long exact sequence of global algebra cohomology assiociated to $X \hookrightarrow Y \hookrightarrow P$,cf.[L1,(3.3.4)], we have an exact sequence

$$0 \longrightarrow H^o(N_{X/Y}) \longrightarrow H^o(N_X) \xrightarrow{\;l^1\;} H^o(N_Y \otimes O_X) \xrightarrow{\;\delta\;}$$
$$A_{X/Y}^2 \longrightarrow H^1(N_X) \xrightarrow{\;l^2\;} H^1(N_Y \otimes O_X) \longrightarrow$$

where $N_{X/Y}$ is the normal sheaf of $X \hookrightarrow Y$, i.e. $N_{X/Y} = \underline{\mathrm{Hom}}_{O_Y}(I_{X/Y},O_X)$ and $I_{X/Y} = \ker\,(O_Y \longrightarrow O_X)$. Concerning $A_{X/Y}^2$; the 2.global algebra cohomology of $X \hookrightarrow Y$ with values in O_X, we have by [L1,(3.2.9)] a spectral sequence $E_2^{p,q}$ giving rise to the exact sequence

$$0 \longrightarrow H^1(N_{X/Y}) \longrightarrow A^2_{X/Y} \longrightarrow H^o(X, \underline{A}^2_{X/Y}) \longrightarrow H^2(N_{X/Y}) \longrightarrow$$

where $\underline{A}^2_{X/Y} = \text{coker}\,(N_X \longrightarrow N_Y \otimes O_X)$, (see (4.4) for a definition of $\underline{A}^2_{X/Y}$ when $\underline{A}^2_X \neq 0$).

(1.9) If $\alpha = \alpha_{X/Y}$ is the composition $H^o(N_Y) \xrightarrow{m^1} H^o(N_Y \otimes O_X) \xrightarrow{\delta} A^2_{X/Y}$ of the maps m^1 and δ appearing in (1.6) and (1.8), then the group

$$A^2(X \subset Y) = \text{coker}\,\alpha$$

contains the obstructions of deforming $t = (X \subset Y \subset P)$ in <u>Sch</u>, cf. [Kl2,A5]. Now once having found the tangent space and an "obstruction space" of D at t, it follows immediately from a general result of Laudal, cf.[L1,(4.2.4)], that

<u>(1.10) Lemma</u>. Let $O_{D,t}$ be the local ring of $D = D(q_1,q_2)$ at the k - point $t = (X \subset Y \subset P)$. If $a^i = \dim A^i(X \subset Y)$, then

$$a^1 - a^2 \le \dim O_{D,t} \le a^1$$

Moreover D is smooth at t iff $\dim O_{D,t} = a^1$.

<u>A fundamental exact sequence relating $A^i(X \subset Y)$ to $H^{i-1}(N_X)$, $i \le 2$.</u>

(1.11) Since N_Y is a local free O_Y - Module by the assumption (1.5i), $0 \longrightarrow N_Y \otimes I_{X/Y} \longrightarrow N_Y \longrightarrow N_Y \otimes O_X \longrightarrow 0$ is exact. Taking cohomology we get by (1.6),

(1.11.1)
$$\ker m^1 = \ker p_1 \approx H^o(N_Y \otimes I_{X/Y}), \quad \text{and}$$
$$\text{coker}\, m^1 = \ker\,[H^1(N_Y \otimes I_{X/Y}) \longrightarrow H^1(N_Y)]\,.$$

Combining with (1.6), (1.8) and (1.9) one proves easily that there is an exact sequence

$$0 \longrightarrow H^o(N_Y \otimes I_{X/Y}) \longrightarrow A^1(X \subset Y) \xrightarrow{p_1} H^o(N_X) \xrightarrow{\beta} \text{coker}\, m^1$$
$$\longrightarrow A^2(X \subset Y) \longrightarrow H^1(N_X) \xrightarrow{1^2} H^1(N_Y \otimes O_X)$$

where the map $\beta = \beta_{X/Y}$ is just the composition of 1^1 and the natural map $H^o(N_Y \otimes O_X) \longrightarrow \text{coker } m^1$, and the map coker $m^1 \longrightarrow A^2(X \subset Y)$ is the map induced by δ onto the cokernels of m^1 and α.

Now suppose $Y=V(F_1,F_2,..,F_r)$, $f_i=\deg F_i$, is a (global) complete intersection (c.i.) in $P = P^n_k$, $r<n$, with Hilbert polynomial q_2. For short, Y is of type $\underline{f}=(f_1,f_2,..,f_r)$. Since $N_Y \approx \Sigma_1^r O_Y(f_i)$ and $H^1(I_X(v)) = \ker [H^1(I_{X/Y}(v)) \longrightarrow H^1(O_Y(v))]$, $v\epsilon Z$, we get

$$H^o(N_Y \otimes I_{X/Y}) \approx \sum_{i=1}^{r} H^o(I_{X/Y}(f_i))$$

$$\text{(1.11.2)} \qquad \text{coker } m^1 \approx \sum_{i=1}^{r} H^1(I_X(f_i))$$

$$H^1(N_Y \otimes O_X) \approx \sum_{i=1}^{r} H^1(O_X(f_i))$$

These expressions inserted in the long exact sequence of (1.11) yields an exact sequence which will be used frequently.

(1.12) Denote by $D(q_1;\underline{f})=D(q_1;f_1,f_2,..f_r)$ the open subscheme of $D(q_1,q_2)$ whose closed points are sequences $(X \subset Y \subset P)$ such that Y is of type $\underline{f}$ in P^n_k. Moreover if dim X=1, then $q_1(x)=dx+1-g$. In this case let

$$H(d,g;n)=H(q_1), \quad H(d,g)=H(d,g;3), \quad D(d,g;\underline{f})=D(q_1;\underline{f})$$

Abusing the language, $pr_i:D(q_1;\underline{f}) \longrightarrow H(q_i)$ denotes the restriction of $pr_i:D(q_1,q_2) \longrightarrow H(q_i)$ to $D(q_1;\underline{f}) \subset D(q_1,q_2)$ as well.

<u>(1.13) Remark</u>. (Vanishing of $A^2(X \subset Y)$).
Let $(X \subset Y \subset P^n)$ be a closed point of $D(q_1;f_1,..,f_r)$ satisfying $\underline{A}^2_X = 0$ and suppose r<n.

a) By (1.11) the obstruction group $A^2(X\subset Y)=0$ iff the map $\beta = \beta_{X/Y}$ $H^o(N_X) \longrightarrow \Sigma_1^r H^1(I_X(f_i))$ is surjective and $1^2 = 1^2_{X/Y}$: $H^1(N_X) \longrightarrow \Sigma_1^r H^1(O_X(f_i))$ is injective.
b) In case dim X = 1 suppose $X \subset P^n$ is reduced and non-special and satisfies $H^1(I_X(f_i))=0$ for i=1,2,...,r. Since one proves easily $H^1(N_X)=0$, cf. [HH, (1.2)] for n=3, we get by a), $A^2(X \subset Y)=0$.
c) If X is a smooth connected curve in P^3 and $Y=V(F_1,F_2)$ where $V(F_1)$ is a smooth hypersurface of degree $f_1 \le 4$, then $A^2(X \subset Y)=0$ provided $H^1(I_X(f_2))=0$ and X is not a complete intersection in P^3. Indeed by [Kl2,13], $A^2(X \subset V(F_1))=0$, and applying a) to $A^2(X \subset V(F_1))$ and $A^2(X \subset Y)$ successively, we conclude easily.

<u>(1.14) Remark</u> (Surjectivity of $1^2_{X/Y}$).
Let $(X\subset Y\subset P^n) \in D(q_1;f_1,..,f_r)$ satisfy $A^2_X=0$. Then the map $1^2 : H^1(N_X) \to \Sigma H^1(O_X(f_i))$ of (1.11) is surjective provided dim X = 1 and $X \hookrightarrow Y$ is a l.c.i. outside a finite number of points. Indeed in this case $\mathrm{Supp}(A^2_{X/Y})$ is finite, cf. (1.8), and we prove easily the surjectivity of 1^2 by splitting $0 \to N_{X/Y} \to N_X \to \Sigma O_X(f_i) \to A^2_{X/Y} \to 0$ into short exact sequences and taking cohomology.

<u>Properties (smoothness etc.) of $pr_1:D(q_1,q_2) \to H(q_1)$ at $(X\subset Y\subset P)$.</u>

(1.15) In [Kl1,(1.3.4)] we proved that pr_1 is smooth at $t=(X\subset Y\subset P)$ provided (1.5i) holds and the map m^1 of (1.6) is surjective. If Y is a c.i. of type $(f_1,..,f_r)$ in $P = P^n$ and $r<n$, then the cokernel of m^1 is $\Sigma H^1(I_X(f_i))$, and we get a part of the following result.

<u>(1.16) Theorem</u>. Let $D = D(q_1;f_1,..,f_r)$ and let $pr_1 : D \to H(q_1)$ be the morphism given by $pr_1((X\subset Y\subset P)) = (X\subset P)$ where $P = P^n$ and $r<n$.

a) The fibers of pr_1 are smooth and geometrically connected of dimension $h^o(N_Y \otimes I_{X/Y}) = \Sigma h^o(I_{X/Y}(f_i))$ at $t=(X\subset Y\subset P) \in D$.

b) Let $t=(X\subset Y\subset P)$ be a closed point of $D(q_1;\underline{f})$ and suppose $H^1(I_X(f_i)) = 0$ for $1\le i\le r$. Then pr_1 is smooth at t.

To give a complete proof of (1.16) and to prove some later results too, we will need

<u>(1.17) Lemma.</u> Let $(A',m_{A'}) \to (A,m_A)$ be any morphism of local artinian k-algebras with residue fields k whose kernel K satisfies $K\cdot m_{A'}=0$. Let $S=\mathrm{Spec}(A) \to S'=\mathrm{Spec}(A')$ be the induced morphism, and let $t=(X\subset Y\subset P^n) \in D(q_1;f_1,..,f_r)$, r<n, be given. Moreover let $X_S \subset Y_S \subset P\times S$, $P = P^n$, be any deformation of t to S and let $X_{S'} \hookrightarrow P\times S'$ (with sheaf ideal $I_{X_{S'}}$) be any deformation of $X_S \hookrightarrow P\times S$ (with sheaf ideal I_{X_S}) to S'. If the natural map

$$r : H^o(I_{X_{S'}}(f_i)) \longrightarrow H^o(I_{X_S}(f_i)),$$

is surjective for i=1,2,..,r, then pr_1 is smooth at t.

<u>Proof.</u> To prove the smoothness of pr_1, i.e. the surjectivity of

$$\underline{D}(S') \longrightarrow \underline{D}(S) \times_{\underline{H}(S)} \underline{H}(S')$$

where $\underline{H}(-) = \mathrm{Mor}(-,H(q_1))$ and $\underline{D}(-) = \mathrm{Mor}(-,D(q_1;\underline{f}))$, we consider a given diagram of deformations

$$\begin{array}{ccc} X_{S'} & \hookrightarrow & P\times S' \\ \uparrow & \circ & \uparrow \\ X_S \hookrightarrow Y_S & \hookrightarrow & P\times S \end{array}$$

of $X\subset Y\subset P$ and $X\subset P$ to S and S' respectively. If I_{Y_S} is the sheaf ideals of $Y_S \hookrightarrow P\times S$ and $\{F_1,..,F_r\}$ is the regular sequence which defines Y in P, then one proves easily that $Y_S=V(H_1,..,H_r)$ for some $H_i \in H^0(I_{Y_S}(f_i)) \subset H^0(I_{X_S}(f_i))$ which maps to F_i via $H^0(I_{Y_S}(f_i)) \twoheadrightarrow H^0(I_Y(f_i))$. (Observe that this map is surjective because $H^1(I_Y(f_i))=0$). By assumption the map τ of (1.17) is surjective , and we can therefore lift $H_i \in H^0(I_{X_S}(f_i))$ to $H_i' \in H^0(I_{X_{S'}}(f_i))$ and thus define a deformation $Y_{S'} = V(H_1',..,H_r') \to P\times S'$ of Y_S containing $X_{S'}$. In fact one proves easily that $Y_{S'}$ is S'-flat by using that $\{F_1,..,F_r\}$ is a regular sequence. This shows the smoothness of pr_1 .

<u>Proof of (1.16)</u>. To prove the smoothness of the fibers of pr_1, we use (1.17) with trivial deformations of X, i.e $X_{S'}= X \times S'$ and $X_S = X \times S$, cf.[M1,Lect 13]. Then we obviously have surjectivity of the map τ and we conclude by (1.17).

Moreover to prove the connectedness of the fibers of pr_1, let $t = (X\subset V(F_1,..,F_r) \subset P)$ and $t'= (X\subset V(G_1,..,G_r) \subset P)$ be given. By [M1,page57] there is an open set $U'\subset B = \operatorname{Spec}[T]$ such that the restriction of

$$Y_B = V(F_1+T(G_1-F_1),..,F_r+T(G_r-F_r)) \subset P \times B$$

to U' is flat over U'. Moreover $T = 0$ and $T = 1$ correspond to the given points t and t' respectively. As remarked in the proof of (1.17), $Y_B \to B$ is flat at t and t' because $\{F_1,..,F_r\}$ and $\{G_1,..,G_r\}$ are regular sequences. Hence if $U = U'\cup\{t,t'\}$, then U is an open set of $B = \mathbb{A}^1_k$ and the restriction of $Y_B \to B$ to U is flat. Clearly $Y_B|_U \supset X\times U$ and the fibers of pr_1 is connected. This argument works for $\operatorname{Spec}(\Omega)$ - points as well where Ω is any overfield of k.

Finally since $p_1:A^1(X\subset Y) \to H^0(N_X)$ is the tangent map of pr_1: $D(q_1;\underline{f}) \to H(q_1)$ at t, the tangent space of the fiber at t is just ker $p_1 \approx H^0(N_Y \otimes I_{X/Y}) \approx \Sigma\, H^0(I_{X/Y}(f_i))$ by (1.11). Combining with the smoothness of the fibers, we have proved a).

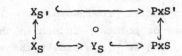

We also have an easy direct proof of b) using (1.17). Indeed taking cohomology of the exact sequence

$$0 \longrightarrow K \otimes I_X(f_i) \longrightarrow I_{X_{S'}} \longrightarrow I_{X_S} \longrightarrow 0$$

and using $H^1(I_X(f_i))= 0$, we get that the map τ of (1.17) is surjective, and we conclude by (1.17).

<u>(1.18) Corollary</u>. Let $(X \subset P^n)$ be a closed point of $H(q_1)$ satisfying $A_X^2 = 0$, cf.(1.5) and let Y be a complete intersection of type $\underline{f} = (f_1,..,f_r)$ containing X. Suppose r<n. If

i) $H^1(I_X(f_i)) = 0$ for $i = 1,2,..,r$, and

ii) $1^2 : H^1(N_X) \hookrightarrow \Sigma H^1(O_X(f_i))$ is injective,

then $H(q_1)$ is smooth at $(X \subset P^n)$.

<u>Proof</u> Since the assumptions of (1.18) imply that $A^2(X \subset Y) = 0$ by (1.13), $D(q_1;\underline{f})$ is smooth at t, cf.(1.10). By (1.16), pr_1 is smooth at t, and we conclude easily.

<u>(1.19) Remark.</u> a) In this proof we used that $H^1(I_X(f_i)) = 0$ for any i implies the smoothness of pr_1, and so the implication

(*) $D(q_1;f)$ is smooth at $(X \subset Y \subset P) \Longrightarrow H(q_1)$ is smooth at $(X \subset P)$

is true. For a converse of (*) under weaker conditions, we observe that the map $A^2(X \subset Y) \longrightarrow H^1(N_X)$ of (1.11) sends obstructions to obstructions, cf. [L1,(4.1.15)] . Hence assuming its injectivity, or equivalently by (1.11), assuming the surjectivity of $\beta : H^o(N_X) \longrightarrow \Sigma H^1(I_X(f_i))$ we have a true implication

$H(q_1)$ is smooth at $(X \subset P)$ $\Longrightarrow$ $D(q_1;\underline{f})$ is smooth at $(X \subset Y \subset P)$

b) The conclusion of (1.18) holds if we replace i) by

i') for each $i = 1,..,r$, either $H^1(I_X(f_i)) = 0$ <u>or</u> $H^1(O_X(f_i)) = 0$

and keep ii). To see this, put $Y = V(F_1,..,F_r)$, $f_i = \deg F_i$, and let S be the intersection of those hypersurfaces $V(F_i)$ for which $H^1(O_X(f_i)) \neq 0$. Applying (1.18) to $(X \subset P)$ and the c.i. $S \supset X$, we see that $(X \subset P)$ is a smooth point of $H(q_1)$ since $1^2_{X/S} = 1^2_{X/Y}$.

<u>(1.20) Remark</u>. We have the following generalization of (1.16). Let $X_H \hookrightarrow P \times H$, $H = H(q_1)$, be the universal object of $\underline{H}(q_1)$, let $\underline{I}$ be the sheaf ideal of $X_H \hookrightarrow P\times H$ and let $\pi : P\times H \longrightarrow H$ be the projection. If $x=(X\subset P) \in H$ and $P=P^n$, we consider the proposition

$$C(i) \begin{cases} R^i\pi_*\underline{I}(f_j) \text{ is locally a free } O_H \text{ - Module in a neighbour-} \\ \text{hood of } x \text{ and the natural map } R^i\pi_*\underline{I}(f_j) \otimes k(x) \longrightarrow \\ H^i(P\times\{x\},\underline{I}(f_j)_x) \approx H^i(I_X(f_j)) \text{ is an isomorphism}, j=1,..,r. \end{cases}$$

Suppose $r<n$. Then by base change theorem [M1,Lect.7] , C(1) implies C(0) and C(0) implies the surjectivety of the map τ of (1.17) for any $t \in pr_1^{-1}(x)$. Hence

> If C(0) <u>or</u> C(1) is satisfied, then pr_1 is smooth at any $t \in pr_1^{-1}(x)$.

In particular by a theorem of Grauert on base change [H,chIII, (12.9)], if V is any reduced irreducible component and if e_i is the minimum value of $H^o(I_X(f_i))$ when $(X\subset P)$ varies in V, then the restriction pr_1: $pr_1^{-1}(U) \longrightarrow U$ is smooth on the open set $U = \{(X\subset P) \in V \mid h^o(I_X(f_i)) = e_i \text{ for } i = 1,2\}$ of V.

(1.21) We can push the argument using Grauert's theorem on base change a bit further. Indeed let V be any locally closed subset of H endowed with the reduced scheme structure. Moreover let $X_V \hookrightarrow P\times V$ be the restriction of the universal object $X_H \hookrightarrow P\times H$ via the natural morphism $i:V \hookrightarrow H$. Now if $\underline{I}$ is the sheaf ideal of $X_V \hookrightarrow P\times V$, $\pi : P\times V \longrightarrow V$ is the projection and $x\in V$, we consider the proposition $C_V(i)$ obtained by replacing H by V in C(i). By base change theorem, $C_V(1)$ implies $C_V(0)$ and $C_V(0)$ implies the surjectivity of the map τ of (1.17) provided the scheme S' of (1.17) allows a morphism $S' \to V$ and $X_{S'} \to P\times S'$ is the pullback of $X_V \to P\times V$ via this morphism. By slightly modifying the proof and the conclusion of (1.17) we get precisely the smoothness of the pullback $(pr_1,i) : D\times V \longrightarrow V$ of pr_1 to V provided $r<n$, i.e.

(1.22) If $C_V(0)$ or $C_V(1)$ is satisfied, then the pullback $(pr_1,i) : D\times V \longrightarrow V$ is smooth at any $t\in (pr_1,i)^{-1}(x)$

<u>(1.23) Proposition.</u> Let V be any irreducible locally closed subset of

the Hilbert scheme $H(q_1)$ endowed with the reduced scheme structure and let $f_1,..,f_r$ be r positive integers. Moreover let $U_V=U_V^0$, resp. U_V^1, be the open set of V consisting of points $(X\subset P^n)$ where the function on V, $h^o(I_X(f_i))$, resp. $h^1(I_X(f_i))$, obtains its minimum values for any $i=1,2,..,r$. Suppose $r<n$. Then

i) $U_V^1 \subset U_V$

ii) the pullback $pr_1^{-1}(U_V) \longrightarrow U_V$ of $pr_1: D(q_1,\underline{f}) \longrightarrow H(q_1)$ is smooth

iii) $pr_1^{-1}(U_V)$ is irreducible

iv) For any $(X\subset Y\subset P) \in pr_1^{-1}(U_V)$

$$\dim pr_1^{-1}(U_V) = \dim U_V + \sum_{i=1}^{r} h^o(I_{X/Y}(f_i))$$

__Proof__. Obviously if $C_V(p)$, cf.(1.21), is satisfied for any x of some open set U of V, then the function $h^p(I_X(f_i))$ is a constant on U for any i. The converse is true by Grauerts theorem [H,chIII,(12.9)]. Hence since $C_V(1)$ implies $C_V(0)$, $U_V^1 \subset U_V$. Moreover ii) follows from (1.22). Since the fibers of pr_1 are geometrically connected, $pr_1^{-1}(U_V)$ is irreducible by [EGA,IV,(4.5.7)]. Then iv) follows from theorem (1.16).

__On the codimension of f-maximal subsets of the Hilbert scheme $H(q_1)$__.

__(1.24) Definition__. A subset W of $H(q_1)$ is called an $(f_1,f_2,..,f_r)$-maximal irreducible subset of $H(q_1)$ provided W is a closed irreducible subset of $H(q_1)$ and there exist an irreducible (non-embedded) component W' of $D(q_1;\underline{f})$ such that $pr_1(W')$ is dense in W.

__(1.25) Remark__. i) If W is $\underline{f}$-maximal and irreducible, then by the irreducibility of $pr_1^{-1}(U_V)$, there is only one irreducible component W' of $D(q_1;\underline{f})$ dominating W.

ii) Moreover if W and V are two $\underline{f}$-maximal irreducible subsets of $H(q_1)$, then it is possible that $W \subsetneq V$. Indeed this phenomenom occurs in H(6,0), cf.(3.3). In general if it happens, we must have $\Sigma\, h^o(I_X(f_i)) > \Sigma\, h^o(I_C(f_i))$ where $X\subset P$, resp. $C\subset P$, is the general curve of W, resp. V, due to the following corollary of (1.23)

__(1.26) Corollary__. Let $W\subset H(q_1)$ be $\underline{f}$-maximal and irreducible with general curve $X\subset P$ and let V be any irreducible subset of $H(q_1)$ properly containing W. If $C\subset P$ is a general curve of V, then either

$$\Sigma\, h^o(I_X(f_i)) > \Sigma\, h^o(I_C(f_i)) \quad ,$$

or C is not contained in a complete intersection of type $\underline{f}$.

Proof Suppose C is contained in a c.i. of type $\underline{f}$ and suppose $\Sigma h^o(I_X(f_i)) \le \Sigma h^o(I_C(f_i))$. By the semicontinuity of $h^o(I_X(v))$, $h^o(I_C(f_i)) = h^o(I_X(f_i))$ for any i, and so if U_V and U_W are the open subsets of V and W of proposition (1.23), then $U_V \supset U_W$. Hence $pr_1^{-1}(U_V) \supset pr_1^{-1}(U_W)$. Since the closure of $pr_1^{-1}(U_W)$ is an irreducible component and since $pr_1^{-1}(U_V)$ is irreducible, $\dim pr_1^{-1}(U_V) = \dim pr_1^{-1}(U_W)$. By (1.23,iv) we have $\dim U_V = \dim U_W$, and this contradicts the assumption $\dim V > \dim W$.

One way of expressing the insight we so for have got into the Hilbert scheme of nests and into its projection morphisems pr_i, is given in the following theorem

(1.27) Theorem. Let W be any $(f_1,f_2,\dots,f_r)$-maximal irreducible subset of the Hilbert scheme $H(q_1)$ whose general member $X \subset P = P_k^n$ satisfies $A_X^2 = 0$, cf.(1.5), and let $V \subset H(q_1)$ be an irreducible (embedded or not) component containing W. Suppose $r<n$ and suppose there is a complete intersection $Y \subset P$ of type $\underline{f}$ containing X such that

$$\text{i) } \beta_{X/Y} : H^o(N_X) \longrightarrow \sum_{i=1}^{r} H^1(I_X(f_i)) \text{ is surjective, and}$$

$$\text{ii) } 1_{X/Y}^2 : H^1(N_X) \longrightarrow \sum_{i=1}^{r} H^1(O_X(f_i)) \text{ is injective}$$

(or weaker, such that $\beta_{X/Y}$ is surjective and $D(q_1;\underline{f})$ is smooth at $(X \subset Y \subset P)$). Then

$$\dim V - \dim W \le \sum_{i=1}^{r} h^1(I_X(f_i)),$$

the inequality is strict iff $H(q_1)$ is singular along W. Furthermore if

$$\text{iii) } h^o(N_X) - h^1(N_X) \le \dim V$$

and i) and ii) are satisfied, then

$$\sum_{i=1}^{r} h^1(I_X(f_i)) - \sum_{i=1}^{r} h^1(O_X(f_i)) \le \dim V - \dim W.$$

(1.28) Remark. The condition iii) is very weak, at least for curves. Indeed if the general point and $C \subset P$ of the set V of (1.27) satisfies $A_C^2 = 0$ and V is non-embedded, then it is known that $h^o(N_C) - h^1(N_C) \le$

dim V, cf.[L1,(5.2.10)]. If the scheme X of (1.27) is a smooth curve in $P = P^n$ or a generic complete intersection curve in $P = P^3$, then $A_X^2 = A_C^2 = 0$, cf.(4.3) and $\chi(N_X) = \chi(N_C)$. So iii) is true in these cases. Finally i) and ii) is equivalent to $A^2(X\subset Y) = 0$ by (1.13a), and the theorem above is therefore a generalization of [Kl2], theorem 10.

__Proof__ The proof of (1.27) is basically the same as for theorem 10 of [Kl2] and we will shortly review it. Indeed by (1.13a), (1.10) and (1.23), dim $W = a^1 - \Sigma h^o(I_{X/Y}(f_i))$ which by (1.11) is equal to $h^o(N_X) - \Sigma h^1(I_X(f_i))$. Since dim $V \le \dim O_{H,X} \le h^o(N_X)$, and dim $V = h^o(N_X)$ iff $H=H(q_1)$ is smooth at $(X\subset P)$, we conclude by subtracting dim W. Finally by ii) and iii)

$$h^o(N_X) - \Sigma\, h^1(O_X(f_i)) \le h^o(N_X) - h^1(N_X) \le \dim V$$

and again, by subtracting dim W, we are done.

__(1.29) Corollary__. In the situation of (1.27), suppose i) and ii).

a) If for each i=1,2..,r ,either $H^1(I_X(f_i)) = 0$ __or__ $H^1(O_X(f_i)) = 0$, then $H(q_1)$ is generically smooth along W and W is of codimension $\Sigma\, h^1(I_X(f_i))$ in $H(q_1)$.

b) If $\Sigma\, h^1(I_X(f_i)) > \Sigma\, h^1(O_X(f_i))$ and (1.27 iii) holds, then W is not an irreducible component of $H(q_1)$.

__Proof__ a) follows from (1.27) provided $H(q_1)$ is smooth at $(X\subset P)$. We therefore conclude by (1.19b). Finally b) is immediate from (1.27).

__(1.30) Example__. To illustrate the results above we consider families of space curves directly linked to two disjoint conics. (We have to involve some results proven in later chapters too). In fact one knows that there is an irreducible component W of the Hilbert scheme H(4,-1) whose general curve Γ is a disjoint union of two conics. If we link Γ (or rather the whole family of two disjoint conics) using a c.i. Y of type (4,k), we get a new family of curves W_k in H(d,g) whose general member X satisfies

$$d=d(X)=4(k-1) \quad ,g=g(X)=2(k-1)^2-3=d^2/8-3$$

Observe that Γ is reduced and non-special, so by (1.13), $A^2(\Gamma\subset Y)=0$ provided $k\ge 3$. Due to the isomorphism $A^2(\Gamma\subset Y) \approx A^2(X\subset Y)$ established in the next section, $A^2(X\subset Y)=0$. The new family W_k is irreducible, cf.(3.4) and (4,k)-maximal, cf.(2.6), and theorem (1.27) applies. Now by standard liaison arguments

$$h^1(I_X(v)) = h^1(I_\Gamma(k-v)) = \begin{cases} 1 & \text{if } v=k-2 \text{ or } k \\ 2 & \text{if } v=k-1 \\ 0 & \text{otherwise} \end{cases}$$

$$h^1(O_X(v)) = h^o(I_{\Gamma/Y}(k-v)) = \begin{cases} 1 & \text{if } v=k-2 \\ 0 & \text{if } v>k-2 \end{cases}$$

Hence $H^1(I_X(4))\neq 0$ iff $4\leq k\leq 6$, and $H^1(O_X(4))=0$ iff $k<6$. Since $H^1(O_X(k))=0$, (1.29a) applies provided $k\neq 6$. In particular if $k\neq 6$, then $H(d,g)$ is non-singular at $(X\subset P)$ and the codimension of W in $H(d,g)$ is $h^1(I_X(4))+h^1(I_X(k))$. Furthermore if k=6, then $h^1(I_X(4))+h^1(I_X(6)) = 2 > h^1(O_X(4))+h^1(O_X(6))=1$ and (1.29b) applies. In this case it is possible to prove that $(X\subset P)$ is a singular point of $H(d,g)=H(20,47)$, cf.(3.16). The codimension of W in H(20,47) is therefore 1.

DEFORMATIONS AND LIAISON OF FAMILIES OF CM - SCHEMES.

(2.1) In this section P is always P^n and talking about an S - point of H(p), D(p,q) or some of its subschemes, we always suppose S is <u>locally noetherian</u>. Now recall that an S - point of D(p,q) is just a sequence of closed embeddings $(X \hookrightarrow Y \hookrightarrow P\times S)$ of flat S - schemes such that for any $s \in S$, the fiber sequence $X_s \hookrightarrow Y_s \hookrightarrow P$ consists of schemes with Hilbert polynomials $p(v) = \chi(O_{X_s}(v))$ and $q(v) = \chi(O_{Y_s}(v))$. Denote by $D(p;\underline{f})_{CM} = D(p;f_1,f_2,..,f_r)_{CM}$ the open subscheme of D(p,q) whose corresponding fiber X_s are Cohen Macaulay (CM) and equidimensional and $Y_s \hookrightarrow P$ are complete intersections (c.i.) of type $\underline{f}$ for any $s \in S$. The main theorem (2.6) of this section states that there is an isomorphism

$$\tau \; : \; D(p;\underline{f})_{CM} \xrightarrow{\sim} D(p';\underline{f})_{CM}$$

defined by liaison. The corresponding tangent and obstruction spaces turn out to be isomorphic, and via the exact sequence of (1.11) this provides an effective tool to compute $h^1(N_X(v))$, $v \in Z$, from similar informations of the linked subscheme. See [K11] and [B] for related results.

<u>Liaison-invariance of the Hilbert scheme $D(p;f)_{CM}$ of nests</u>.

<u>(2.2) Definition</u>. Let $(X \hookrightarrow Y \hookrightarrow P\times S)$ be an S - point of $D(p,\underline{f})_{CM} \subset D(p,q)$ and suppose deg p = deg q, i.e. dim X_s = dim Y_s for any $s \in S$.

i) Then we define the linked subscheme $X' \hookrightarrow Y$ of $X \hookrightarrow Y$ by letting its sheaf ideal $I_{X'/Y}$ in Y be

$$I_{X'/Y} = O_X^{\vee} \quad \text{where } (-)^{\vee} = \underline{\mathrm{Hom}}_{O_Y}(-,O_Y).$$

We also say $X' \hookrightarrow Y$ is linked to $X \hookrightarrow Y$ or that X and X' are linked by Y. By proposition (2.4), $(X')'=X$ in Y, and the relation "liaison of families", i.e. linkage of schemes over S, is therefore symmetric.

ii) If for any $x \in \mathrm{Ass}(O_X)$, $O_{X,x} = O_{Y,x}$, we say $X' \hookrightarrow Y$ is geometrically linked to $X \hookrightarrow Y$.

(2.3) Let $K \supset k$ be an overfield and let $(X \subset Y \subset P)$ be a K - point of $D(p,\underline{f})_{CM}$ with dim X = dim Y. By [PS,(1.3)], the linked scheme X' by Y is Cohen Macaulay and equidimensional, dim X = dim Y and $(X')' = Y$. In particular $(X' \subset Y \subset P)$ is a K - point of $D(p';\underline{f})_{CM}$ where $p'(v) = \chi(O_{X'}(v))$. In order to clarify the concept of liaison of families and to prove the main theorem of this chapter we need the following result, cf.[KM,(4.1)] for a special case.

<u>(2.4) Proposition</u>. Let $(X \to Y \to P \times S)$ be an S - point of $D(p;\underline{f})_{CM}$, let $X' \to Y$ be linked to $X \to Y$ and let $(X')_s \to Y_s$ be the induced morphism of $X' \to Y$ on the fibers $(X')_s$ and Y_s at $s \in S$. Then $(X')_s \to Y_s$ is linked to $X_s \to Y_s$. Moreover $X' \to S$ is S - flat and the dobly linked subscheme $(X')' \to Y$ is just $X \to Y$.

<u>Proof.</u> Let $s \in S$ and let m be the maximal ideal of $O = O_{S,s}$. Put $O_i = O/m^i$, $J_i = \ker(O_i \to O_{i-1})$ and

$$X_i = X \underset{S}{\times} \mathrm{Spec}(O_i) \hookrightarrow Y_i = Y \underset{S}{\times} \mathrm{Spec}(O_i).$$

Thus $O_1 = k(s)$ and $X_1 = X_s \hookrightarrow Y_1 = Y_s$. Moreover let $O_{X_i}{}^{\vee} = \underline{\mathrm{Hom}}_{O_{Y_i}}(O_{X_i},O_{Y_i})$ and $O_X{}^{\vee} = \underline{\mathrm{Hom}}_{O_Y}(O_X,O_Y)$. We first claim that

$$O_{X_i}{}^{\vee} \underset{O_i}{\otimes} O_{i-1} \xrightarrow{\sim} O_{X_{i-1}}{}^{\vee}$$

is an isomorphism. To see this we observe that the S - flatness of $Y \to S$ gives us an exact sequence

$$0 \longrightarrow O_{Y_i} \otimes J_i \longrightarrow O_{Y_i} \longrightarrow O_{Y_{i-1}} \longrightarrow 0$$

We therefore get a commutative diagram of exact horizontal sequences

$$\begin{array}{ccccccc} & O_{X_i}^{\vee} \otimes_{O_i} J_i & \longrightarrow & O_{X_i}^{\vee} & \longrightarrow & O_{X_i}^{\vee} \otimes_{O_i} O_{i-1} & \longrightarrow 0 \\ *) & \downarrow & \circ & \wr\downarrow & \circ & \downarrow & \\ 0 \longrightarrow & \underline{Hom}(O_{X_i}, O_{Y_i} \otimes J_i) & \longrightarrow & \underline{Hom}(O_{X_i}, O_{Y_i}) & \longrightarrow & \underline{Hom}(O_{X_i}, O_{Y_{i-1}}) & \longrightarrow 0 \end{array}$$

Indeed recall that J_i is a $k(s)$ - module and $Y_1 = Y_s$ is Gorenstein, so

$$\underline{Ext}^1_{O_{Y_i}}(O_{X_i}, O_{Y_i} \otimes J_i) \approx \underline{Ext}^1_{O_{Y_1}}(O_{X_1}, O_{Y_1}) \underset{k(s)}{\otimes} J_i = 0$$

Now the vertical arrow to the right in the diagram *) is surjective for _any_ i. In particular $O_{X_i}^{\vee} \otimes k(s) \longrightarrow\!\!\!> O_{X_1}^{\vee}$ is surjective, and since J_i is a $k(s)$- module, the vertical arrow to the left in the diagram *) is surjective. This in turn implies the injectivity of the vertical arrow to the right, i.e. the claim is proved.

Next we prove that

$$O_X^{\vee} \underset{O}{\otimes} k(s) \xrightarrow{\sim} O_{X_s}^{\vee}$$

is an isomorphism. Note that $I_{X'/Y} = O_X^{\vee}$ and $I_{(X_s)'/Y_s} = O_{X_s}^{\vee}$, so by tensoring $0 \longrightarrow I_{X'/Y} \longrightarrow O_Y \longrightarrow O_{X'} \longrightarrow 0$ by $k(s)$ over O, we see that the isomorphism above implies that $X' \to S$ is flat at $s \in S$ and that the fiber morphism $(X')_s \to Y_s$ of $X' \to Y$ at $s \in S$ coincides with $(X_s)' \to Y_s$, i.e. that $(X_s)' = (X')_s$. Now let $x \in X$ map to $s \in S$ via the structure morphism $X \to S$. Then $x \in Y$ and abusing the language, $x \in X_i$ and $x \in Y_i$ as well. Put

$$A = O_{X,x}\ ,\ B = O_{Y,x}\ ,\ A_i = O_{X_i,x}\ ,\ B_i = O_{Y_i,x}\ ,$$

$$\hat{B} = \varprojlim B_i,\ \hat{A} = \varprojlim A_i,\ A^{\vee} = \mathrm{Hom}_B(A,B) \text{ and } A_i^{\vee} = \mathrm{Hom}_{B_i}(A_i, B_i).$$

Then it will be sufficient to show $A^{\vee} \underset{B}{\otimes} B_1 \xrightarrow{\sim} A_1^{\vee}$. Since we already know

$$A_i^{\vee} \underset{B_i}{\otimes} B_{i-1} = A_i^{\vee} \underset{O_i}{\otimes} O_{i-1} \approx A_{i-1}^{\vee}$$

we deduce [EGA, O_I, (7.2.9)] $(\varprojlim A_i^{\vee}) \underset{\hat{B}}{\otimes} B_1 \approx A_1^{\vee}$. Moreover one knows that [EGA, O_I, (7.2.10)]

$$\varprojlim A_i^{\vee} = \mathrm{Hom}_{\hat{B}}(\hat{A}, \hat{B}) = A^{\vee} \underset{B}{\otimes} \hat{B}$$

and we conclude as expected.

Finally it remains to show that $X''=X$ where $X''=(X')'$. Since $I_{X''/Y}=\check{O}_{X'} = \check{\check{I}}_{X/Y}$, $X''\subset X$. So let

$$0 \longrightarrow I_{X''/X} \longrightarrow O_X \longrightarrow O_{X''} \longrightarrow 0$$

be exact. Since X'' is S-flat and $(X'')_s = (X_s)'' = X_s$, $I_{X''/X} \otimes k(s) = 0$, and so by Nakayamas lemma, $X''=X$, and the proof is complete.

(2.5) Let $(X \hookrightarrow Y \hookrightarrow P)$ be a Spec(k)-point of $D(p,\underline{f})_{CM}$. If $X \hookrightarrow Y$ and $X' \hookrightarrow Y$ are linked and if $p'(v)$ is the Hilbert polynomial of X' in P, then by [AK,I,(2.3)]

$$I_{X/Y}=\underline{\mathrm{Hom}}_{O_Y}(O_{X'},O_Y)=\underline{\mathrm{Hom}}_{O_Y}(O_{X'},\Omega_Y) \otimes \Omega_Y^{-1} \approx \Omega_{X'} \otimes \Omega_Y^{-1}$$

Moreover using $\Omega_Y \approx O_Y(\Sigma f_i-n-1)$ and duality for X', we get

$$X(O_X(v))=X(O_Y(v))-X(I_{X/Y}(v))=X(O_Y(v))-(-1)^{n-r}X(O_{X'}(\Sigma f_i-n-1-v))$$

Hence

(2.5.1) $$p(v) = q(v) - (-1)^{n-r}p'(\Sigma f_i -n-1-v)$$

<u>(2.6) Theorem</u>. Let q be the Hilbert polynomial of a (global) complete intersection of type $f=(f_1,f_2,..,f_r)$ in $P = P^n$ and let p and p' be polynomials satisfying (2.5.1). Then there is an isomorphism

$$\tau \ : \ D(p;\underline{f})_{CM} \xrightarrow{\simeq} D(p';\underline{f})_{CM}$$

which on S-points is defined by sending $(X \hookrightarrow Y \hookrightarrow P\times S)$ onto $(X' \hookrightarrow Y \hookrightarrow P\times S)$ where X and X' are linked by Y.

<u>Proof</u> Let $S = D(p;\underline{f})_{CM}$ and let $(X \to Y \to P\times S)$ be the restriction of the universal object of $D(p,q)$ to S. By (2.4) $X'\to S$ is flat and the fibers $(X')_s \to Y_s$ of the linked subscheme $X'\to Y$ are linked to $X_s \to Y_s$. Hence $(X')_s$ is Cohen Macaulay and equidimensional of dimension dim Y_s by [PS,(1.3)], cf.(2.3). This implies that $(X' \to Y \to P\times S)$ is an S-point of $D(p';\underline{f})_{CM}$, i.e.

$$D(p;\underline{f})_{CM} = S \longrightarrow D(p';\underline{f})_{CM}$$

Starting with $S = D(p';\underline{f})_{CM}$ and using (2.4) as above, we have an inverse and the proof is complete.

(2.7) If we are working with non - projective schemes, we do not have the representability of the Hilbert functor of nests, i.e the scheme D(p,q) does not necessarily exist. However the liaison - isomorphism of theorem (2.6) is still true for the corresponding local deformation functor. We can also generalize (2.6) by weakening the Cohen Macaulay assumptions on the schemes X and X'. To be precise let $\underline{l}$ be the category of affine schemes S = Spec (A) where A is an artinian local k - algebra with residue field k. If $X \hookrightarrow Y \hookrightarrow Z$ is a given sequence of closed embeddings of locally noetherian k - schemes, we define its local deformation functor $Def_{X\subset Y}$ on $\underline{l}$ as follows. The set $Def_{X\subset Y}(S)$, S = Spec(A), consists of sequences $X_S \hookrightarrow Y_S \hookrightarrow Z_S$ of S - flat schemes whose pullback via A ->>k is the given sequence $X \hookrightarrow Y \hookrightarrow Z$. Then we have

<u>(2.8) Theorem.</u> Let $X \hookrightarrow Y \hookrightarrow Z$ be a given sequence of closed embeddings of locally noetherian k-schemes, and let X' be the subscheme of Y whose sheaf ideal is $I_{X'/Y} = O_X^{\vee}$. Suppose

i) $$I_{X/Y} = O_{X'}^{\vee}$$

ii) $$\underline{Ext}^1_{O_Y}(O_X, O_Y) = 0 \text{ and } \underline{Ext}^1_{O_Y}(O_{X'}, O_Y) = 0$$

Then there is a natural isomorphism

$$\tau : Def_{X\subset Y} \xrightarrow{\sim} Def_{X'\subset Y}$$

<u>(2.9) Remark.</u> Indeed if $(X_S \hookrightarrow Y_S \hookrightarrow Z\times S) \in Def_{X\subset Y}(S)$, the corresponding objekt $(X'_S \hookrightarrow Y_S \hookrightarrow Z\times S) \in Def_{X'\subset Y}(S)$ defined by τ is given by $I_{X'_S/Y_S} = O_{X_S}^{\vee}$. Moreover observe that there is a projection p_2 : $Def_{X\subset Y} \longrightarrow Hilb_Y$ onto the local Hilbert functor of $Y \hookrightarrow Z$, $Hilb_Y$, defined by sending $(X_S \to Y_S \to Z\times S)$ onto $(Y_S \to Z\times S)$. Obviously the isomorphism τ commutes with p_2 and p_2' : $Def_{X'\subset Y} \longrightarrow Hilb_Y$.

<u>(2.10) Remark.</u> If $X \hookrightarrow Y \hookrightarrow Z$ are equal to Spec(A) $\hookrightarrow$ Spec(B) $\hookrightarrow$ Spec(R) where R and B are local Gorenstein rings and A is a local ring of pure height, then the condition (2.8i) is always true by [S,(2.2)]. Moreover by [S,(4.1)] and Gorenstein duality, (2.8ii) is equivalent to the claim that A <u>and</u> A' = B/Hom(A,B) satisfies the Serre condition S_2.

<u>Proof</u> As in the first part of the proof of (2.4), one shows $I_{X'_S/Y_S} \otimes_A k$

$= I_{X'/Y}$ from which the isomorphism $X_S' \otimes \mathrm{Spec}(k) \cong X'$ and the flatness of $X' \to S = \mathrm{Spec}(A)$ are easily deduced. Since $(X_S')' = X_S$ in Y_S by the final part of the proof of (2.4), we are done.

Liaison-invariant tangent and obstruction groups.

(2.11) Let $X \hookrightarrow Y \hookrightarrow P$ and $X' \hookrightarrow Y \hookrightarrow P$ be linked, cf.(2.2). Then for any open $U \subset P$, $X\cap U \hookrightarrow Y\cap U \hookrightarrow U$ and $X'\cap U \hookrightarrow Y\cap U \hookrightarrow U$ satisfies the conditions of (2.8). Hence

(2.11,1) $$\mathrm{Def}_{X\cap U \subset Y\cap U} \xrightarrow{\simeq} \mathrm{Def}_{X'\cap U \subset Y\cap U}$$

If we define the sheaf $\underline{A}^1(X\subset Y)$ by the cartesian diagram

(2.11.2) $$\begin{array}{ccc} \underline{A}^1(X\subset Y) & \xrightarrow{\underline{p}_2} & N_Y \\ \downarrow & \square & \downarrow \\ N_X & \longrightarrow & N_Y \otimes O_X \end{array}$$

then $\Gamma(U,\underline{A}^1(X\subset Y))$ is the tangent space of $\mathrm{Def}_{X\cap U\subset Y\cap U}$ by (1.7). Combining with (2.9), i.e. using $p_2' \cdot \tau = p_2$, the isomorhism of (2.11.1) induces an isomorphism of tangent spaces, fitting into a commutative diagram

$$\begin{array}{ccc} \Gamma(U,\underline{A}^1(X\subset Y)) & \xrightarrow{\simeq} & \Gamma(U,\underline{A}^1(X'\subset Y)) \\ & \searrow \quad \circ \quad \swarrow & \\ & \Gamma(U,N_Y) & \end{array}$$

We have therefore proved

(2.12) Corollary. Let X and X' be equidimensional CM-schemes linked by a complete intersection Y in P^n. Then there is a sheaf isomorphism $\underline{A}^1(X\subset Y) \approx \underline{A}^1(X'\subset Y)$ fitting into a commutative diagram

$$\begin{array}{ccc} \underline{A}^1(X\subset Y) & \xrightarrow{\underline{p}_2} & \\ \wr\downarrow & \circ & N_Y \\ \underline{A}_1(X'\subset Y) & \xrightarrow{\underline{p}_2'} & \end{array}$$

the global sections of which give us precisely the tangent spaces and the tangent maps of

$$\begin{array}{ccc} D(p;\underline{f})_{CM} & \xrightarrow{pr_2} & \\ \wr\downarrow & \circ & H(q) \\ D(p';\underline{f})_{CM} & \xrightarrow{pr_2'} & \end{array}$$

at $(X \subset Y \subset P^n)$ and $(X' \subset Y \subset P^n)$. In this diagram $H(q)$ is the Hilbert scheme to which $(Y \subset P^n)$ belongs.

(2.13) At least if the linkage is <u>geometric</u>, we also have an isomorphism of obstruction spaces $A^2(X \subset Y) \approx A^2(X' \subset Y)$ of $D(p;\underline{f})$ and $D(p';\underline{f})$ at $(X \subset Y \subset P)$ and $(X' \subset Y \subset P)$ respectively. To see this, observe that in this case $(I_{X/Y})_x = 0$ for any $x \in \mathrm{Ass}(O_X)$. Hence

$$N_{X/Y} = \underline{\mathrm{Hom}}(I_{X/Y}, O_X) = 0.$$

Moreover if the cotangent sheaf $\underline{A}^2_X$ vanishes, then there is an exact sequence $0 \to N_X \to N_Y \otimes O_X \to \underline{A}^2_{X/Y} \to 0$ by the definition (1.8) of $\underline{A}^2_{X/Y}$. Combining with (2.11.2), we see that

$$(2.13.1) \qquad 0 \longrightarrow \underline{A}^1(X \subset Y) \xrightarrow{\underline{p}_2} N_Y \longrightarrow \underline{A}^2_{X/Y} \longrightarrow 0$$

is exact. By (1.9) $A^2(X \subset Y)$ is the cokernel of $H^0(N_Y) \to H^0(\underline{A}^2_{X/Y})$. More generally we define $A^i(X \subset Y)_v$, for $1 \le i \le 2$ by

$$(2.13.2) \qquad \begin{aligned} A^1(X \subset Y)_v &= H^0(Y, \underline{A}^1(X \subset Y)(v)) \\ A^2(X \subset Y)_v &= \mathrm{coker}\,[H^0(N_Y(v)) \longrightarrow H^0(\underline{A}^2_{X/Y}(v))] \end{aligned}$$

Then $A^2(X \subset Y) = A^2(X \subset Y)_0$ and by (2.13.1), $A^2(X \subset Y)_v \approx \ker H^1(Y, \underline{p}_2(v))$ for any integer v. Hence

<u>(2.14) Corollary.</u> Let X and X' be equidimensional CM - schemes, geometrically linked by a complete intersection Y in P^n. If the cotangent sheaves $\underline{A}^2_X$ and $\underline{A}^2_{X'}$ vanish, then

$$A^2(X \subset Y)_v \xrightarrow{\simeq} A^2(X' \subset Y)_v$$

for any $v \in \mathbb{Z}$. In particular for $v = 0$ we get an isomorphism $A^2(X \subset Y) \approx A^2(X' \subset Y)$ of obstruction spaces.

<u>Proof</u> Immediate from $A^2(X \subset Y)_v \approx \ker H^1(Y, \underline{p}_2(v))$, corollary (2.12) and the injectivity of $\underline{p}_2$ and $\underline{p}_2'$.

<u>(2.15) Remark.</u> In a later section we prove $\underline{A}^2_X \approx \underline{A}^2_{X'}$ for geometrically linked equidimensional CM - schemes. So any $X \subset P^n$ which is in the liaison class of a local complete intersection, satisfies $\underline{A}^2_X = 0$. In

particular if X is a curve and a generic complete intersection in P^3, then $\underline{A}_X^2 = 0$.

(2.16) Remark. Concerning the use of theorem (1.27) we have by (2.15), (2.14) and (1.13a) that the condition: "$\underline{A}_X^2 = 0$, (i) and (ii)" of (1.27) is invariant under geometric linkage. The remaining condition (iii) is trivially satisfied for geometrically linked curves in P^n due to the following result

(2.17) Corollary. Let $(X \subset P) \in H(d,g)_{CM}$, $(X' \subset P) \in H(d',g')_{CM}$ be curves, geometrically linked by a c.i. Y in P^n. If $\underline{A}_X^2$ and $\underline{A}_{X'}^2$ vanish and if $\chi(N_X)=(n+1)d+(n-3)(1-g)$ (true if X is a smooth curve), then

$$\chi(N_{X'}) = (n+1)d' + (n-3)(1-g')$$

(2.18) Before giving the proof we want to establish some relations between a curve $X \subset Y = V(F_1,F_2,\ldots,F_{n-1})$ in P^n and its linked curve $X' \subset Y$. We have already seen $I_{X/Y} \approx \Omega_{X'}(n+1-f)$, cf.(2.5). This and (2.5.1) gives immediately

$$\text{(2.18.1)}\qquad \begin{aligned} d+d' &= f_1 f_2 \cdots f_{n-1} = \pi f_i \\ g-g' &= (d-d')(f-n-1)/2 \qquad , f = \sum_{i=1}^{r} f_i \\ h^0(I_{X/Y}(v)) &= h^1(O_{X'}(f-n-1-v)) \\ h^1(O_X(v)) &= h^0(I_{X'/Y}(f-n-1-v)) \end{aligned}$$

Moreover twisting the sequence $0 \longrightarrow \Omega_X(n+1-f) \longrightarrow O_Y \longrightarrow O_{X'} \longrightarrow 0$ by $s = f-n-1-v$ and taking cohomology , we get an exact sequence

$$H^0(O_Y(s)) \xrightarrow{h'} H^0(O_{X'}(s)) \longrightarrow H^1(\Omega_X(-v)) \xrightarrow{h} H^1 O_Y(s))$$

The cokernel of h' is obviously $H^1(I_{X'}(f-n-1-v))$ and dualizing $H^1(\Omega_X(-v)) \xrightarrow{h} H^1(O_Y(f-n-1-v))$ we get $(\text{coker } h)^{\vee} \approx H^1(I_X(v))$ as well. It follows that

$$\text{(2.18.2)}\qquad h^1(I_X(v)) = h^1(I_{X'}(f-n-1-v))$$

For details see [S].

Proof To see that $\chi(N_X)=(n+1)d+(n-3)(1-g)$ for smooth curves, just take $\chi(-)$ of the two exact sequences

$$0 \longrightarrow T_X \longrightarrow T_P|_X \longrightarrow N_X \longrightarrow 0$$

$$0 \longrightarrow O_X \longrightarrow O_X(1)^{\oplus n+1} \longrightarrow T_P|_X \longrightarrow 0$$

where T_X, resp. $T_P|_X$, is the tangent sheaf of X, resp. of P restricted to X. Of course we have to use that $\chi(T_X) = \chi(\Omega_X^{-1}) = 3-3g$

Next, and this is the main point, $\chi(A^i) = \dim A^1(X\subset Y) - \dim A^2(X\subset Y)$ is invariant under geometric linkage by (2.12) and (2.14). Combining with (1.11), we see that $\chi(N_X)+\Sigma\chi(I_X(f_i))$, or equivalently $\chi(N_X)-\Sigma\chi(O_X(f_i))$, is invariant under direct linkage too. By assumption

$$\chi(N_X) - \Sigma\chi(O_X(fi)) = (n+1)d + (n-3)(1-g) - \Sigma(df_i+1-g)$$

The right hand side is simply $2g-2-d(f-n-1)$ where $f = \Sigma f_i$. Since

$$2g-2-d(f-n-1) = 2g'-2-d'(f-n-1)$$

by (2.18.1), we conclude easily.

(2.19) Under the assumptions of (2.14) we can use the liaison - invariance of $A^i(X\subset Y)_v$ for i=1,2, to compute $h^1(N_{X'}(v))$ for any integer v provided we know $h^1(N_X(v))$. If we will do this for $v \neq 0$ we need to generalize the exact sequence of (1.11). Indeed using (2.11.2) and (2.13.2) one proves in exactly the same way as in (1.11) that there is an exact sequence

$$\begin{aligned} &0 \longrightarrow \Sigma H^0(I_{X/Y}(f_i+v)) \longrightarrow A^1(X\subset Y)_v \longrightarrow H^0(N_X(v)) \longrightarrow \\ (2.19.1)\quad & \\ &\Sigma H^1(I_{X/Y}(f_i+v)) \longrightarrow A^2(X\subset Y)_v \longrightarrow H^1(N_X(v)) \xrightarrow{1_v^2} \Sigma H^1(f_i+v)) \end{aligned}$$

cf.[P]. The case $v = o$ is just the sequence of (1.11), and 1_v^2 is surjective under the assumptions of (1.14).

<u>(2.20) Example.</u> Let $X\subset P^3$ be any smooth curve of maximal rank in H(8,g) with g<5. Since d>2g-2, $H^1(O_X(1))=0$. So $H^1(N_X(v))=0$ for $v\geq 0$. Since $\chi(I_X(3))=g-5$ and $\chi(I_X(4))=g+2$, it follows that $H^0(I_X(v))=0$ for $v\leq 3$ and $H^1(I_X(v))=0$ for $v\geq 4$ by maximal rank. Let X and X' be geometrically linked by a c.i. Y of type (f_1,f_2). Then $A^2(X\subset Y)_v=0$ for $v\geq 0$ by the long exact sequence of (2.19). Combining (2.14), (2.15) and (2.19) we get

$$h^1(N_{X'}(v)) = \Sigma\, h^1(O_{X'}(f_i+v)) = \Sigma\, h^0(I_{X/Y}(f_i-4-v)) \ , \ v \geq 0$$

In particular for any fixed $v \geq 0$, $h^1(N_{X'}(v))=0$ provided $f_i - 4 - v \leq 3$, i.e. $f_i \leq 7+v$ for $i=1,2$.

<u>(2.21) Remark.</u> The arguments used in (2.20) shows easily that if $X \subset P^3$ is any curve of maximal rank satisfying $H^1(N_X)=0$ and if X and X' are geometrically linked by a c.i. Y of type (f_1, f_2), then

$$H^1(N_{X'}(v)) \approx \Sigma\, H^1(O_{X'}(f_i + v)), \qquad v \geq 0.$$

IRREDUCIBILITY AND SMOOTHNESS OF THE LINKED FAMILY.

(3.1) If $\{X_t \subset P \mid t \in U\}$ is a family of subschemes of $P = P^n$, $U \subset H(p)_{CM}$, and if each member X_t is contained in a c.i. $Y_t \subset P$ of type $\underline{f}=(f_1, f_2, \ldots, f_r)$, the type doesn't vary with $t \in U$, then let $U' = \{X'_t \subset P\}$ $H(p')_{CM}$ be the total family of subschemes obtained by liaison, see (3.2) for a precise definition. In this section we will consider the following questions, arising naturally in the study of the Hilbert scheme using the technique of liaison:

i) Is U' irreducible if U is irreducible?
ii) Is U' open in $H(p')_{CM}$ provided $U \subset H(p)_{CM}$ is open?
iii) Is $H(p')$ smooth along U' if $H(p)$ is smooth along U?

There are in fact counterexamples to all these questions, and so, in this section, we will make explicite conditions under which i), ii) and iii) are true. All this will in fact be consequences of the liaison isomorphism $\tau : D(p;\underline{f})_{CM} \xrightarrow{\sim} D(p';\underline{f})_{CM}$ of (2.6) and of the general results of Section 1.

<u>(3.2) Definition.</u> If $U \subset H(p)_{CM}$ is a locally closed subset (contained in $pr_1(D(p;\underline{f})_{CM})$), then the family of $\underline{f}$-linked subschemes U' is by definition

$$U' = pr'_1(\tau(pr_1^{-1}(U)))$$

where

$$\begin{array}{ccc} D(p;\underline{f})_{CM} & \xrightarrow[\sim]{\tau} & D(p';\underline{f})_{CM} \\ \downarrow{\scriptstyle pr_1} & & \downarrow{\scriptstyle pr'_1} \\ H(p)_{CM} & & H(p')_{CM} \end{array}$$

In this case we write : $U' \sim U$ via $\underline{f}$

Irreducibility of the f-linked family.

We now turn to the question of the irreducibility of U'. Even if U is irreducible, the linked family U' need not be, as the following example shows.

(3.3) Example. Start with the two classically well-known families of curves in $H(6,0)_S$[1] whose general curves sit on smooth cubic surfaces F_1 and F_2 and whose corresponding invertible sheaves are given by $(2,0,...,0)$ and $(3,2,1,0,..,0)$ via the usual isomorphism $Z^{\oplus 7} \xrightarrow{\sim}$ $Pic(F_i)$, cf.[H,V,§4]. One knows that $H(6,0)_S$ is irreducible and so is the open set $U=\{(X \subset P) \epsilon H(6,0)_S | h^o(I_X(2))=0\}$ of $H(6,0)_S$. By computing $X(I_X(v)) = \binom{v+3}{3} - (6v+1)$ for different v's, one proves easily that any curve $X \subset P$ of U is contained in a c.i. of type $(3,4)$. Thus $U \subset$ $pr_1(D(6,0;3,4))$ and the linked family $U' \subset H(6,0)$ turns out to consist of two irreducible subsets, one of which contains smooth non-connected curves. Indeed the invertible sheaves of the general curves of these two families are $(2,0,..,0)$ and $(3,1,1,1,1,0,-1)$ via $Z^{\oplus 7} \approx Pic(F_i)$. Moreover $U' \subset H(6,0)$ is open, cf. (3.8), so the closure of these two irreducible families of U' form two irreducible components of $H(6,0)$.

However if we combine theorem (2.6) with the study of the projection morphism $pr_1:D = D(p;f)_{CM} \longrightarrow H = H(p)_{CM}$ from Section 1, we can at least say the following about the irreducibility of U'.

(3.4) Proposition. Let a_i, $1 \leq i \leq r$, be r integers, let $D = D(p;\underline{f})_{CM}$ and let $U \subset pr_1(D)$ be any locally closed irreducible subset consisting of schemes $X \subset P^n$ satisfying $h^o(I_X(f_i)) = a_i$ for $1 \leq i \leq r$. Then the $\underline{f}$-linked family $U' \subset H(p')_{CM}$ is irreducible.

Proof. $pr_1^{-1}(U)$ is irreducible by (1.23) and so is $\tau(pr_1^{-1}(U))$ by (2.6). Hence $U'=pr'_1(\tau(pr_1^{-1}(U))$ is irreducible.

(3.5) Remark. Proposition (3.4) remains true if we replace the condition "$h^o(I_X(f_i))=a_i$ for $1 \leq i \leq r$ " by "$h^1(I_X(f_i)) = a_i$ for $1 \leq i \leq r$", cf.$(1.23i)$.

(3.6) Corollary. Let f_i, a_i and a'_i for $1 \leq i \leq r$ be non-negative

---

[1] $H(d,g)_S$ is the subscheme of $H(d,g)$ of smooth connected curves

integers, let $f = \sum_{i=1}^{r} f_i$ and suppose $r<n$. If U is the set

$$\{(X\subset P) \in pr_1(D) \mid h^1(I_X(f_i))=a_i \text{ and } h^{n-r}(I_X(f-f_i-n-1)=a_i', 1\le i\le r\}$$

and if U' is the $\underline{f}$-linked family, then the number of irreducible components of U and U' coincides and $U''= U$ where $U''=(U')'$. Indeed if U_i is an irreducible component of U, then U_i' is an irreducible component of U' and $U_i=U_i''$. Moreover

$$\dim U_i + \sum_{i=1}^{r} h^o(I_{X/Y}(f_i)) = \dim U_i' + \sum_{i=1}^{r} h^o(I_{X'/Y'}(f_i))$$

for any $(X\subset Y\subset P) \in pr_1^{-1}(U_i)$ and $(X'\subset Y'\subset P) \in pr_1'^{-1}(U_i')$.

<u>Proof</u> If X' is linked to X by a c.i. of type $\underline{f}$, then

$$(3.6.1) \qquad h^i(I_{X'}(v)) = h^{n+1-r-i}(I_X(f-n-1-v)) \ , \quad v \in Z$$

for $0<i\le n-r$, cf.[S,(5.3)] or (2.18.2). Therefore by (3.2) U' is equal to

$$\{(X'\subset P)\in pr_1(D') \mid h^1(I_{X'}(f_i))=a_i' \text{ and } h^{n-r}(I_{X'}(f-n-1-f_i))=a_i, \forall i\}$$

where $D'= D(p';\underline{f})_{CM}$. Interpreting U'' similarly, we get $U''=U$.

Suppose $U=\bigcup_{i=1}^{e} U_i$ is an irreducible decomposition of U. It is straightforward from the definition (3.2) to see $U_i\subset U_i''$ and $U'=\bigcup_{i=1}^{e} U_i'$. By (3.5) U_i' is irreducible and so is its closure $\overline{U_i'}$ in U'. Up to exclusion of those $\overline{U_i'}$ which satisfy $\overline{U_i'}\subset \overline{U_j'}$ for some $j\ne i$,

$$U' = \bigcup_{i=1}^{e} \overline{U_i'}$$

is an irreducible decomposition of U'. So to prove that the number of irreducible components of U and U' coincides, we will show a contradiction, supposing $\overline{U_{i_o}'} \subset \overline{U_{j_o}'}$ for $i_o \ne j_o$. Now by (3.6.1) $h^1(I_{X'}(f_i))$ is a constant on $\overline{U_j'}$ for any j because $\overline{U_j'}\subset U'$. The $\underline{f}$-linked family $(\overline{U_j'})'$ is therefore irreducible by (3.5) and we get

$$U'' = \bigcup_{i=1}^{e} (\overline{U_i'})' \quad \text{and} \quad U_i \subset U_i'' \subset (\overline{U_i'})'$$

Since U_i is an irreducible component of $U=U''$, $U_i=U_i''=(\overline{U_i'})'$. It follows that $U_{i_o} \subset U_{j_o}$ which is a contradiction.

Since we have proved $U_i = (\overline{U_i'})'$, it follows that $U_i' = (\overline{U_i'})''$. In general $U_i' \subset \overline{U_i'} \subset (\overline{U_i'})''$ and we get that U_i' is closed in U', i.e. U_i' is an irreducible component of U'.

Finally by (1.23)

$$\dim pr_1^{-1}(U_i) = \dim U_i + \Sigma\, h^o(I_{X/Y}(f_i))$$

for any $(X \subset Y \subset P) \in pr_1^{-1}(U_i)$. We have a corresponding formula relating $pr_1'^{-1}(U_i')$ and U_i' and since $\tau(pr_1^{-1}(U_i)) \subset pr_1'^{-1}(U_i')$, cf.(3.2) , we get $pr_1^{-1}(U_i) \subset \tau^{-1} pr_1'^{-1}(U_i') \subset pr_1^{-1}(U_i'')$. But now $U_i'' = U_i$ implies that $\dim pr_1^{-1}(U_i) = \dim pr_1'^{-1}(U_i')$ and the dimension formula relating $\dim U_i$ and $\dim U_i'$ in (3.6) is proved.

Along the same lines we can treat linkage of specializations, i.e.

<u>(3.7) Proposition.</u> Let $(X \subset P)$ and $(C \subset P)$, $P = P^n$, be two points of $H(p)_{CM}$, $(X \subset P)$ a specialization of $(C \subset P)$, and suppose there is a c.i. Y_o of type $(f_1, f_2, .., f_r)$ containing X such that for $1 \le i \le r$,

$$h^o(I_X(f_i)) = h^o(I_C(f_i)) \text{ and } h^{n-r}(O_X(s_i)) = h^{n-r}(O_C(s_i))$$

where $s_i = \Sigma f_j - n - 1 - f_i$. Then there is a c.i. Y of type $\underline{f}$ containing C such that if C', resp. X', is the subscheme of P linked to C by Y, resp. to X by Y_o, then $(X' \subset P)$ is a specialization of $(C' \subset P)$.

The proof is a consequence of (3.4), (1.23) and the fact that smooth morphisms map generic points onto generic points, the details of which we leave to the reader.

<u>Openness and smoothness of the f-linked family</u>.

We now turn to the question of the openness and smoothness of the $\underline{f}$-linked family U'.

<u>(3.8) Proposition.</u> Let $D = D(p; f_1, f_2, .., f_r)_{CM}$, $1 \le r < n$, and let U be any <u>open subset of $pr_1(D)$</u> consisting of schemes $X \subset P^n$ satisfying $h^{n-r}(I_X(\Sigma fj - n - 1 - f_i)) = 0$ for $1 \le i \le r$. Then the $\underline{f}$-linked family U' is an <u>open subset of $H(p')_{CM}$</u>.

<u>Proof.</u> Since $U \subset pr^1(D)$ is open, $\tau(pr_1^{-1}(U))$ is open in $D(p'; \underline{f})_{CM}$, cf.(3.2). Using (3.6.1) and (1.16 b) the openness of U' follows

readily from the smoothness of $pr_1' : D(p';\underline{f})_{CM} \longrightarrow H(p')_{CM}$ restricted to $\tau(pr_1^{-1}(U))$.

(3.9) Proposition. Let a_i and f_i ,$1 \le i \le r$, be non-negative integers and let $U \subset H(p)_{CM}$ consist of schemes $X \subset P^n$ satisfying

$$h^0(I_X(f_i)) = a_i \quad \text{and} \quad h^{n-r}(I_X(\Sigma f_j - n-1-f_i)) = 0$$

for $1 \le i \le r$. Suppose $r<n$ and that U is open is $H(p)_{CM}$. Then $H(p')_{CM}$ is smooth along the $\underline{f}$-linked family U' provided $H(p)_{CM}$ is smooth along U. Moreover U' is irreducible if U is.

Proof. Consider U as a subscheme of $H=H(p)_{CM}$ with scheme structure given by $O_{U,x} = O_{H,x}$ for any $x \in U$. Then U is smooth by assumption and since the pullback $pr_1^{-1}(U) \longrightarrow U$ is smooth by (1.23ii), the composition $pr_1^{-1}(U) \to U \to Spec(k)$ is smooth. Hence $D(p';\underline{f})_{CM}$ is smooth at any point of $\tau(pr_1^{-1}(U))$ and by the smoothness of pr_1' over $pr_1'^{-1}(U') \supset \tau(pr_1^{-1}(U))$, cf. (1.16), U' is smooth over k. The irreducibility of U' follows readily from (3.4).

(3.10) Corollary. Let $(X \subset P^n) \in H(p)_{CM}$ be non-obstructed and suppose there is a c.i. Y of type $(f_1, f_2, .., f_r), r<n$, containing X and linking X to X'. If for any $i=1,2,..,r$ and any generization $(C \subset P^n)$ of $(X \subset P^n)$ in $H(p)_{CM}$,

$$h^{n-r}(I_X(\Sigma f_j-n-1-f_i)) = 0 \text{ and } h^0(I_X(f_i)) = h^0(I_C(f_i))$$

then $(X' \subset P^n)$ is non-obstructed.

(3.11) The proof is an immediate consequence of (3.9). Note that in (3.9) the assumption that U is <u>open</u> in $H(p)_{CM}$ is fullfilled provided we have

$$H^1(I_X(f_i)) = 0 \text{ for any } i \text{ and any } (X \subset P) \in U$$

cf. (1.23,i). The openness condition on U in (3.8) is considerably weaker, and in terms of generizations it corresponds the claim that any generization $(C \subset P)$ of $(X \subset P)$ in $H(p)_{CM}$ which sits on a c.i. of type $\underline{f}$ satisfies $h^0(I_X(f_i)) = h^0(I_C(f_i))$. It would have been desireable to prove (3.10) assuming this weaker claim, but such a generalization is false, cf. (3.21). We can, however, prove the following variant of (3.10)

<u>(3.12) Proposition</u>. Let $(X \subset P^n)$ be non-obstructed and let Y and X' be as in (3.10). If the map

$$\beta_{X/Y} : H^o(N_X) \longrightarrow \Sigma H^1(I_X(f_i))$$

of (1.11) is <u>surjective</u> and the corresponding map $\beta_{X'/Y}$ of the linked subscheme is the <u>zero</u> map, then $(X' \subset P^n)$ is non- obstructed. In particular $(X' \subset P^n)$ is non-obstructed provided

$$H^1(I_X(f_i)) = 0 = H^{n-r}(I_X(\Sigma f_j - n - 1 - f_i))$$

for any $i = 1,2,..,r$.

<u>Proof.</u> By (1.19), $D(p;\underline{f})_{CM}$ is smooth at $(X \subset Y \subset P)$ and so is $D(p';\underline{f})_{CM}$ at $(X' \subset Y \subset P)$ by (2.6). Consulting (1.11) and combining with $\beta_{X'/Y}=0$, we get that the tangent map p_1': $A^1(X' \subset Y) \longrightarrow H^o(N_{X'})$ of pr_1' is surjective and we conclude by [EGA IV,(17.11.1)]. For the conclusion in the final sentence of (3.12) we have to use (3.6.1).

<u>(3.13) Example.</u> Observe that for the subset $U=U(f_1,f_2)$ $H(d,g)_{CM}$ given by

$$\{(X \subset P^3) \in pr_1(D(d,g;f_1,f_2)_{CM}) \mid H^1(I_X(f_i))=0=H^1(I_X(f_i-4)), i=1,2\}$$

we have by (3.6), (3.8) and (3.12):

1) U is irreducible iff U'is, and U"=U
2) U and U' are open in H(d,g) and H(d',g') respectively
3) H(d,g) is smooth at $(X \subset P) \in U$ iff H(d',g') is smooth at a corresponding linked curve $(X' \subset P) \in U'$.

In [Kl3] we applied this to Buchsbaum curves with 1 dimensional Rao module $M(X) = \Sigma H^1(I_X(v))$, concentrated in degree $c=c(X)$, in the following way. Let

$$e(X) = \max\{t \mid H^1(O_X(t)) \neq 0\} \text{ and } s(X) = \min\{t \mid H^o(I_X(t)) \neq 0\}$$

and let

$$U_1(d,g) = \{ (X \subset P^3) \in H(d,g)_{CM} \mid M(X) \approx k,\ e(X) < c(X) < s(X) \}$$

Then the functions $e(X)$, $c(X)$ and $s(X)$ defined on $U_1(d,g)$ do not vary with $(X \subset P)$. Thus _any_ curve $(X \subset P)$ of $U_1(d,g)$ is contained in a c.i. of type (f_1,f_2) where $f_1=s(X)$ and $f_2=c(X)+2$. Due to (2.18) the linked curves X' of $U_1(d,g)'$ satisfies $e(X')<c(X')<s(X')$ as well, thus allowing a further linkage using c.i. of type $(s(X'),c(X')+2)$ for _any_ X' of $U_1(d,g)'=U_1(d',g')$. In this way we make a sequence of liaisons which by [Kl3] terminates with $U_1(2,-1)$. Now observe that

$$U_1(d,g) \subset U(f_1,f_2)$$

and so 1), 2) and 3) above applies for the sets $U_1(d,g)$ and $U_1(d,g)'$ and correspondingly for any such pair of sets in the whole liaison sequence. One knows that $H(2,-1)_{CM}$ is smooth and irreducible, and it follows that $U_1(d,g)$ is _irreducible_ and consists of _non-obstructed_ points.

The irreducibility of $U_1(d,g)$ proven above is now a very special case of the more general irreducibility criterion, recently proven by Bolondi [Bo]. It applies particularly nicely to families of Buchsbaum curves and its classification, cf.[BM1] ,[BM2] and [BM3].

Criterions for obstructedness of a curve.

We will finish this chapter by using the irreducibility criterion to give a criterion for a linked curve $(X' \subset P^n)$ to be a singular point of its Hilbert scheme $H(d',g';n)_{CM}$. The proof plays on the fact that two non-singular points $(X' \subset P^n)$ and $(C' \subset P^n)$ of $H(d',g';n)$ which belong to a common irreducible component $V' \subset H(d',g';n)$ must satisfy $h^o(N_{C'}) = h^o(N_{X'})$. Our result is a generalization of [EF,th.III.6], and it is possible to generalize the proof further so that it works for subschemes of P^n by introducing additional conditions.

(3.14) Proposition. Let $(C \subset P)$ and $(X \subset P)$, $P = P^n$, be two points of $H(d,g;n)_{CM}$, $(X \subset P)$ a specialization of $(C \subset P)$[2], and suppose there is a c.i. $Y \supset C$ of type $(f_1,..,f_{n-1})$ linking C geometrically to C'. Let $s_i = \Sigma_{j=1}^{n-1} f_j - f_i - n - 1$ for $i=1,..n-1$, and suppose the cotangent sheaves $A_X^2 = 0$, $A_C^2 = 0$ and that $A^2(C \subset Y) = 0$. If

i) $\qquad h^o(I_X(f_i)) = h^o(I_C(f_i)) \qquad$ for $i=1,..,n-1$ and

---

[2] i.e.there is irred.$V \subset H(d,g;n)$ with general point $(C \subset P)$,and $(X \subset P) \in V$

ii) $\sum_{i=1}^{n-1} h^o(I_X(s_i)) > \sum_{i=1}^{n-1} h^o(I_C(s_i))$

then any curve $(X'\subset P)$ geometrically linked to $(X\subset P)$ by some c.i. of type $(f_1,f_2,..,f_{n-1})$ is a singular point of $H(d',g';n)_{CM}$. Moreover if

iii) $h^1(I_C(s_i)) = 0$ for $i=1,..,n-1$ <u>or</u>

iv) $h^o(I_{C/Y}(s_i)) = 0$ for $i=1,..,n-1$,

then $H(d',g';n)$ is non-singular at $(C'\subset P)$. In the case iv), $H^o(N_{C'}) = 0$. Furthermore the unique irreducible component which contains $(C'\subset P)$ contains also $(X'\subset P)$.

<u>Proof</u>. We first claim that $h^1(N_{X'}) > h^1(N_{C'})$. Indeed since X and X' are <u>geometrically</u> linked by some c.i. Z of type $(f_1,..,f_{n-1})$, the map $1^2_{X/Z} : H^1(N_{X'}) \longrightarrow \Sigma H^1(O_{X'}(f_i))$ is surjective, cf.(1.14). So

$$h^1(N_{X'}) \geq \Sigma h^1(O_{X'}(f_i)) = \Sigma h^o(I_{X/Z}(s_i))$$

by (2.18.1). Moreover by assumption and (2.14), $A^2(C'\subset Y) = 0$. The map $1^2_{C'/Y}$ is therefore injective, cf. (1.13), and it follows that

(*) $$h^1(N_{C'}) \leq \Sigma h^1(O_{C'}(f_i)) = \Sigma h^o(I_{C/Y}(s_i))$$

So it suffices to prove $\Sigma h^o(I_{X/Z}(s_i)) > \Sigma h^o(I_{C/Y}(s_i))$. To show this consider the exact sequence $0 \to I_Y \to I_C \to I_{C/Y} \to 0$ and the corresponding sequence for $X\subset Z$. Since $h^o(I_Y(s_i)) = h^o(I_Z(s_i))$ and $h^1(I_Y(v)) = 0 = h^1(I_Z(v))$ for any v, the claim follows readily from ii).

To prove that $(X'\subset P)$ is a singular point of $H(d',g';n)$, suppose the converse. Let V be an irreducible, locally closed subset of $H(d,g;n)$ containing $(X\subset P)$ and with general point $(C\subset P)$, and let U be the subset of $pr_1(D(d,g;\underline{f})_{CM})\cap V$ whose points $(X_1\subset P)$ satisfy $h^o(I_{X_1}(f_i)) = h^o(I_C(f_i))$ for $i=1,..,n-1$. Since U contains the "generic point" of V, U is dense in V. It follows that U is irreducible. By (3.4), U' is irreducible, and it contains $(C'\subset P)$ and $(X'\subset P)$. So there is an irreducible component $T\subset H(d',g';n)_{CM}$ containing U' and we get

$$h^o(N_{X'}) = \dim T \leq \dim O_{H',(C'\subset P)} \leq h^o(N_{C'}),$$

using that $H'=H(d',g';n)_{CM}$ is non-singular at $(X'\subset P)$. Hence $h^o(N_{X'}) \leq$

$h^o(N_{C'})$ which is equivalent $h^1(N_{X'}) \le h^1(N_{C'})$ by (2.17). This contradicts the proven claim $h^1(N_{X'}) > h^1(N_{C'})$ above.

Next suppose iii). Then $H^1(I_{C'}(f_i))=0$ by (2.18.1), and since $A^2(C' \subset Y)=A^2(C \subset Y)=0$, $H(d',g';n)$ is non-singular at $(C' \subset P)$ by (1.18). Finally suppose iv). Consulting (*) we get immediately that $H^1(N_{C'})=0$.

<u>(3.15) Remark.</u> a) If the condition $A^2(C \subset Y)=0$ is not satisfied, then $(X' \subset P)$ is still a singular point of $H(d',g';n)_{CM}$ provided we suppose $\underline{A}^2_X=\underline{A}^2_C=0$, i) and

$$\text{ii')} \quad \Sigma\, h^o(I_X(s_i)) > \Sigma\, h^o(I_C(s_i)) + \dim A^2(C \subset Y)$$

b) Another more important variant of (3.14) is the following. Instead of assuming "$A^2(C \subset Y)=0$" we assume "$\beta_{X'/Y}$ is surjective" for some specialization $(X \subset Z)$ of $(C \subset Y)$ linking X geometrically to X' (and we keep the assumptions i), ii) and $\underline{A}^2_X = \underline{A}^2_C = 0$ of (3.14)). Then $(X' \subset P)$ is obstructed. To prove this, we proceed as we did in (3.14) with one extra observation, namely the semicontinuity of $A^2(C' \subset Y)$. In fact we claim

$$(*) \qquad \dim A^i(C' \subset Y) \le \dim A^i(X' \subset Z)$$

for $i=1,2$. Indeed (*) for $i=2$ follows from (*) for $i=1$ because for curves $\dim A^1(C' \subset Y) - \dim A^2(C' \subset Y)$ depends only on $d'=d(C')$, $g'=g(C')$ and $(f_1,\ldots,f_{n-1})$ (easily seen from (1.11) and the surjectivity of 1^2 (1.14), cf. [K11, (2.2.14)] for n=3). Since $\dim A^1(X' \subset Z) - \dim A^2(X' \subset Z)$ is given by the same expression, (*) holds for $i=2$ iff (*) holds for $i=1$. Moreover (*) for $i=1$ is true because it is well known that the embedding dimension of scheme of finite type is upper semicontinuous , cf. [H, III], exercise 12.1.

<u>(3.16) Example.</u> One knows that $H(8,5)_S$ is irreducible and contains curves $(C \subset P)$ and $(X \subset P)$ satisfying

A) $h^1(I_C(2)) = 2$ and $H^1(I_C(v)) = 0$ for $v \neq 2$

B_1) $h^1(I_X(2))=2$, $h^1(I_X(1))=h^1(I_X(3))=1$ and $H^1(I_X(v))=0$ for $v \notin \{1,2,3\}$

B_2) $h^1(I_X(2))=2$, $h^1(I_X(3))=1$ and $H^1(I_X(v))=0$ for $v \notin \{2,3\}$

cf. [GP]. The curve $(C \subset P)$ in the class (A) represents the "generic" curve, and since $H(8,5)_S$ is irreducible, the curves in (B_1) and (B_2) are specializations. Making liaison, starting with two disjoint conics, resp. a twisted cubic and a line, one may find ($X_1 \subset P$) in the class (B_1) resp. ($X_2 \subset P$) in the class (B_2), sitting on a smooth quartic surface. The same conclusion holds for ($C \subset P$) by Castelnuovo-Mumfords lemma [M1, lect.14]. Therefore there are linkages using c.i. of type (4,7) so that their linked curves ($C' \subset P$), ($X_1 \subset P$) and ($X_2 \subset P$) are smooth and connected and belongs to $H(20,47)_S$.

Now we apply (3.14) to ($C \subset P$) and ($X_i \subset P$) for i=1,2. Its assumptions are easily verified. Indeed $A^2(C \subset Y)=0$ by (1.13c), and from (A) and (B_i) above one verifies i) to iv). So $H(20,47)_S$ is non-singular at ($C' \subset P$) and singular at ($X_1 \subset P$) and at ($X_2 \subset P$). Observe that this proves that $H(20,47)_S$ is singular along the set W_6 of (1.30) because one may link a general curve of W_6 so that its linked curve belongs to (B_1).

(3.17) Inspired by this example we now give a criterion for a doubly linked curve to be obstructed. Indeed the curves ($X_i \subset P$) $\in H(20,47)_S$, i=1,2, admit a double link to curves in H(4,-1), and starting in H(4,-1), then the first link, using c.i. of type (4,3), give curves in $H(8,5)_S$ which are proper specializations of the generic curve of $H(8,5)_S$. And this indicates the idea of our proof, namely we first link a family (with generic curve X) such that the linked family has positive codimension in H(d',g') by (1.27), i.e. X' is a proper specialization of some other curve. Then we link once more in such a way that we can use (3.14). This gives

<u>(3.18) Theorem.</u> Let $(X \subset P^3) \in H(p)_{CM}$ be a curve satisfying $H^1(N_X) = 0$, and let $Y \supset X$ be a c.i. of type (f,g), linking X geometrically to X', such that

i) $H^1(I_X(f)) = H^1(I_X(g)) = 0$

ii) $H^0(I_{X/Y}(f-4)) = 0$, $H^1(I_X(f-4)) \neq 0$ and $H^1(I_X(f-8)) = 0$

iii) either $H^0(I_{X/Y}(g-4)) = 0$ or $H^1(I_X(g-4)) = 0$

Let h be any integer such that

iv) $h \geq f$ and $H^1(I_X(f+g-4-h)) = 0$,

and let $Y' \supset X'$ be any c.i. of type (h,g+4) linking X' geometrically to X". Then $X" \subset P^3$ is obstructed.

<u>(3.19) Corollary.</u> Let $X \subset P^3$ be a curve satisfying $H^1(N_X) = 0$ and

suppose there is a surface of degree f containing X such that $f < s(X)+4$ and

$$H^1(I_X(f)) = H^1(I_X(f-8)) = 0 \text{ and } H^1(I_X(f-4)) \neq 0$$

Then for any double link (geometric linkage) using successively c.i. of type (f,g) and $(f,g+4)$, $g>>o$, we get a curve $X'' \subset P^3$ which is obstructed.

<u>Proof.</u> Immediate from (3.18). See [SE] and [EF] for examples.

<u>Proof of (3.18).</u> We first claim that the linked curve $X' \subset P$, $P=P^3$ is a specialization of some other curve $C \subset P$ for which we have

(*) $$h^0(I_C(g)) < h^0(I_{X'}(g))$$

(We need this extra information (*) later in order to apply (3.14)). Indeed by i), the assumption $H^1(N_X)=0$ and (1.13a), we get $A^2(X \subset Y)=0$ which in turn implies $A^2(X' \subset Y)=0$ by (2.14). Now observe that either $H^1(I_{X'}(f))=0$ or $H^1(O_{X'}(f))=0$ by iii) and (2.18), and corresponding $H^1(O_{X'}(g))=0$ by ii). So $X \subset P$ is non-obstructed by (1.19b). Now instead of proving (*) using (1.27) and (1.26) we get an easier proof using (1.23) directly. In fact by non-obstructedness $(X' \subset P)$ belongs to a unique integral component V of $H(p')_{CM}$, say with "generic" point $(C \subset P)$. Then we must have (*) because otherwise we have equality and then (1.23ii) leads to the smoothness of $pr_1': D(p';g)_{CM} \longrightarrow H(p')_{CM}$ at $(X' \subset V(G) \subset P)$ where $Y=V(F,G)$, $g=\deg G$. This in turn gives a surjective tangent map p_1', cf. (1.6), which via the exact sequence of (1.11) implies that $\beta_{X'/V(G)}=0$. Since $A^2(X' \subset Y)=0$ implies $\beta_{X'/Y}$ surjective by (1.13a) and since there is a commutative diagram

$$\begin{array}{ccc} H^0(N_{X'}) & \xrightarrow{\ \beta_{X'/Y}\ } & H^1(I_{X'}(f)) \oplus H^1(I_{X'}(g)) \\ & \searrow^{\beta_{X'/V(G)}} \quad \circ & \downarrow \\ & & H^1(I_{X'}(g)) \end{array}$$

where the vertical arrow is surjective, we get $H^1(I_{X'}(g)) = 0$, contradicting the assumption $H^1(I_X(f-4)) \neq 0$.

We now apply (3.14) or rather (3.15b) to $(C \subset P)$ and the specialization $(X' \subset P)$, using a c.i. $Z \supset X'$ of type $(h, g+4)$. In fact the cotangent sheaves vanish by (2.15), and (3.14i) follows from

$$(**)\quad \begin{aligned} h^1(I_{X'}(h)) &= h^1(I_X(f+g-4-h)) = 0 \\ h^1(I_{X'}(g+4)) &= h^1(I_X(f-8)) = 0. \end{aligned}$$

Moreover (**) together with (1.16) ensures that C is contained in a c.i. of type (h,g+4) as well, allowing a geometric linkage to another curve. Since (3.14ii) is an immeditate consequence of (*), the final point to check is surjectivity of $\beta_{X''/Z}$. Due to (**) $\beta_{X'/Z}$ is obviously surjective, and since we have already proved $A^2(X' \subset Y)=0$, we conclude by (4.7) of the next section, and the proof is complete.

(3.20) Of course the criterion (3.18) applies very widely and it implies that non-obstructedness is no longer a rare property. Moreover it is possible to generalize (3.18) to curves in P^n for $n \geq 4$ using the same proof. It seems however impossible to use it to get obstructed curves of maximal rank, and we will therefore close this section by giving two examples of obstructed curves (of degree 4 and 5 respectively), one of which is a (non-reduced) curve of maximal rank.

<u>(3.21) Example.</u> As the example about H(8,5) in [GP] seems to indicate, there exist smooth curves $X \subset P$ in H(8,5) which belong to both (B_1) and the closure of the family (B_2) defined in the preceding example. If S is the cubic surface containing one such X, then the description above should show that D(8,5;3) is singular at $t=(X \subset S \subset P)$. (Anyway it is possible to prove the obstructedness of t by using (4.10.1) and the resolution of I_X appearing in [GP] to conclude that $\dim H^2(R,A,A) = 1$. Comparing the exact sequences of (4.10,*) and (1.11) one gets $\dim A^2(X \subset S) = 1$. Since $h^1(I_X(3)) = 1$ and $h^1(N_X) = 0$ it follows from (1.11) that $\beta_{X/S} = 0$. So one gets a surjective tangent map p_1: $A^1(X \subset S) \to H^0(N_X)$. If t is non-obstructed, the projection pr_1: $D(8,5;3)_S \to H(8,5)_S$ is smooth at t, and this is clearly false by (3.16).) Now any such curve X is easily seen to be contained in a c.i. Y of type (3,4). So if $U \subset H(8,5)_S$ contains any curve as in (B_1) and (B_2), then $pr_1^{-1}(U) = D(8,5;3,4)_S$ consists of (at least) two irreducible components whose intersection contains the curves X above. In fact $pr_1^{-1}(U) = D(8,5;3,4)_S$ by [GP]. Now by (3.8) the linked family U' is <u>open</u> in $H(4,-1)_{CM}$ and $(X' \subset P)$ sits in the intersection of two irreducible components and is therefore obstructed.

Finally we claim that $h^1(N_{X'})=1$. Indeed by (1.11)

$$(*)\qquad H^0(N_X) \longrightarrow H^1(I_X(3)) \oplus H^1(I_X(4)) \longrightarrow A^2(X \subset Y) \longrightarrow H^1(N_X)$$

is exact and $\beta_{X/Y}$ which is not surjective by (1.19a), must vanish because $h^1(I_X(3)) + h^1(I_X(4)) = 1$. It follows that $\dim A^2(X\subset Y) = 1$. The sequence (*) for the linked curve, together with $A^2(X\subset Y) \approx A^2(X'\subset Y)$, shows the claim.

(3.22) Example. The obstructed curve $(X'\subset P^3) \in H(4,-1)_{CM}$ of (3.21) sits on a reduced irreducible surface S of degree 3 because $(X\subset P^3) \in H(8,5)_S$ does. Moreover by (2.18), $h^1(I_{X'}(v)) = 0$ for $v \notin \{0,1,2\}$ and $h^o(I_{X'}(3)) = 6$, cf.(3.16,B_1). So there exists a c.i. Y' of type (3,3) containing X' and the linked curve $X''\subset P^3$ in $H(5,0)_{CM}$ is obstructed by (3.12) or by 3) of (3.13).

FURTHER LIAISON INVARIANTS.

(4.1) We have already proved the liaison invariance of $A^i(X\subset Y)$, and among other things noticed its importance in determining $\dim H^1(N_X)$ from informations about its linked curve. Unfortunately $A^i(X\subset Y)$ depends on Y. Moreover if the index of speciality e(X), see (3.13), is large compared with s(X), then $h^1(N_X)$ and so $\dim A^2(X\subset Y)$ might be large, and the computation of $h^1(N_X)$ via (2.19.1) and liaison is non-trivial. In this case it is not clear if we can use theorem (1.27) either. The purpose of this section is to remedy this defect, i.e. we will introduce a liaison invariant subgroup $C(X\subset Y)$ of $A^2(X\subset Y)$ which is "more independent" of Y and does not "grow" with e(X). In fact if $R=k[X_o,..,X_n]$ and $A \approx R/I$ is the minimal cone of $X\subset P = P^3$, then, under some assumptions (maximal rank of X is enough), the independence of Y follows from the isomorphism $C(X\subset Y)\approx {}_oH^2(R,A,A)$ where ${}_oH^2(R,A,A)$ is the graded piece of degree zero of the 2. algebra cohomology group (or cotangent group, cf.[LS]) associated to R -> A.

The liaison-invariance of the algebra cohomology group $H^2(k,A,A)$.

Let us start with the case of local rings. We define the notion of linkage according to the rules of (2.2) even if A_i is not Cohen-Macaulay. Then we can prove

(4.2) Theorem. Let R -> B -> A_i for i=1,2 be surjections of local k-algebras and suppose R is k-smooth and ker(R ->B) is generated by an R-regular sequence. If A_1 and A_2 are geometrically linked by B and if for each i=1,2 , A_i satisfies the Serre condition S_3,then there is a canonical isomorphism

$$H^2(k,A_1,A_1) \approx H^2(k,A_2,A_2)$$

<u>Proof.</u> There is a long exact sequence of algebra cohomology groups

$$(4.2.1) \quad \to H^1(R,A_i,A_i) \to H^1(R,B,A_i) \xrightarrow{\delta} H^2(B,A_i,A_i) \to H^2(R,A_i,A_i)$$

cf. [LS], [L1,(3.3)] or [An,(18.2)] . By assumption R is smooth and $H^2(R,B,-)=0$ and it follows that

$$\operatorname{coker} \alpha_i \approx H^2(R,A_i,A_i) \approx H^2(k,A_i,A_i)$$

where α_i is the composition of the "restriction map" $m:H^1(R,B,B) \to H^1(R,B,A_i)$ and the connecting homomorphism δ appearing in (4.2.1). So it suffices to prove the existence of an isomorphism

$$(4.2.2) \qquad \theta : H^2(B,A_1,A_1) \approx H^2(B,A_2,A_2)$$

fitting into a commutative diagram

$$(4.2.3) \qquad \begin{array}{ccc} & & H^2(B,A_2,A_2) \\ & \overset{\alpha_1}{\nearrow} & \\ H^1(R,B,B) & \circ & \downarrow \theta \\ & \underset{\alpha_2}{\searrow} & \\ & & H^2(B,A_2,A_2) \end{array}$$

To prove that θ is an isomorphism, we first claim that

$$(4.2.4) \qquad \operatorname{Ext}^j{}_B(A_i,B) = 0 \qquad \text{for } i=1,2 \text{ and } j=1,2$$

Indeed by Gorenstein duality this is equivalent to the vanishing of the local cohomology group $H^j_m(A_i)$ for $\dim B-3<j<\dim B$; $i=1,2$. We conclude by [S,(4.1)] which states that A_i satisfies S_3 iff $H^j_m(A_{3-i})=0$ for $\dim B-3<j<\dim B$.

Next we claim that $H^2(B,A_i,A_i) \approx \operatorname{Ext}^1_B(K_i,A_i)$ for $i=1,2$ where $K_i = \ker(B \to A_i)$. Indeed since the liaison is geometric, $B_p \approx (A_i)_p$ for any $p \in \operatorname{Ass}(A_i)$ and we get

$$\operatorname{Hom}_{A_i}(H_2(B,A_i,A_i),A_i) = 0 \quad , \text{ for } i=1,2$$

$$\operatorname{Hom}_{A_i}(\operatorname{Tor}^B_1(K_i,A_i),A_i) = 0 \quad , \text{ for } i=1,2$$

Therefore by well known spectral sequences, cf. [An,(16.1)]

$$H^2(B,A_i,A_i) \approx \mathrm{Ext}^1_{A_i}(K_i/K_i^2,A_i) \approx \mathrm{Ext}^1_B(K_i,A_i) \quad ,i=1,2$$

as claimed. Now we define a morphism

$$\Phi : \mathrm{Ext}^1_B(K_1,A_1) \longrightarrow \mathrm{Ext}^1_B(K_2,A_2)$$

by dualizing extensions, i.e by sending

$$(H^\cdot) : \qquad 0 \longrightarrow A_1 \longrightarrow H \longrightarrow K_1 \longrightarrow 0$$

onto

$$(H^\cdot)^\vee : \qquad 0 \longrightarrow K_1{}^\vee \longrightarrow H^\vee \longrightarrow A_1{}^\vee \longrightarrow 0$$

recalling that $K_1{}^\vee = \mathrm{Hom}_B(K_1,B) = A_2$ and $A_1{}^\vee = K_2$ follows from $\mathrm{Ext}^1_B(A_i,B) = 0$. The right exactness of $(H^\cdot)^\vee$ is a consequence of $\mathrm{Ext}^2_B(A_i,B) = \mathrm{Ext}^1_B(K_i,B)=0$. An inverse of Φ is defined in the same way and θ is an isomorphism.

Finally we prove the commutativity of (4.2.3) by making α_i explicite in terms of extensions and then we prove that the two extensions which, via α_1 and α_2, come from the same $\mu \in H^1(R,B,B) = \mathrm{Hom}_R(I_B,B)$, $I_B=\ker(R\to B)$, are isomorphic. We leave the verification to the reader.

There is a different proof of (4.2) by B. Ulrich and R.O.Buchweitz in [BU] for complete local CM-algebras.

<u>(4.3) Corollary.</u> Let $X \subset P^n$ be an equidimensional CM-scheme. If X and X' are geometrically linked by a c.i. $Y \subset P^n$, then the algebra cohomology sheaves (or cotangent sheaves) $\underline{A}^2_X = T^2(O_X/k,O_X)$ and $\underline{A}^2_{X'}$, are isomorphic. In particular if X is a curve and a generic complete intersection in P^3, then $\underline{A}^2_X = 0$.

(4.4) This follows immediately from (4.2) and the fact that generic complete intersection curves in P^3 can be linked in several steps to smooth curves, cf.[R]. Moreover observe that the sequence (4.2.1) under the general assumptions of (4.3) can be written as

$$(4.4.1) \qquad 0 \longrightarrow N_{X/Y} \longrightarrow N_X \longrightarrow N_Y \otimes O_X \longrightarrow \underline{A}^2_{X/Y} \longrightarrow \underline{A}^2_X \longrightarrow 0$$

where $\underline{A}^2_{X/Y}$ is the 2. algebra cohomology sheaf of $X \to Y$ with values in

O_X. This coincides with the definition (1.8) since we there required $\underline{A}_X^2 = 0$.

The liaison-invariant subgroup $C(X \subset Y)$ of $A^2(X \subset Y)$.

In the following let $X \subset P = P_k^n$ be an equidimensional CM scheme and let X' be its linked subscheme via a c.i. Y containing X. Now if we apply the arguing in (4.2) to the graded minimal cones of X and X' in $R=k[X_o,..,X_n]$, letting $C(X\subset Y)$ and $C(X'\subset Y)$ be the cokernels of the corresponding maps α_i in (4.2.3), we get the liaison invariant groups we are aiming at.

(4.5) Theorem. Let $X \hookrightarrow P=P^{r+1}$ be a curve and suppose $X \subset P$ and $X' \subset P$ are geometrically linked by a complete intersection $Y \subset P$ of type $(f_1,f_2,.,f_r)$. Let $I = \Sigma_t H^o(I_X(t))$ and suppose $\underline{A}_X^2 = 0$. Then there are subgroups $C(X\subset Y)_v$ and $C(X'\subset Y)_v$ of $A^2(X\subset Y)_v$ and $A^2(X'\subset Y)_v$ respectively and an isomorphism

$$C(X\subset Y)_v \xrightarrow{\sim} C(X'\subset Y)_v$$

Moreover there is an exact sequence

$$\longrightarrow A^1(X\subset Y)_v \longrightarrow H^o(N_X(v)) \xrightarrow{v\beta X/Y} \sum_{i=1}^{r} H^1(I_X(f_i+v)) \longrightarrow$$

$$C(X\subset Y)_v \longrightarrow H^1(N_X(v)) \xrightarrow{v l_{gr}^2} {}_v\mathrm{Hom}(I,\Sigma_t H^1(O_X(t)))$$

and of course a corresponding sequence replacing X by X'.

(4.6) Remark. In [Kl1,(2.3.14)] there is another proof of (4.5) for curves in P^3 where we also proved that coker ${}_v l_{gr}^2 \approx {}_v\mathrm{Ext}_R^2(I,\Sigma_t H^1(I_X(t)))$, and that this cokernel is liaison-invariant.

Proof. Let B, A_1 and A_2 be the minimal cones in $R=k[X_o,..,X_n]$ of $Y \subset P$, $X \subset P$ and $X' \subset P$ respectively. If $K_i = \ker(B \to A_i)$, $\overline{A}_1 = \Sigma H^o(O_X(t))$ and $\overline{A}_2 = \Sigma H^o(O_{X'}(t))$, we claim that

(4.6.1) $$A_i^v \approx \overline{A}_i^v \approx K_{3-i} \quad \text{and} \quad K_i^v \approx \overline{A}_{3-i}$$

for i=1,2 where $(-)^v = \mathrm{Hom}_B(-,B)$. Indeed since $\mathrm{depth}_m B=2$, m the irrelevant maximal ideal of B, we get $\mathrm{depth}_m M^v=2$ for any B-module M of finite type, cf.[KL,(2.2.2)]. It follows that $\mathrm{Hom}_B(M,B) \approx \Sigma_t H^o(\underline{\mathrm{Hom}}_{O_Y}(\widetilde{M},O_Y)(t))$, and since (4.6.1) is true locally outside m by the definition of liaison, the claim is proved.

Next we claim that (4.2.2) and (4.2.3) hold provided we replace $H^2(B,A_i,A_i)$ by $H^2(B,A_i,\overline{A}_i)$ and α_i by the composition

$$(4.6.2) \qquad H^1(R,B,B) \to H^1(R,B,A_i) \to H^1(R,B,\overline{A}_i) \xrightarrow{\delta} H^2(B,A_i,\overline{A}_i)$$

of natural maps and of a connecting homomorphism δ, cf. (4.2.1) and replace A_i by $\overline{A}_i$. Indeed by Gorenstein duality on B, $\mathrm{Ext}^2_B(A_i,B)=0$, i.e. $\mathrm{Ext}^1(K_i,B)=0$. This together with (4.6.1) is sufficient for showing $\mathrm{Ext}^1_B(K_1,\overline{A}_1) \approx \mathrm{Ext}^1_B(K_2,\overline{A}_2)$ by dualizing extensions as in the proof of (4.2). It follows that $H^2(B,A_1,\overline{A}_1) \approx H^2(B,A_2,\overline{A}_2)$ by the spectral sequences of (4.2). Since the corresponding diagram of (4.2.3) is still commutative, the claim follows.

If $\overline{C}(X\subset Y) = \mathrm{coker}\,\alpha_1$ and $\overline{C}(X'\subset Y) = \mathrm{coker}\,\alpha_2$, we have by the claim above a degree-preserving isomorphism $\overline{C}(X\subset Y) \approx \overline{C}(X'\subset Y)$. Moreover letting m_i be the composition of the first two arrows in (4.6.2) we get an exact sequence

$$(4.6.3) \qquad H^1(R,A_i,\overline{A}_i) \to \mathrm{coker}\,m_i \to \mathrm{coker}\,\alpha_i \to H^2(R,A_i,\overline{A}_i) \to 0$$

essentially because of (4.2.1) and the fact $H^2(R,B,\overline{A}_i) = 0$. Furthermore $H^1(R,B,-)$ is right exact since $H^2(R,B,-) = 0$. Applying $H^1(R,B,-)$ to the exact sequence $B \to \overline{A}_1 \to H^1_m(A_1) \to 0$, letting $H^1_m(A_1) = \Sigma_t H^1(I_X(t))$ and $I_B = \ker(R \to B)$, we get

$$(\mathrm{coker}\,m_1)_v \approx {}_vH^1(R,B,H^1_m(A_1)) \approx {}_v\mathrm{Hom}_R(I_B,\Sigma_t H^1(I_X(t))) \approx \sum_{i=1}^{r} H^1(I_X(f_i+v))$$

for the elements of degree v. The exact sequence of (4.5) follows now from (4.6.3) and (4.7), provided we let $C(X\subset Y)_v = (\overline{C}(X\subset Y))_v$.

Finally comparing this exact sequence with the corresponding exact sequence involving $A^1(X\subset Y)_v$, cf. (2.19.1), it follows that $C(X\subset Y)_v$ is a sub k-vectorspace of $A^2(X\subset Y)_v$ and the proof is complete.

<u>(4.7) Lemma.</u> Let $f: X \hookrightarrow P = P^n$ be an equidimensional CM scheme with minimal cone A in $R=k[X_o,..,X_n]$ and let $\overline{A} = \Sigma_t H^o(O_X(t))$. If $\underset{X}{\overset{2}{A}} = 0$ and dim X > 0, then ${}_vH^1(R,A,A) \approx H^o(X,N_X(v))$ and there is an exact sequence of graded R-modules which in degree v looks like

$$0 \to {}_vH^2(R,A,\overline{A}) \longrightarrow H^1(N_X(v)) \xrightarrow{v^1 g r^2} {}_v\mathrm{Hom}_R(I,\Sigma H^1(O_X(t)) \to {}_vH^3(R,A,\overline{A})$$

Proof. Due to Laudal [L1,(3.2)] there exists a local algebra cohomology group $H_m^\bullet(R,A,M)$, M graded of finite type, fitting into a long exact sequence

$$(4.7.1)\quad \to {}_vH_m^i(R,A,M) \to {}_vH^i(R,A,M) \to A^i(k,f,\widetilde{M}(v)) \to {}_vH_m^{i+1}(R,A,M) \to$$

where $A^i(k,f,\widetilde{M}(v))$ is the global algebra cohomology of $f:X \hookrightarrow P^n$ introduced by Illusie and Laudal. Moreover this sequence is governed by two spectral sequences

$$(4.7.2)\quad \begin{aligned} E_2^{p,q} &= H^p(R,A,H_m^q(M)) \Longrightarrow H_m^{p+q}(R,A,M) \\ 'E_2^{p,q} &= H^p(X,\widetilde{H^q(R,A,M(v))}) \Longrightarrow A^{p+q}(k,f,\widetilde{M}(v)) \end{aligned}$$

where $H_m^q(-)$ is the local cohomology with support V(m), cf. [Kl, Section 3] which treats this situation in detail. Now let $M = \bar{A}$. Since $H_m^q(\bar{A})=0$ for q<2, we get $H_m^i(R,A,\bar{A}) = 0$ for i=1,2 and $H_m^3(R,A,\bar{A})=\mathrm{Hom}_R(I,H_m^2(\bar{A}))= \mathrm{Hom}_R(I,\Sigma H^1(O_X(t)))$. Using $\widetilde{H^2}(R,A,\bar{A}) = \underset{\sim}{A}_X^2 = 0$, we conclude as claimed.

(4.8) Corollary. Let X, X' and Y be as in (4.5) and suppose $\beta_{X/Y}$ is surjective. Then $\beta_{X'/Y}$ is surjective provided there exists a c.i. Z containing X such that $A^2(X\subset Z) = 0$.

Proof. Since $C(X\subset Z)_o$ is a subspace of $A^2(X\subset Z)_o=A^2(X\subset Z)$ (true also when $X \hookrightarrow Z$ is generically non-isomorphic), $C(X\subset Z)_o=0$. It follows from the exact sequence of (4.5) that ${}_ol_{gr}^2$ is injective, and combining with the surjectivity of $\beta_{X/Y}$, we get $C(X\subset Y)_o=0$ using (4.5) once more. Hence $C(X'\subset Y)_o=0$ again by (4.5), and in particular we get that $\beta_{X'/Y}$ is surjective.

(4.9) Remark. (Comparing $C(X\subset Y)$ with graded algebra cohomology). From (4.5) and (4.7) we get an exact sequence

$$0 \to \mathrm{coker}\,\beta_{X/Y} \to C(X\subset Y) \to {}_oH^2(R,A,A) \to 0$$

where the term on the right side of $C(X\subset Y) := C(X\subset Y)_o$ is independent of Y. This is certainly not the case for $A^2(X\subset Y)$ where the corresponding sequence, deduced from (2.19.1), depends on Y "on both sides." Now suppose

$$(*)\quad c(X) < s(X) \quad \text{where} \quad c(X)= \max\{t \mid H^1(I_X(t)) \neq 0\}$$

(including arithmetically Cohen Macaulay curves as the special case: "$c(X) = -\infty$"). Then we claim

$$_0H^i(R,A,A) \xrightarrow{\simeq} {}_0H^i(R,A,\bar{A}) \qquad \text{for } i=1,2$$

Indeed by the long exact sequence of algebra cohomology associated to the sequence $0 \to A \to \bar{A} \to H^1_m(A) \to 0$ of A-modules, the claim follows if we can prove $_0H^i(R,A,H^1_m(A))=0$ for i=1,2. Recalling $H^1_m(A) \approx \Sigma H^1(I_X(t))$, we get $_0H^1(R,A,H^1_m(A)) \approx {}_0\mathrm{Hom}(I,H^1_m(A))=0$ from (*), and using the well known description of $H^2(R,A,-)$ in [LS] or [SGA, exp VI] which gives a surjection map $_0\mathrm{Hom}(E/K,-) \twoheadrightarrow {}_0H^2(R,A,-)$ where E, resp. K, is the R-module of "relations", resp. "trivial relations", among the minimal generators of I, we get $_0H^2(R,A,H^1_m(A))=0$ as well. In conclusion, supposing (*) we get coker $\beta_{X/Y} = 0$, and it follows that

$$C(X\subset Y) \approx {}_0H^2(R,A,A)$$

<u>A vanishing criterion for $C(X\subset Y)$ and $H^1(N_X)$.</u>

(4.10) Before finishing this section with an example using the liaison invariance of $C(X\subset Y)$ we will include a vanishing criterion for $C(X\subset Y)$, which, under the extra assumption "$e(X)<s(X)$" leads to a vanishing criterion of $H^1(N_X)$. In what follows X is a generic complete intersection curve in P^3. Let $\{H_1,\ldots,H_a\}$ be a minimal set of generators of the homogeneous ideal I of X in $R=k[X_0,\ldots,X_3]$, and let $h_i = \deg H_i$ and $A=R/I$. In [Kl1,(2.2.8)] we proved that

$$(4.10.1) \qquad {}_vH^2(R,A,A)^\vee \approx {}_{-v-4}\mathrm{Hom}_R(I,H^1_m(A)) \subset \sum_{i=1}^{a} H^1(I_X(h_i-v-4))$$

where $(-)^\vee = \mathrm{Hom}_k(-,k)$. Moreover let $r(X) = \min\{t \mid E_t \neq 0\}$, cf. (4.9), be the minimal degree of a relation among the H_i's, and suppose $c(X)<r(X)$. Then by the argument in (4.9), $_0H^2(R,A,H^1_m(A)) = 0$. Letting $M = A$ in (4.7.1) and using (4.7.2), we get an exact sequence

$$*) \to H^0(N_X) \to {}_0\mathrm{Hom}(I,H^1_m(A)) \to {}_0H^2(R,A,A) \to H^1(N_X) \xrightarrow{{}^{o}\!\mathrm{gr}^{2}_{1}} {}_0\mathrm{Hom}(I,\Sigma H^1(O_X(t)))$$

closely related to that of (4.5). Comparing we deduce

$$(4.10.2) \qquad {}_0H^2(R,A,A)=0 \Longrightarrow C(X\subset Y)=0$$

provided the c.i. $Y=V(F_1,F_2)$ of type (f_1,f_2) containing X is chosen such that the natural map ${}_0\mathrm{Hom}(I,H^1_m(A)) \longrightarrow \Sigma_{i=1}^{2} H^1(I_X(f_i))$ is surjective. This is true provided

$$(4.10.3)\quad \begin{cases} \text{either } H^1(I_X(f_i))=0 \quad \text{for } i=1,2, \ \underline{\text{or}} \\ F_1 \text{ is a minimal generator of } I \text{ and } H^1(I_X(f_2))=0, \ \underline{\text{or}} \\ \{F_1,F_2\} \text{ is part of a minimal set of generators of } I. \end{cases}$$

We therefore have the following result

(4.11) Proposition. In the situation of (4.10), we suppose (4.10.3), $c(X)<r(X)$ and $e(X)\leq \min\{t|H^1(I_X(t))\neq 0\}$. Moreover suppose the Rao-module $\Sigma H^1(I_X(t))$ is non-vanishing in at most two consecutive degrees (i.e $\mathrm{diam}(X)\leq 2$). Then

$$C(X\subset Y) = 0 \qquad \text{and} \qquad H^1(N_X) \approx \sum_{i=1}^{a} H^1(O_X(h_i))$$

The conclusion applies in particular to arithmetically Cohen Macaulay curves provided $e(X)<r(X)$.

Proof. One knows that $h_i\leq \max\{c(X)+2,e(X)+3\}$ for any i by [K13,remark2] or by [GM,(3.15)]. Hence $H^1(I_X(h_i-4))=0$ by the assumption on the diameter and on $e(X)$. It follows that ${}_0H^2(R,A,A)=0$ by (4.10.1) and we conclude by (4.10.2) and the exact sequence of (4.5).

(4.12) Corollary. Let X be a generic complete intersection curve in P^3 and suppose $\mathrm{diam}(X) \leq 2$ and $e(X) < s(X)$. If $\mathrm{diam}(X) \neq 0$, suppose $e(X) \leq \min\{t|H^1(I_X(t))\neq 0\}$ and $c(X) \leq s(X)$. Then

$$H^1(N_X) = 0.$$

(4.13) Example. Take a curve $C\subset P^3$ in $H(14,25)_S$ sitting on a smooth cubic hypersurface F, satisfying

$$\sum_{v=1}^{\infty} h^1(I_C(v)) = h^1(I_C(4)) = 1$$

(for instance take C to be a curve directly linked to a curve Γ in $H(4,0)_S$ by a c.i. of smooth surfaces of degree 3 and 6). Computing $\chi(I_C(v))=\binom{v+3}{3}-(dv+1-g)$ for different v's, one may see that the homogeneous ideal is generated by 5 generators of degree 3,6,6,6 and 6. Moreover $h^1(O_C(3))=1$ and $h^1(O_C(v))=0$ for $v\geq 4$. Taking cohomology of $0 \to N_{C/F} \to N_C \to O_C(3) \to 0$, using that the normal bundle of $C\subset F$ is $N_{C/F} \approx \Omega(1)$, we get $h^1(N_C)=h^1(O_C(3))=1$. Now link C geometrically to a curve X by a c.i. Y of type (f_1,f_2) and say we want to compute $h^1(N_X)$ for any possible value of $f_i \geq 6$. By (2.19.1), (2.14) and (1.14) we have exact horizontal sequences

$$(*)\quad \begin{array}{l} H^o(N_C) \longrightarrow \Sigma H^1(I_C(f_i)) \longrightarrow A^2(C\subset Y) \longrightarrow H^1(N_C) \longrightarrow\!\!\!\!\to \Sigma H^1(O_C(f_i)) \\ \qquad\qquad\qquad\qquad\qquad\qquad\qquad \wr| \\ H^o(N_X) \xrightarrow{\beta} \Sigma H^1(I_X(f_i)) \longrightarrow A^2(X\subset Y) \longrightarrow H^1(N_X) \longrightarrow\!\!\!\!\to \Sigma H^1(O_X(f_i)) \end{array}$$

The upper one gives immediately dim $A^2(C\subset Y)=h^1(N_C)=1$. If $\beta_{X/Y}$ is surjective (true if $f_i \neq 8$ for $i=1,2$ because in this case $H^1(I_X(f_i)) \approx H^1(I_C(f_1+f_2-4-f_i))^V=0$) then the lower one implies that

$$(**)\qquad h^1(N_X) = 1 + \Sigma h^1(O_X(f_i)) = 1 + \Sigma h^o(I_{C/Y}(f_i-4))$$

and $h^1(N_X)$ is readily found. For instance if $(f_1,f_2)=(6,9)$, then $h^1(N_X)=1+h^o(I_C(2))+h^o(I_C(5))=11$. If, however, $f_i = 8$ for some i, we can not compute $h^1(N_X)$ from (*) unless we know dim coker $\beta_{X/Y}$. Using the invariance of $C(X\subset Y)$ under geometric linkage we can overcome this difficulty. In fact we claim that $\beta_{X/Y}$ is surjective, thus proving (**) in any case. To see this we observe that $C(C\subset Y)=0$ by (4.11). By (4.5), $C(X\subset Y)=0$ and the surjectivity of $\beta_{X/Y}$ follows from the exact sequence of (4.5) (One may also prove the surjectivity of $\beta_{X/Y}$ by combining (4.8) and (1.13c)).

As a final observation we see that theorem (1.27) applies to the family W of linked curves X obtained by varying $(C\subset P) \in H(14,25)$ as above and the c.i. $Y\supset C$ of type (f_1,f_2). Indeed even though $A^2(X\subset Y) \neq 0$, we find that the parenthetical conditions of (1.27) are satisfied because the vanishing of $C(C\subset Y)$ and the non-obstructedness of $(C\subset P) \in H(14,25)_S$, cf.(1.18), lead to the non-obstructedness of $(X\subset Y\subset P) \in D(d',g';f_1,f_2)$ by (1.19a).

REFERENCES.

[AK] Altman, A. and Kleiman, S., Introduction to Grothendieck Duality Theory. Springer Lecture Notes 146 (1970).

[An] André, M., Méthode Simpliciale en Algèbre Homologique et Algèbre Commutative. Springer Lecture Notes 32 (1967).

[Bo] Bolondi, G., Irreducible families of curves with fixed cohomology. Preprint, Camerino, 1987.

[BM1] Bolondi, G. and Migliore J., Classification of maximal rank curves in the liaison class L_n", (to appear in Math.Ann).

[BM2] Bolondi and Migliore, Buchsbaum liaison classes. Preprint 1987, Trento. Italy

[BM3] Bolondi and Migliore, The Lazardsfeld - Rao and Zeuten problems for Buchsbaum curves, Preprint 1987, Trento.

[B] Buchweitz.R.O: Contributions à la theorie des singularités. Univ. Paris VI (Thesis)

[BU] Buchweitz.R.O. and Ulrich B., Homological properties which are invariant under linkage. Preprint 1983

[BH] Brun J.et Hirschowitz A. Le problèm de Brill Noether pour les idéaux de P^2. Ann.scient.éc.norm.sup.,4.ser.,t.20,1987, 171-200

[E] Ellingsrud, G., Sur le schéma de Hilbert des variéteé de codimension 2 dans P^e a cône de Cohen-Macaulay. Ann.scient.éc.norm.sup., 4.sér., t.8 (1975), 423-432.

[EF] Ellia Ph.,Fiorentini M.,Défaut de postulation et singularités du schéma de Hilbert. Ann.Univ.Ferrara Nuova Ser.Sez.VII, 30, (1984),185-198

[E] Ellia Ph., D'autre composantes non réduites de Hilb P^3. Math.Ann 277, 433 - 446 (1987)

[EGA] Grothendieck, A. And Dieudonné, J., Éléments de Géométrie Algébrique. Publ.Math. I.H.E.S. 4 (1960), 8 (1961), 11 (1961), 17 (1963), 20 (1964), 24 (1965), 28 (1966), 32 81967).

[GP] Gruson, L. and Peskine, C., Genre des courbes de l'espace projectif. Proc. Tromsø, Conference on alg. geom.(1977),32-59.

[H] Hartshorne, R., Algebraic Geometry. Springer Verlag (1977).

[HH] Hartshorne R. and Hirschowitz A., Smoothing Algebraic Space Curves, Proc.Alg.Geom.,Sitges 1983, Springer Lecture Notes 1124, 98-131

[Ha] Halphen,G., Mémoire sur la classification des courbes gauches algébriques. Oeuvres completes t.III.

[Kl] Kleppe, J.O., Deformations of Graded Algebras. Math.Scand. 45 (1979), 205 - 231.

[Kl1] Kleppe, J.O. The Hilbert-flag scheme, its properties and its connection with the Hilbert scheme. Applications to curves in 3-space. Preprint Ser.(1981), Univ. of Oslo (part of thesis).

[Kl2] Kleppe. J.O., Non-reduced components of the Hilbert scheme of smooth space curves, Proc., Rocca di Papa 1985. Lecture Notes in Math, Springer-Verlag, 1266

[Kl3] Kleppe, J.O., Singularities in Codimension 1 of the Hilbert scheme. An example. Preprint 1983, College of Engineering, Oslo

[KL] Kleiman, S.L. and Landolfi, J., Geometry and Deformation of Special Schubert varieties. Comp.Math.23, Fasc. 4 (1971), 407-434.

[KM] Kustin A.R. and Miller M. Deformation and linkage of Gorenstein algebras, Trans. Amer. Math. Soc., Vol 284, No 2, 1984, 501-534.

[L1] Laudal, O.A., Formal Moduli of Algebraic Structures. Springer Lecture Notes 754 (1979).

[L2] Laudal, O.A., A generalized trisecant lemma. Proc. Tromsø, Conference on alg. geometry (1977), 112-149.

[LS] Lichtenbaum, S. and Schlessinger M., The cotangent complex of a morphism. Trans. Amer. Math. Soc. 128 (1967), 41-70.

[M1] Mumford, D,. Lectures on Curves on an Algebraic Surface. Annals of Math. Studies 59, Princeton Univ. Press (1966).

[M2] Mumford, D., Further Pathologies in Algebraic Geometry. Amer. J. Math. 84 (1962), 642-648.

[N] Noether, M., Zur Grundlegung der Theorie der algebraischen Raumcurven. J. Reine u. Angew. Math. (1882), 271-318 (or Steinerische Preisschrift, Verlag d. König. Akad. D. Wiss., Berlin (1883)).

[P] Perrin D. Courbes passant par m point généraux de P^3. (These). Memoire de la Soc. Math de France n 28/29 Supplement an Bull S.M.F., Tome 115, 1987, Fase 3.

[PS] Peskine, C. and Szpiro, L., Liaison des variétés algébriques. Inv. Math.26 (1974), 271-302.

[R] Rao, P. Liaison Among curves in P . Inv. Math. 1979, 205-217

[R1] Lazardsfeld R. and Rao P., Linkage of General Curves of Large Degree, Proc.Ravello 1982, Springer Lecture Notes, 997

[S] Schenzel P., Notes on Liaison and Duality, J.Math of Kyoto Univ. 22 (1982), 485-498

[S1] Schlessinger, M., Functor of Artin rings, Trans Amer. Math. Soc. 130 (1968), 208-222.

[Se] Sernesi, E., Un esempio di curve obstruita in P^3 in "Seminario di variabili complesse,Bologna,1981", CNR, Università degli Studi di Bologna (1982), 223-231

[SGA2] Grothendieck, A., Cohomologie locale des faisceaux cohérents et Theoremes de Lefschetz locaux et globaux. North-Holland, Amsterdam (1968).

[SGA7] Grothendieck, A., Raynaud, M. and Rim, D.S., Groupes de Monodromie en Géométrie Algébrique. Springer Lecture Notes 288 (1972) and 340(1973).

[U] Ulrich B,Liaison and deformations, J. Pure and Applied Alg.39 (1986), 165-175, North-Holland

ON CURVES ON K3 SURFACES

G. Martens
Mathematisches Institut, Universität
Erlangen-Nürnberg, Bismarckstr. $1\frac{1}{2}$,
D-852o Erlangen

§ 1. Introduction

We deal with smooth irreducible curves on a complex projective K3 surface X. Our method is based on Green's and Lazarsfeld's result on the constancy of the Clifford index in linear systems on X (cf. (2.1)).

We show that X contains no smooth plane curves of genus > 10 (cf. (3.1)) and we study the Clifford index of smooth curves on X in greater detail ((3.3, (3.5)). In the last section we discuss the question of the constancy of the gonality in linear systems on X originally asked by J. Harris and Mumford in 1982.

Notation: We work over the complex numbers. C always denotes a smooth irreducible projective curve of genus g. A g^r_n on C is a linear system of C of degree n and of projective dimension r (a pencil if $r = 1$).

§ 2. Method

The Clifford index c of C is defined by

$$c := \min\{\deg A - 2(h^0(A)-1) \mid A \in \mathrm{Pic}(C):\ h^0(A) \geq 2 \text{ and } h^1(A) \geq 2\}$$

if $g \geq 4$.

A line bundle A on C is said to compute the Clifford index c of C if $h^0(A) \geq 2$, $h^1(A) \geq 2$ and if $\mathrm{cliff}(A) := \deg A - 2(h^0(A)-1)$ is equal to c.

We first recall the main result of [GL].

(2.1) Theorem: Let X be a complex projective K3 surface and let

$C \subset X$ <u>be a smooth irreducible curve of genus</u> $g \geq 4$ <u>whose Clifford index</u> c <u>is strictly less than the generic value</u> $\left[\frac{g-1}{2}\right]$. <u>Then there is a line bundle</u> L <u>on</u> X <u>such that</u> $L \otimes \mathcal{O}_{C'}$ <u>computes the Clifford index of any smooth curve</u> C' <u>in</u> $|C|$. ∎

(2.2) <u>**Lemma:**</u> <u>In Theorem (2.1) we may assume</u> $h^0(L \otimes \mathcal{O}_{C'}) = h^0(L)$ <u>for any smooth curve</u> C' <u>in</u> $|C|$.

<u>Proof:</u> Let A be a line bundle on C of degree $\deg A \leq g-1$ which computes the Clifford index $c < \left[\frac{g-1}{2}\right]$ of C. Starting with A, Green and Lazarsfeld ([GL], §3) construct a line bundle $L \in \operatorname{Pic} X$ and a locally free $\mathcal{O}_X$-module F such that

$$h^0(L \otimes \mathcal{O}_{C'}) \geq h^0(L) \geq 2$$

and

$$h^1(L \otimes \mathcal{O}_{C'}) \geq h^0(\det F) \geq 2.$$

If $c_1(F)^2 > 0$ they conclude

$$g+1-c \leq h^0(L) + h^0(F) \leq h^0(L) + h^0(\det F) \leq h^0(L \otimes \mathcal{O}_{C'}) + h^1(L \otimes \mathcal{O}_{C'})$$
$$= g+1-\operatorname{cliff}(L \otimes \mathcal{O}_{C'}) .$$

Thus $\operatorname{cliff}(L \otimes \mathcal{O}_{C'}) = c$, and we have equality at each step. In particular, $h^0(L \otimes \mathcal{O}_{C'}) = h^0(L)$.

If $c_1(F)^2 = 0$ we may proceed analogously (using the bundle $\omega_C \otimes A^*$, cf. [GL], (3.5)) unless we are in a certain case in which we may choose $L = \mathcal{O}_X(E)$ for some smooth elliptic curve E on X. But in that case Green and Lazarsfeld show that

$$c \geq s \cdot \operatorname{cliff}(L \otimes \mathcal{O}_{C'}) \geq s \cdot c$$

where $s := h^1(A)-1 \geq 1$. In particular, $\operatorname{cliff}(L \otimes \mathcal{O}_{C'}) = c$.

Let $c \neq 0$. Then we have $s = 1$. Hence $\deg A \geq g-1$. Since $\deg A \leq g-1$ by hypothesis, we obtain $\deg A = g-1$ and $h^0(A) = h^1(A) = 2$. We see

$$g-3 = \deg A - 2 = c < \left[\frac{g-1}{2}\right]$$

whence the contradiction $g \leq 3$.

So let $c = 0$, i.e. C a hyperelliptic curve. Then there is a pencil of smooth elliptic curves on X cutting on any smooth $C' \in |C|$ its unique g^1_2 unless $g = 5$ and $C' \in |2B|$ where B is a smooth genus 2 curve on X ([SD], 5.8.3). In the latter case, of course, $|B|_{C'}|$ is the g^2_4 of C'. ∎

(2.3) <u>**Corollary: Let L be the line bundle of (2.2) and let M be L or $L^*(C)$. Then we have:**</u>

(1) $M\otimes\mathcal{O}_C$ <u>computes the Clifford index</u> c <u>of</u> C.

(2) $h^o(M\otimes\mathcal{O}_C) = h^o(M)$.

(3) $M = \mathcal{O}_X(H)$ <u>for some smooth irreducible curve</u> H <u>on</u> X <u>of genus</u> $g(H) = h^o(M\otimes\mathcal{O}_C) - 1$.

(4) <u>If</u> $h^o(M\otimes\mathcal{O}_C) \geq 3$ <u>holds for both</u> $M = L$ <u>and</u> $M = L^*(C)$ <u>the inequality</u>

$$g \leq \frac{(\deg(M\otimes\mathcal{O}_C))^2}{4(h^o(M\otimes\mathcal{O}_C) - 2)} + 1$$

<u>is a nontrivial condition for</u> g (<u>the same for</u> $M = L$ <u>and</u> $M = L^*(C)$).

<u>**Proof:**</u> (1) is clear. If C is hyperelliptic (2) and (3) are easy consequences of the closing lines in the proof of (2.2). To prove (2) and (3) in the nonhyperelliptic case we may assume that $|M|$ has no base components (note that $M\otimes\mathcal{O}_C$ is base point free since it computes c). Thus we have $M^2 \geq o$ for the self-intersection number of M on X. Considering first $M = L$ we distinguish two cases:

Let $L^2 > o$. Then we clearly have $L = \mathcal{O}_X(H)$ for a smooth irreducible curve H on X of genus $g(H) = \frac{1}{2}H^2 + 1 = h^o(L) - 1 = h^o(L\otimes\mathcal{O}_C) - 1$, by (2.2). (Cf. [SD], 2.6, 5.8, 6.1.)

Let $L^2 = o$. In this case, $L = \mathcal{O}_X(kE)$ for some (smooth) elliptic curve E on X and $k = h^1(L) + 1 = h^o(L) - 1$ ([SD], 2.6). Of course,

$$\mathcal{O}_X(C)\cdot L = kC\cdot E = kn$$

where $n := C\cdot E > o$, and the pencil $|E|$ cuts on C a g^1_n. From (2.2) and the minimality of the Clifford index c of C we obtain

$$k(n-2) = \text{cliff}(L\otimes\mathcal{O}_C) = c \leq \text{cliff}(g^1_n) \leq n-2$$

whence $k = 1$ or $n = 2$. Since we assume here that C is not hyperelliptic we have $k = 1$ and $L = \mathcal{O}_X(H)$ with $H = E$.

We thus have proved (3) for $M = L$. Since $h^1(H) = o$ we deduce from (2.2) and from the exact sequence

$$o \to \mathcal{O}_X(H-C) \to \mathcal{O}_X(H) \to \mathcal{O}_C(H) \to o$$

that $h^o(H-C) = h^1(H-C)$. Therefore, by Riemann-Roch,

$$h^o(L^*(C)) = h^o(C-H) = \chi(C-H) = \frac{1}{2}(C-H)^2 + 2 = g - 1 - C\cdot H + h^o(L\otimes\mathcal{O}_C) =$$
$$= h^1(L\otimes\mathcal{O}_C) = h^o(L^*(C)\otimes\mathcal{O}_C) .$$

This proves (2). To see (3) for $M = L^*(C)$ we can argue as we did for L.

Finally, if $h^o(M\otimes\mathcal{O}_C) \geq 3$ we conclude from (2) and (3) that $M^2 = 2(h^o(M\otimes\mathcal{O}_C) - 2) > o$. By the Hodge index theorem, then,

$$C^2 M^2 \leq (\mathcal{O}_X(C)\cdot M)^2.$$

This is the inequality (4). ∎

§ 3. Applications

We want to give two applications of the results in § 2.

I. Smooth plane curves on K3 surfaces

By (2.3) (4) we have

(3.1) **Theorem:** *A complex projective K3 surface contains no curve isomorphic to a smooth plane curve of degree* $d > 6$.

Proof: Let C be a smooth plane curve of degree $d > 6$ on a K3 surface X. C has only two line bundles computing its Clifford index $c = d-4$: The line bundle A corresponding to the unique g_d^2 on C and the bundle $\omega_C \otimes A^*$. Cf. [M], Satz 4.
By § 2 there is a line bundle L on X such that $L\otimes\mathcal{O}_C = A$ and $h^o(L) = h^o(A) = 3$. The inequality (2.3) (4) says then

$$\tfrac{1}{2}(d-1)(d-2) = g \leq \tfrac{1}{4}d^2 + 1$$

whence the contradiction $d \leq 6$. ∎

I learnt that theorem (3.1) is known to Morrison. For $d > 8$ it is an immediate consequence of Reid's result ([R]) stating that any complete and base point free pencil g_n^1 on a smooth irreducible curve C of genus $g > \frac{1}{4}(n+2)^2 + 1$ on a K3 surface X is cut out on C by a pencil of elliptic curves on X. In fact, since X has an at most countable collection of such pencils Reid observes

(3.2) **Theorem:** *A smooth irreducible curve C of genus $g > \frac{1}{4}(n+2)^2 + 1$ having infinitely many complete and base point free pencils of degree n cannot lie on a K3 surface.* ∎

In particular, a K3 surface contains no smooth curve having a plane model of degree d with e double points if $o \leq e < \frac{1}{4}(d-1)(d-7)-2$.

II. On the Clifford index of curves on K3 surfaces

An obvious consequence of (2.3) is

(3.3) Theorem: Let X be a K3 surface containing no nonrational smooth curves of genus $< r$ $(r \geq 2)$. Then any smooth curve C on X of genus $g \geq 4$ and Clifford index $c < \left[\frac{g-1}{2}\right]$ has a line bundle A computing c such that $\deg A \leq g-1$ and $h^0(A) \geq r+1 \geq 3$. ∎

For instance, a smooth curve C on a general quartic surface in $\mathbb{P}_3$ is a complete intersection, and it is easy to see that $\mathcal{O}_C(1)$ computes the Clifford index of C unless C is a hyperplane section.

(3.4) Example: On a smooth trigonal curve C of genus $g \geq 5$ the unique g^1_3 is the only linear system of degree $\leq g-1$ computing c, cf. [M], Satz 4. Thus, if C is contained in a K3 surface X the g^1_3 is cut out on C by a pencil of elliptic curves on X. Note however that on a smooth hyperelliptic curve $C \subset X$ of genus 5 the g^1_2 need not be cut out by an elliptic pencil. ∎

According to (2.3) the genus of the curve C in (3.3) is bounded. In fact, we have

(3.5) Theorem: Let C be a smooth curve of Clifford index c and genus $g > \frac{1}{4}(c+4)^2 + 1$ on a K3 surface X. Then there is an elliptic curve E on X such that $|E|$ cuts a g^1_{c+2} (computing c) on any smooth curve C' in $|C|$. Moreover, every g^1_{c+2} on C is cut out by a pencil of elliptic curves on X. If C is not hyperelliptic and if $g > \frac{1}{4}(c+5)^2 + 1$ then every linear series of degree $\leq g-1$ of C which computes c is a pencil g^1_{c+2}.

Proof: If $g > \frac{1}{4}(c+4)^2 + 1$ any smooth C' in $|C|$ has a g^1_{c+2}, by [M], Kor. 1 of Satz 4 and by (3.1). From the main result of [R] (quoted before (3.2)) it follows that every g^1_{c+2} of C' is cut out on C' by a pencil of elliptic curves.
Assume that C is not hyperelliptic. By (3.2), C cannot be a double covering of an elliptic curve. By [M], Satz 4, then, any linear series of C computing c and of degree $\leq g-1$ which is not a pencil is a base point free net g^2_{c+4}. Let $f: C \to \mathbb{P}_2$ denote the corresponding morphism and let m be the degree of f. If $\frac{c+4}{m} \geq 3$ the plane curve $f(C)$ of degree $\frac{c+4}{m}$ carries infinitely many complete and base point

free pencils of degree $\frac{c+4}{m} - 1$ whence C has infinitely many such pencils of degree $c+4-m \leq c+3$. For $g > \frac{1}{4}(c+5)^2 + 1$ this contradicts (3.2). If $\frac{c+4}{m} = 2$ (i.e. $f(C)$ a conic) C has a pencil of degree $m = \frac{c}{2} + 2$. Since $m-2 \geq c$ (by definition of c), C must be hyperelliptic. This case has been excluded. ∎

§ 4. On the Harris-Mumford-conjecture

(4.1) Let us call a curve C an exceptional curve if the Clifford index c of C cannot be computed by a pencil (i.e. if there is no g^1_{c+2} on C).

There are (up to now) only two types of exceptional curves of Clifford index c known (cf. [ELMS]):

(4.1.1) smooth plane curves (degree $c+4$, genus $g = \frac{1}{2}(c+2)(c+3)$)

(4.1.2) for odd c: certain projectively normal and half canonically embedded curves of degree $2c+3$ in $\mathbb{P}_{\frac{1}{2}(c+3)}$ (genus $g = 2c+4$).

It is conjectured that there are no other exceptional curves.

We call a curve C d-gonal (and d its gonality) if C has a pencil g^1_d but no g^1_{d-1}. For example, the curves of type (4.1.1) and (4.1.2) are known to be (c+3)-gonal.

Let X be a K3 surface and $C \subset X$ be a smooth irreducible curve of Clifford index c. Clearly, by (2.1), if the linear system $|C|$ on X contains no exceptional curve the gonality of smooth curves in $|C|$ is constant: all smooth curves in $|C|$ are (c+2)-gonal though there need not be a pencil of elliptic curves on X cutting on C a g^1_{c+2}. Consider the following examples of hyperelliptic K3 surfaces and curves of Clifford index $c = 2$ (cf. [SD], 5.9):

(4.2) Example: Let $f: X \to \mathbb{P}_1 \times \mathbb{P}_1$ be a double covering of $\mathbb{P}_1 \times \mathbb{P}_1$ branched along a smooth curve of bidegree $(4,4)$ on $\mathbb{P}_1 \times \mathbb{P}_1$. If $E \subset \mathbb{P}_1 \times \mathbb{P}_1$ is a smooth curve of bidegree $(2,2)$ then $C_o := f^{-1}(E)$ is a double covering of genus 9 of an elliptic curve whereas a general curve C in $|C_o|$ maps isomorphically by f to a smooth curve of bidegree $(4,4)$. C_o and C are 4-gonal but C_o has infinitely many g^1_4 and C has only two such pencils. ∎

(4.3) <u>Example</u> (Donagi): Let $f:X \to \mathbb{P}_2$ be a double covering of $\mathbb{P}_2$ branched along a smooth plane sextic. If $E \subset \mathbb{P}_2$ is a smooth plane cubic then $C_o := f^{-1}(E)$ is a double covering of genus 10 of an elliptic curve whereas a general curve C in $|C_o|$ maps isomorphically by f to a smooth plane sextic. Thus C_o is 4-gonal and C is 5-gonal. ∎

Note that C in Donagi's example (4.3) is exceptional and that (according to (3.1) and (3.2)) 10 is the maximum genus for a double covering of an elliptic curve and also for a smooth plane curve to lie on a K3 surface. Of course, there is a curve H of genus 2 (resp. of genus 3 in (4.2)) such that $|H|$ cuts a g^2_6 (resp. a g^3_8) computing c on every smooth curve in $|C_o|$.

I learnt that J. Harris and Mumford discussed the question of the constancy of the gonality in linear systems on K3 surfaces during their joint work on the moduli space of curves (Invent. math. 67 (1982)). After Donagi found the counterexample (4.3) this "conjecture" was modified and proved in the form (2.1), cf. [GL].
It is the aim of this § to prove the following

(4.4) <u>Theorem: Let C be an exceptional curve of Clifford index c on X. If C is of type (4.1.1) resp. (4.1.2) then so too is every smooth curve in $|C|$ unless we are in Donagi's example (4.3).</u>

In particular, if C in (4.4) is not a smooth plane sextic, all smooth curves in $|C|$ are $(c+3)$-gonal.

For a proof we first note the following consequence of the base point free pencil trick.

(4.5) <u>Lemma: Let C be a smooth irreducible curve of genus $g = 4r-2$ $(r \geq 2)$ admitting a line bundle A such that $\deg A = g-1$, $h^o(A) = r+1$ and A computes the Clifford index of C. Then C is $2r$-gonal.</u>

<u>Proof:</u> Since C has Clifford index $\deg A - 2(h^o(A)-1) = 2r-3$ there is no pencil of degree $\leq 2r-2$ on C. Assume that B is a base point free line bundle of degree $2r-1$ on C. Then the base point free pencil trick says

$$h^o(A\otimes B) \geq 2\,h^o(A) - h^o(A\otimes B^*) .$$

Since $\deg(A\otimes B^*) = (4r-3) - (2r-1) = 2r-2$ we have $h^o(A\otimes B^*) \leq 1$ and therefore

$$h^0(A\otimes B) \geq 2h^0(A) - 1 = 2r+1 .$$

By Riemann-Roch, then

$$h^0(\omega_C \otimes A^* \otimes B^*) = h^0(A\otimes B) - \deg B \geq (2r+1) - (2r-1) = 2 .$$

But $\omega_C \otimes A^* \otimes B^*$ again is a line bundle of C of degree $2r-2$, and we have a contradiction.

Since the gonality of every smooth curve of genus g is $\leq \frac{g+3}{2}$ the result follows. ∎

To prove theorem (4.4) we have to distinguish two cases:

(4.6) $C \subset X$ is of type (4.1.2).

If $r := \frac{1}{2}(c+3)$, C is of genus $g = 4r-2$ and degree $g-1$ in $\mathbb{P}_r$. Let C' be a smooth curve in $|C|$. By (2.1) there is a line bundle L on X such that $L\otimes\mathcal{O}_{C'}$ computes the Clifford index $c = 2r-3$ of C'. We have $\mathcal{O}_X(C')\cdot L = \mathcal{O}_X(C)\cdot L = g-1$ since every line bundle of C computing c has degree $g-1$ ([ELMS]). Thus, by (4.5), C' is exceptional.

Let A be a line bundle of C' computing c with minimal $h^0(A)$. The image of C' under the induced morphism $C' \longrightarrow \mathbb{P}_s$ (where $s = h^0(A) - 1 \geq 2$) is a smooth curve of degree $\deg A$ without any $(2s-2)$-secant $(s-2)$-planes ([ELMS]). Since $g = 2(c+2)$ a crucial result of [ELMS] (concerning vanishing properties of Castelnuovo's polynomial counting these planes) implies then that $\deg A = g-1$, and from

$$g - 1 - 2s = \deg A - 2s = c = g - 1 - 2r$$

it follows that $s = r$. Consequently, C' is again of type (4.1.2).

(4.7) $C \subset X$ is a smooth plane curve of degree $d = c+4$ (i.e. has a very ample net g^2_d). Then $d \leq 6$, by (3.1).

If $d = 4$ and if there is a hyperelliptic curve C' in $|C|$ then $|C|$ is birational very ample whereas $|C'|$ is not ([SD]), a contradiction.

For $d = 5$ every smooth curve in $|C|$ is a smooth plane quintic; note that this has already been proved in (4.6) since the types (4.1.1) and (4.1.2) coincide for $c = 1$.

So let $d = 6$. Since the unique g^2_6 on C is the only linear series of degree $\leq g-1$ on C computing c ([M], Satz 4) we conclude from (2.3) that there is a smooth genus 2 curve H on X (such that $|H|$ cuts on C the g^2_6). Clearly, $|H|$ exhibits X as a double plane (cf. [SD], 5.1).

Any smooth curve C' in $|C|$ has genus 1o, Clifford index 2 and a g^2_6. Thus if C' is not a smooth plane sextic it has to be a double covering of an elliptic curve, and we are in Donagi's case (4.3).

Finally, note that smooth plane quintics and sextics can only lie on degree 2 K3 surfaces (i.e. double planes). For the K3 surfaces containing curves of type (4.1.2) cf. [ELMS]; they contain a line.

References

[ELMS] D. Eisenbud, H. Lange, G. Martens, F.-O. Schreyer: The Clifford dimension of a projective curve. Preprint, to appear

[GL] M. Green, R. Lazarsfeld: Special divisors on curves on a K3 surface. Invent. math. 89 (1987), 357-37o

[M] G. Martens: Funktionen von vorgegebener Ordnung auf komplexen Kurven. J. reine angew. Math. 32o (198o), 68-85

[R] M. Reid: Special linear systems on curves lying on a K3 surface. J. London Math. Soc. (2) 13 (1976), 454-458

[SD] B. Saint-Donat: Projective models of K3 surfaces. Amer. J. Math. 96 (1974), 6o2-639

GONALITY AND HILBERT SCHEMES OF SMOOTH CURVES

by

Emilia Mezzetti and Gianni Sacchiero

Introduction.

Let $I_{d,g,n}$ be the open subset of the Hilbert scheme of curves of degree d and genus g in $\mathbf{P}^n$ (1) parametrizing irreducible, smooth, non degenerate curves.
It is well-known that in the Brill-Noether range (i.e. $\rho(d,g,n) \geq 0$), there exists a unique irreducible generically smooth component M of $I_{d,g,n}$ which dominates the moduli space M_g.
In particular, in the non special range ($d \geq g+n$), all the points representing non special curves belong to M. It is natural to ask if M coincides with $I_{d,g,n}$. Severi ([Se]) claimed it, but with an incomplete proof. Clearly, if it happens, $I_{d,g,n}$ is irreducible. Recently L. Ein showed that $I_{d,g,n} = M$ when $n = 3, 4$ ([E1], [E2]) but J. Harris exhibited an example of another component W_3 of $I_{d,g,n}$ containing a family of trigonal curves, for $n \geq 6$.
We have tried to clarify the example of Harris, asking for the general curve of the component W_3. We prove in fact that it is a trigonal curve. Furthermore, our proof works also in a more general case, showing the existence of irreducible generically smooth components of the Hilbert scheme, whose general curve is an m-secant curve on a rational ruled surface. The method we use is standard: on one hand, it is easy to compute the dimension of the family $W^m_{d,g,n} \subseteq I_{d,g,n}$ of m-secant curves lying on a rational ruled surface. On the other hand, we show that $dim\ W^m_{d,g,n} = h^0(\mathcal{N})$, where $\mathcal{N}$ is the normal bundle of a general curve of the family; this is the more difficult part. More precisely, from the Segre formula, it follows that a smooth curve X of $W^m_{d,g,n}$ satisfies the equality:

$$(*) \qquad 2g-2 = (m-1)(2d - m(r-1)) - 2m$$

where $r = h^0(\mathcal{O}_X(1)) - 1$. Our main result is the following:

Theorem.

Let

$$d_m(n,g) := \frac{(m+1)(n+1)-4m}{(m-1)(n+1)} g + \frac{n+4m+1}{n+1}$$

Then for any d, g, n, r fullfilling (*) and the inequalities:

$$\gamma_m(r,n) < d \leq d_m(n,g),^{(2)}$$

$W^m_{d,g,n}$ is an irreducible, generically smooth component of $I_{d,g,n}$.

One can easily see that, for example, if $n \geq 5$ (resp. $n \geq 6$), there exist irreducible components of trigonal curves in the Brill-Noether range (resp. in the non special range). A similar statement holds in the case of 4–gonal curves if $n \geq 6$ (resp. $n \geq 8$). And so on.

---

(1) $\mathbf{P}^n = \mathbf{P}^n{}_k$, k an algebrically closed field with char $k = 0$.
(2) For the definition of $\gamma_m(r,n)$, see §2

The paper consists of 3 sections: the first one contains some preliminary results used in the second one, devoted to the proof of the theorem. The third section includes conjectures and open problems.

§1. Preliminary facts.

Let us fix the notations:

$\mathcal{F} = \bigoplus_{i=1}^{s} \mathcal{O}_{\mathbf{P}^1}(a_i), a_i > 0$, is a vector bundle of rank s on $\mathbf{P}^1$; $r+1 = dim\ H^0(\mathbf{P}^1, \mathcal{F})$; $V \subseteq H^0(\mathbf{P}^1, \mathcal{F})$ is a linear subspace of dimension $n+1 \geq s+1$.

Consider the morphism $\varphi_V : \mathbf{P}(\mathcal{F}) \to \mathbf{P}(V) = \mathbf{P}^n$; denoting by F the image of $\mathbf{P}(\mathcal{F})$, let $\pi : F \to \mathbf{P}^1$ be the structural morphism and f_x be the fiber over $x \in \mathbf{P}^1$; if dim $F = s$, then F is called a rational scroll of dimension s.

If $s = 2$, then F is a rational surface; in this case, we will write R instead of F. Let us remark that, if $\mathcal{F} = \mathcal{O}_{\mathbf{P}^1}(a_1) \oplus \mathcal{O}_{\mathbf{P}^1}(a_2)$ and $R = \mathbf{P}(\mathcal{F})$, then $r = a_1 + a_2 + 1$.

a) The Galois group of a curve X lying on a ruled surface R.

We denote by H_R the unisecant divisor on R defined by $\mathcal{O}_R(H_R) = \mathcal{O}_{\mathcal{F}}(1)$.

It is a well-known fact that the fiber f and H_R generate the Picard group of R. Fixing $m \geq 3$ an integer, let $X \subseteq R$ be an m-secant curve of degree d in a very ample linear system on R; the very ampleness is equivalent to the condition: $d > m[\frac{r}{2}]$ (see [H]). Hence

(1) $$X \sim mH_R + (d - m(r-1))f.$$

The set of m-secant curves of degree d is a linear system on R(†); its dimension is

$$N(m,d,r) = (m+1)(d+1) - \binom{m+1}{2}(r-1) - 1.$$

The morphism $\pi|_X : X \to \mathbf{P}^1$ is finite of degree m.

Proposition 1.

Let X be a smooth curve, which is a general element of the linear system $|mH_R + (d - (m(r-1))f|$.

Then in each ramification fiber, there is only one point of simple ramification.

Proof.

Consider the natural embedding:

$$\phi_{|X|} : R \to \phi(R) = S \subset \mathbf{P}^N,$$

$N = N(m,d,r)$. The image of any line of R is a rational normal curve of degree $m : S_x = \phi_{|X|}(f_x)$.

Consider the following subvarieties of $\check{\mathbf{P}}^N$:

$$X_{S_x} = \{H \in \check{\mathbf{P}}^N \mid \exists P_1, P_2 \in S_x \text{ such that } (H \cdot S_x) \geq 2P_1 + 2P_2\},$$

(†) In fact it can be thought as the linear variety of invertible quotients of degree d of $Sym^m(\mathcal{F})$.

$$Y_{S_x} = \{H \in \check{\mathbf{P}}^N \mid \exists P \in S_x \text{ such that } (H \cdot S_x) \geq 3P\}.$$

It is known that X_{S_x} and Y_{S_x} have codimension two in $\check{\mathbf{P}}^N$. So $X_S = \bigcup_{x \in \mathbf{P}^1} X_{S_x}$ and $Y_S = \bigcup_{x \in \mathbf{P}^1} Y_{S_x}$ have codimension one.

Let H be a hyperplane in $\check{\mathbf{P}}^N \setminus \{X_S \cup Y_S \cup \check{S}\}$.

It is clear that $H \cap S$ is the image of an m-secant curve X such that the morphism $\pi|_X$ has the required property.

Corollary.

Let X be a general m-secant curve. Then the Galois group G of $\pi|_X$ is the full symmetric group on m letters.

Proof.

The morphism $\pi|_X$ is simply ramified by Prop. 1 and this implies that G is the full symmetric group (see, for instance, [N], Lemma 5.22).

b) Bound for the genus of a curve on a s-scroll.

Let X be an m-secant curve of degree d and genus g on a rational normal scroll F of dimension s in $\mathbf{P}^r$.

We say that X is "in uniform position " on F if every s-tuple of points of X on a general fiber f_x generates all the fiber.

From the Segre formula (cfr.[G-S]), if X is in uniform position then:

$$g \leq \Sigma_s^m(d,r) := \frac{m-1}{s(s-1)}(s(d-s+1) - m(r-s+1)).$$

Proposition 2.

Let $X \subseteq \mathbf{P}^r$ be a curve of degree d and genus g having a g^1_m, such that the associated Galois group is the full symmetric group. If $g > \Sigma_s^m(d,r)$ and $d > \frac{m}{s+1}(2r-s+1)$, then X lies on a rational scroll of dimension at most $s-1$.

Proof.

Let D be a divisor of the g^1_m. We are interested in computing dim $< D >$; for, we want to prove that $\Sigma_{s+i}^m(d,r)$, $i \in \mathbf{N}$, is a decreasing function of i. In fact: the difference

$$\Sigma_s^m(d,r) - \Sigma_{s+i}^m(d,r) = \frac{(m-1)i}{(s-1)(s+i-1)}\left(d - m\frac{(r+1)(2s+i-1) - s(s+i)}{s(s+i)}\right)$$

is positive if $d > \frac{m}{s(s+i)}((r+1)(2s+i-1) - s(s+i))$.

The right-hand side term of the above inequality is a decreasing function of i. By assumption, the inequality is verified for $i = 1$, so it is always true. Being $g > \Sigma_s^m(d,g)$, the linear span $< D >$ has dimension at most $s-2$, when $D \in g^1_m$.

c) The 2-dual variety of a ruled surface R.

Let us recall the definition of higher order dual varieties of a given variety $Z \subseteq \mathbf{P}(V) = \mathbf{P}^n([PS])$.

For each $q \geq 0$, consider the homomorphism

$$a^q : V_Z \to P^q_Z(1),$$

where $P^q_Z(1)$ denotes the sheaf of principal parts of order q of $\mathcal{O}_Z(1)$. The fibers of a^q determine the osculating spaces to Z: for $z \in Z$, let $I_q(z)$ denote the image of the fiber $a^q(z) : V \to P^q_Z(1)(z)$ of a^q. Then $P(I_q(z))$ is the q-th order osculating space to Z at z. The dimension of these spaces does not depend, in general, on the point z. So, define the q-dual variety $\check{Z}_q \subset \check{\mathbf{P}}^n$ of Z as the closure of the set of hyperplanes containing a q-osculating space to Z.
Now consider $\check{R}_2 \subset \check{\mathbf{P}}^n$, the 2-dual variety of a ruled surface R. Let $\mathcal{F}^*_2(R)$ be the vector bundle on $\mathbf{P}^1$ corresponding to $\check{R}_2$.
It is know that: $rk(\mathcal{F}^*_2(R)) = n-3$.
Now let X be a smooth linearly normal curve on R and suppose that X is an m-secant curve of degree d.
Denoting by $\mathcal{N}_X$ the normal bundle of X in $\mathbf{P}^r$, we get

$$\mathbf{P}(\mathcal{N}_X(-1)) \subseteq X \times \mathbf{P}^r$$

(see for instance [S2]).
Moreover, if θ is the second proiection, then $\check{X} = \theta(\mathbf{P}(\mathcal{N}_X(-1)))$ is the dual variety of X.
Set $\Gamma = f \cdot X$ and $\mathcal{L} = \mathcal{O}_X(1)$.

Lemma.
With the above notations, if R is linearly normal, then R induces the following quotient:

$$\mathcal{N}_X \to \mathcal{L}(2\Gamma)^{\oplus(r-3)} \to 0.$$

Moreover $\theta(\mathbf{P}(\mathcal{L}(2\Gamma)^{\oplus(r-3)}))$ coincides with $\check{R}_2$.

Proof.
Consider an unisecant curve U contained in R, of degree $r-3$. The linear projection centered in $< U >$ maps R to a conic, so we get a hypersurface S_U of degree 2, which contains R, and has equation $G_U = 0$. Let us consider the following sequence (induced by S_U):

$$0 \to \mathcal{O}_R \xrightarrow{\cdot G_U} \mathcal{I}_R/\mathcal{I}^2_R(2)$$

where $\mathcal{I}_R$ is the ideal sheaf of R. Let us dualize and restrict to X the previous sequence.
Since $(\cdot G_U)\check{}$ is not surjective on U, we twist by $\mathcal{O}_X(-U \cdot X)$ and get:

$$(\mathcal{I}_X/\mathcal{I}^2_X)\check{}(-2) \to \mathcal{O}_X(-U \cdot X) \to O.$$

From being $H \sim U + 2f$ we get

$$\mathcal{N}_X \to \mathcal{L}(2\Gamma) \to O.$$

As the unisecant curves of degree $r-3$ form a linear system of dimension $r-4$, the first claim is proved. The second part immediately follows from the definition of $\check{R}_2$, noting that a tangent hyperplane to S_U is a 2–osculating one to R. Then $\theta(\mathbf{P}(\mathcal{L}(2\Gamma)^{\oplus(r-3)})) \subseteq \check{R}_2$. Since they have the same dimension, equality holds.

Corollary.
Under the assumptions of the lemma, we have:

$$\mathcal{F}_2^*(R) = \mathcal{O}_{\mathbf{P}^1}(2)^{\oplus(r-3)}.$$

Let now $L \subset G(r-n-1, r)$ be a linear subspace of dimension $r-n-1$ in $\mathbf{P}^r$, such that $L \cap R = \emptyset$. Consider the linear projection $h_L : \mathbf{P}^r --> \mathbf{P}^n$ and set $R_L = h_L(R)$.

Proposition 3.
Let $R \subset \mathbf{P}^r$ be a rational normal ruled smooth surface. Then, for a general projection, $\mathcal{F}_2^*(R_L)$ is balanced, i.e. $\mathcal{F}_2^*(R_L) = \overset{n-3}{\underset{i=1}{\oplus}} \mathcal{O}(a_i)$, with $\Sigma_{i=1}^{n-3} a_i = 2(r-3)$ and $|a_i - a_j| \leq 1$, for all i, j.

Proof.
It is easy to see that the required property is open, so it sufficies to construct an example of a rational ruled surface $R_L \subseteq \mathbf{P}^n$ with this property.
Let $Y \subset R$ be a hyperplane section of R. By the lemma, R induces the following quotient:

$$\mathcal{N}_Y(-1) \to \mathcal{O}_{\mathbf{P}^1}(2)^{\oplus(r-3)} \to 0. \tag{2}$$

Let us remark that (2) factorizes through $\mathcal{N}_{R|Y}$.
Note that $\mathcal{N}_Y(-1) \simeq \mathcal{O}_{\mathbf{P}^1} \oplus \mathcal{O}_{\mathbf{P}^1}(2)^{\oplus(r-3)}$, because Y is a rational normal curve contained in a hyperplane of $\mathbf{P}^r$ (cfr. [S1]). We are going to prove that the quotients of the form (2) correspond bijectively to the rational normal ruled surfaces having Y as hyperplane section. Clearly, if R and R' are two different surfaces of this type containing Y, then the obtained quotients are different. On the other hand, we will show that those ruled surfaces form an irreducible variety of dimension $r-3$, which is exactly the projective dimension of the quotients (2). In fact, let $H_{r-1,o,r}$ be the Hilbert scheme of rational curves of degree $r-1$ in $\mathbf{P}^r$; it is irreducible of dimension $(r+1)^2-4$. Let $\mathcal{H}_{r-1,r}$ be the Hilbert scheme of surfaces of degree $r-1$ in $\mathbf{P}^r$, if $r \neq 5$ (resp. the irreducible component containing rational normal scrolls if $r=5$). In both cases, it is irreducible of dimension $(r+3)(r-1)-3$ (cfr. [C]). Let

$$\Gamma = \{(Y,R) | Y \subset R\} \subset H_{r-1,o,r} \times \mathcal{H}_{r-1,r}$$

and let p_1, p_2 be the projections on $H_{r-1,o,r}$ and $\mathcal{H}_{r-1,r}$, respectively. If $R \in \mathcal{H}_{r-1,r}$ then $p_2^{-1}(R) \simeq \mathbf{P}^r$. So Γ is irreducible and $\dim p_1^{-1}(Y) = r-3$.
Let Y' be a smooth rational curve of degree $r-1$, generating a hyperplane of $\mathbf{P}^n$, such that $\mathcal{N}_{Y'}(-1) \simeq \mathcal{O}_{\mathbf{P}^1} \oplus \mathcal{O}_{\mathbf{P}^1}(2) \oplus \left(\overset{n-3}{\underset{i=1}{\oplus}} \mathcal{O}_{\mathbf{P}^1}(a_i)\right)$, with: $\Sigma_{i=1}^{n-3} a_i = 2(r-3)$ and $|a_i - a_j| \leq 1, \quad \forall i,j$ (see [S1] for the existence of such a curve Y'). Then Y' is a projection of a curve $Y \subset \mathbf{P}^r$; looking

at this commutative diagram:

$$
\begin{array}{ccccccccc}
 & & & & 0 & & 0 & & \\
 & & & & \downarrow & & \downarrow & & \\
 & & & & \mathcal{O}_{\mathbf{P}^1}^{r-n} & = & \mathcal{O}_{\mathbf{P}^1}^{r-n} & & \\
 & & & & \downarrow & & \downarrow & & \\
0 & \longrightarrow & \mathcal{O}_{\mathbf{P}^1} \oplus \mathcal{O}_{\mathbf{P}^1}(2) & \longrightarrow & \mathcal{N}_Y(-1) & \longrightarrow & \mathcal{O}_{\mathbf{P}^1}(2)^{\oplus(r-3)} & \longrightarrow & 0 \\
 & & \| & & \downarrow & & \downarrow \psi & & \\
0 & \longrightarrow & \mathcal{O}_{\mathbf{P}^1} \oplus \mathcal{O}_{\mathbf{P}^1}(2) & \longrightarrow & \mathcal{N}_{Y'}(-1) & \longrightarrow & \bigoplus_{i=1}^{n-3} \mathcal{O}_{\mathbf{P}^1}(a_i) & \longrightarrow & 0 \\
 & & & & \downarrow & & \downarrow & & \\
 & & & & 0 & & 0 & &
\end{array}
$$

we are done.

§2. Proof of the theorem.

Throughout this section, we set $R = \mathbf{P}(\mathcal{F})$, $\mathcal{F} = \mathcal{O}_{\mathbf{P}^1}(a_1) \oplus \mathcal{O}_{\mathbf{P}^1}(a_2)$, with $|a_1 - a_2| \leq 1$. Moreover let us denote: $\alpha = \left[\frac{2(r-3)}{n-3}\right]$, $\beta = 2(r-3) - \alpha(n-3)$, and

$$
i = \begin{cases} 1 & \text{if } \beta > 0 \\ 0 & \text{if } \beta = 0. \end{cases}
$$

Put

$$
\gamma_m(r,n) = \begin{cases} \frac{(m+1)(r-1)}{2} + \alpha + i, & \text{if } r \text{ is odd} \\ \frac{(m+1)r}{2} - 2 + \alpha + i, & \text{if } r \text{ is even.} \end{cases}
$$

Proposition 4.

Let $X \subset R$ be a general m-secant curve, $\mathcal{L} = \mathcal{O}_x(1)$ and $\delta = h^1(X, \mathcal{L})$. If

$$
\delta \geq \frac{2(r-3)(m-2)}{n-3},
$$

$$
d > \gamma_m(r,n)
$$

and L is general in $G(r-n-1, r)$, $n \geq 4$, then :

$$
h^1(\mathcal{N}_{X_L}) \leq (n-3)\delta + \frac{m-3}{m-1}g - 2(r-2)(m-2) + 2.
$$

Proof.

Let us recall the relation:

$$
K_X \sim (m-2)H_X + (d - (m-1)(r-1) - 2)\Gamma, \tag{3}
$$

where K_X, H_X and Γ are the canonical divisor of X, the hyperplane section and the divisor $f \cdot X$, respectively.

Consider the following exact sequence:

$$0 \to \mathcal{N}_{X|R} \to \mathcal{N}_X \to \mathcal{N}_{R|x} \to 0. \tag{4}$$

By (1) and (3), we get:
$\mathcal{N}_{X|R} = \mathcal{O}_X(X) = \omega_X \otimes \mathcal{L}^2(-(r-3)\Gamma)$; so being $d > \frac{m}{2}(r-3)$, it is $h^1(\mathcal{N}_{X|R}) = 0$. Then: $h^1(\mathcal{N}_{X|\mathbf{P}^r}) = h^1(\mathcal{N}_{R|x})$.

The lemma gives the following quotient:

$$\mathcal{N}_{R|X}(-1) \to \mathcal{O}_X(2\Gamma)^{\oplus(r-3)} \to 0;$$

since $c_1(\mathcal{N}_{X|\mathbf{P}^r}) = (r+1)H_X + K_X$, by (4) we get the exact sequence:

$$0 \to \mathcal{L}^2(-(r-3)\Gamma) \to \mathcal{N}_{R|x} \to \mathcal{L}(2\Gamma)^{\oplus(r-3)} \to 0.$$

The projection $h_L : \mathbf{P}^r - - > \mathbf{P}^n$ produces the commutative diagram:

$$\begin{array}{ccccccc}
 & & 0 & & 0 & & \\
 & & \downarrow & & \downarrow & & \\
 & & \mathcal{M} & = & \mathcal{M} & & \\
 & & \downarrow & & \downarrow & & \\
0 \longrightarrow & \mathcal{L}^2(-(r-3)\Gamma) & \longrightarrow \mathcal{N}_{R|X} & \longrightarrow & \mathcal{L}(2\Gamma)^{\oplus(r-3)} & \longrightarrow & 0 \\
 & \| & \downarrow & & \downarrow \psi & & \\
0 \longrightarrow & \mathcal{L}^2(-(r-3)\Gamma) & \longrightarrow h_L^*(\mathcal{N}_{R_L|X_L}) & \longrightarrow & \mathcal{L}(\alpha\Gamma)^{\oplus(n-3-\beta)} \oplus \mathcal{L}((\alpha+1)\Gamma)^{\oplus\beta} & \longrightarrow & 0 \\
 & & \downarrow & & \downarrow & & \\
 & & 0 & & 0 & &
\end{array}$$

where $c_1(\mathcal{M}) = (r-n)H_X$. Considering the embedding:

$$\phi_1 : R \xrightarrow{|2H-(r-3)f|} \phi_1(R) = S_1 \subseteq \mathbf{P}^8,$$

we note that $((\phi_{1|x}^* \mathcal{O}_{\mathbf{P}^8}(1) = \mathcal{L}^2(-(r-3)\Gamma)$. Being S_1 a conic bundle then $\dim < \phi_1(\Gamma) >= 2$. Let us prove that the curve $\phi_1(X)$ is linearly normal. If not, $\phi_1(X)$ would be a projection of a curve Y, lying on a rational scroll of dimension at least 3 in $\mathbf{P}^9$.

In this case we get by (3) that $g > \Sigma_3^m(2d-(r-3)m, 9)$: which contradicts Proposition 2. Hence, by Riemann-Roch theorem we deduce:

$$h^1(\mathcal{L}^2(-(r-3)\Gamma)) = (m-3)\left(\frac{g}{m-1} - 2\right).$$

Likewise, considering the embedding:

$$\phi_{\alpha+i} : R \xrightarrow{|H+(\alpha+i)f|} \mathbf{P}^{r+2(\alpha+i)}, \qquad i = 0, 1,$$

the condition $d > \gamma_m(r,n)$ implies that $\phi_{\alpha+i}(X)$ is linearly normal. In fact , by Lemma 2.4 ([H], Ch. V) and Serre duality, we get:

$$H^1(R, \mathcal{O}_R(-X)(1)) = 0.$$

Hence:

$$h^1(\mathcal{L}(\alpha+i)\Gamma)) = \delta - (\alpha+i)(m-2).$$

Let $W^m_{d,g,n}$ be the family of smooth curves of genus g and degree d in $\mathbf{P}^n$, which are m-secant on a rational ruled surface of degree $r-1$. By the Segre formula, we obtain:

$$2g-2 = (m-1)(2d - m(r-1)) - 2m. \tag{5}$$

Theorem.

Let

$$d_m(n,g) := \frac{(m+1)(n+1)-4m}{(m-1)(n+1)}g + \frac{n+4m+1}{n+1}.$$

Then, for any d such that (5) holds and

$$\gamma_m(r,n) < d \leq d_m(n,g),$$

$W^m_{d,g,n}$ is an irreducible, generically smooth component of $I_{d,g,n}$.

Proof.

With the notations as in §1, we have:

$$\begin{aligned}\dim W^m_{d,g,r} &= \dim \mathcal{H}_{r-1,r} + N(m,d,r) = \\ &= (r+3)(r-1) - 3 + (m+1)\left(d+1-\frac{m(r-1)}{2}\right) - 1 = \\ &= (m+1)(d+1) + (r-1)\left(r+3-\binom{m+1}{2}\right) - 4.\end{aligned}$$

We first consider the case $r = n$.

From Proposition 4 and from $\delta = r+g-d$ (Riemann-Roch), we get:

$$h^0(\mathcal{N}_{X|\mathbf{P}^r}) \leq \mathcal{X}(\mathcal{N}_{X|\mathbf{P}^r}) + h^1(\mathcal{L}^2(-(r-3)\Gamma) + (r-3)h^1(\mathcal{L}(2\Gamma)) = \dim W^m_{d,g,r} \leq h^0(\mathcal{N}_{X|\mathbf{P}^r});$$

so the equality holds and the theorem is true.

In the case $r > n$, note that each curve of $W^m_{d,g,n}$ is a projection of some curve of $W^m_{d,g,r}$; hence:

$$\begin{aligned}\dim W^m_{d,g,n} &= \dim W^m_{d,g,r} - \dim PGL(r+1,k) + \dim G(r-n-1,r) + \\ &\quad + \dim PGL(n+1,k) = \\ &= \dim W^m_{d,g,r} - (r+1)^2 + (n+1)(r-n) + (n+1)^2 = \\ &= \dim W^m_{d,g,r} - (r+1)(r-n).\end{aligned}$$

On the other hand, let $X \subseteq \mathbf{P}^r$ project to $X_L \subseteq \mathbf{P}^n$. The condition $d \leq d_m$ implies $\delta \geq \frac{2(r-3)(m-2)}{n-3}$, then by Proposition 4, we have:

$$\begin{aligned} h^0(\mathcal{N}_{X|\mathbf{P}^r}) - h^0(\mathcal{N}_{X_L|\mathbf{P}^n}) &= (r-n)(d-g+1) + h^1(\mathcal{N}_{X|\mathbf{P}^r}) - h^1(\mathcal{N}_{X_L|\mathbf{P}^n}) \geq \\ &\geq (r-n)(d-g+1) + (r-3)(\delta - 2m + 4) - (n-3)\delta + \\ &\quad + 2(r-3)(m-2) = \\ &= (r-n)(r+1). \end{aligned}$$

So we conclude that : $h^0(\mathcal{N}_{X_L|\mathbf{P}^n}) = \dim W^m_{d,g,n}$.

Remarks .

1. The function $d_m(n,g)$ is a decreasing function of m, as one can observe by looking at its derivative. So, if n and g are fixed, $d_3(n,g)$ is the maximum integer such that there exists an irreducible component of the type described in the theorem.
Hence, if $d \leq d_3(n,g) = \frac{2(n-2)g+n+13}{n+1}$, then $I_{d,g,n}$ is not always irreducible, as Harris noticed first (see [E2]).

2. In the case $m = 3$, from the fact $h^1(\mathcal{L}^2(-(r-3)\Gamma)) = 0$, we get the following statement: $W^3_{d,g,n}$ is an irreducible component of $I_{d,g,n}$ if and only if $\mathcal{X}(d,g,n) \leq \dim W^3_{d,g,n}$, where $\mathcal{X}(d,g,n) = \mathcal{X}(\mathcal{N}_{X|\mathbf{P}^n})$ is the Euler-Poincaré characteristic of the normal bundle of a smooth curve X in $\mathbf{P}^n$.

3. For any m, there exists a component $W^m_{d,g,n}$ in the non special range (resp. in the Brill-Noether range); in fact, one can easily find a function $n_m(g)$ (resp. $\bar{n}_m(g)$) such that, if $n \geq n_m(g)$ (resp. $n \geq \bar{n}_m(g)$), then $d_m(n,g) > g+n$ (resp. $d_m(n,g) \geq \frac{n}{n+1}g + n$).
For example, if $m = 3$: $n_3(g) = 6$, $\bar{n}_3(g) = 5$.
If $m = 4$ and g is large enough: $n_4(g) = 8$, $\bar{n}_4(g) = 6$.

§3.Conjectures.

Let us set:
$D_m(n,g) := \max\ \{d|$ there exists an irreducible component $W \subset I_{d,g,n}$ such that the general curve in W is m-gonal $\}$.
One can easily see that: $D_3(n,g) = d_3(n,g)$, while in the general case $D_m(n,g) \geq d_m(n,g)$.
Let us consider, for example, the case $m = 4$. Since the curves we are dealing with are special, the linear span of the 4-gonal divisor is a line or a plane.
In the first case, we can define a function $d_4^2(n,g)$ which corresponds to the maximum degree d such that the family of 4-gonal curves, each of them lying on a rational scroll surface, is a full component of $I_{d,g,n}$. So comparing dim $W^4_{d,g,n}$ with $\mathcal{X}(d,g,n)$ and applying the theorem, we get:

$$\frac{5n-11}{3(n+1)}g + \frac{n+17}{n+1} \leq d_4^2(n,g) \leq \frac{5n-9}{3(n+1)}g + \frac{n+13}{n+1}.$$

In the second case, we can define the function $d_4^3(n,g)$ corresponding to 4−gonal curves on rational scrolls of dimension 3. An upper bound for $d_4^3(n,g)$ can be found in the following way.

Let $\mathcal{L} = \omega_X(-\alpha\Gamma)$ be the invertible sheaf that gives the embedding of X in $\mathbf{P}^r$ as a linearly normal curve. In this case, the dimension of the family, say $\bar{W}_4(r,g)$ is exactly:

$$\dim \bar{W}_4(r,g) = \dim M^1_{g,4} + \dim PGL(r+1,k),$$

where $M^1_{g,4}$ is the moduli space of 4-gonal curves of genus g. Since $\dim M^1_{g,m} = 2g + 2m - 5$, we get:

$$\dim \bar{W}_4(r,g) = 2g + 2 + (r+1)^2.$$

As before, projecting in $\mathbf{P}^n$ and comparing with $\mathcal{X}(d,g,n)$ we obtain:

$$d_4^3(n,g) \leq \frac{2(n-3)}{n+1} g + \frac{2(n+13)}{n+1}.$$

Suppose the equality holds; then for g large enough, we have:

$$D_4(n,g) = \begin{cases} d_4^3(n,g) & \text{if } n \geq 9 \\ d_4^2(n,g) & \text{otherwise.} \end{cases}$$

It leads us to state the following:

Conjecture 1: $D_m(n,g)$ is a decreasing function of m (at least for $\rho \geq 0$).

Conjecture 1 implies:

Conjecture 2: If $d \geq d_3(n,g)$, then $I_{d,g,n}$ is irreducible. This has been recently proved for $n \geq 9$ ([S3]).

Moreover, by taking into account the Remark 3 (§2), Conjecture 1 implies:

- If n=3,4 and $\rho(d,g,n) \geq 0$, then $I_{d,g,n}$ is irreducible.

- If n=5, $d \geq g+5$, then $I_{d,g,n}$ is irreducible.

Let us now consider the Brill-Noether number $\rho(d,g,n)$, for curves of $W^m_{d,g,n}$ in the linearly normal case. From (5), we find that it is negative if

$$d > \frac{n(m+1)}{2} - \frac{n(m-1)(m-3)}{2(mn-2n-1)}.$$

This condition is always satisfied under the hypothesis of the theorem. Furthermore, if we estimate $d_m^{m-1}(n,g)$ as before, we obtain that $\rho\left(d_m^{m-1}(n,g),n,g\right)$ is also negative, for linearly normal curves. So, denoting by $I^{LN}_{d,g,n}$ the union of the irreducible components of $I_{d,g,n}$ such that the general curve is linearly normal, it seems sensible the following:

Conjecture 3: $I^{LN}_{d,g,n}$ is irreducible, if $\rho(d,g,n) \geq 0$.

We are interested in generalizing the construction we have done studying the components $W^m_{d,g,n}$. Consider a component $W \subset I_{d,g,n}$: if $W \not\subset I^{LN}_{d,g,n}$, it is clear that we can find a "linearly normal" component $W_r \subset I^{LN}_{d,g,r}$ such that all curves of W are projections of curves of W_r.
Let us start from a component $W_r \subset I^{LN}_{d,g,r}$; and let $W_n \subset I_{d,g,n}$ be the family obtained by projecting curves of W_r. We ask whether W_n is a component of $I_{d,g,n}$. It is clear that a necessary condition is:

(6) $$\dim W_n \geq \mathcal{X}(d,g,n).$$

On one hand we have (cfr. §2):

$$\dim W_r - \dim W_n = (r-n)(r+1)$$

On the other hand:

(7) $$\mathcal{X}(d,g,r) - \mathcal{X}(d,g,n) = (r-n)(r+1-\delta).$$

So it follows that (6) is not always satisfied.

Problem 1: Find suitable assumptions under which (6) is also a sufficient condition.

Let us restrict to the case of generically smooth components. We have the following fact:

– Let W_r be an irreducible generically smooth (i.g.s.) component. Then W_n is i.g.s. if and only if

$$h^1(\mathcal{N}_{X|\mathbf{P}^n}) - h^1(\mathcal{N}_{X_L|\mathbf{P}^n}) = (r-n)\delta$$

(recall that X_L is the projection of X from a linear subspace L of $\mathbf{P}^r$).

As an immediate consequence, a necessary condition in order to get a component $W_n \subset I_{d,g,n}$ is: $h^1(\mathcal{N}_{X|\mathbf{P}^r}) \geq (r-n)\delta$.
Observe that always: $h^1(\mathcal{N}_{X|\mathbf{P}^r}) - h^1(\mathcal{N}_{X_L|\mathbf{P}^n}) \geq (r-n)\delta$, as we can see by considering the following exact sequence:

$$0 \to \mathcal{L} \to \mathcal{N}_{X|\mathbf{P}^r} \to h_p^*\mathcal{N}_{X_P|\mathbf{P}^{r-1}} \to 0,$$

where h_p is the projection from a point $P \in L$ to $\mathbf{P}^{r-1}$; now, take $P_1 \in h_p(L)$ and get a similar sequence. Repeating this procedure $r-n$ times, we reach $\mathbf{P}^n$.

Problem 2: Find additional conditions so that W_r i.g.s. and $h^1(\mathcal{N}_{X|\mathbf{P}^r}) \geq (r-n)\delta$ imply W_n i.g.s.. This is equivalent to giving conditions in order that:

$$h^1(\mathcal{N}_{X|\mathbf{P}^r}) - h^1(\mathcal{N}_{X_L|\mathbf{P}^n}) = (r-n)\delta.$$

References.

[C] C. Ciliberto: On the Hilbert scheme of curves of maximal genus in a projective space, Math. Zeit. 194 (1987), 351-363.

[E1] L. Ein: Hilbert scheme of smooth space curves, Ann. Scient. Ec. Norm. Sup., 4 ser., 19 (1986) 469-478.

[E2] L. Ein: The irreducibility of the Hilbert scheme of smooth space curves, preprint.

[GS] F. Ghione - G.Sacchiero: Genre d'une courbe lisse tracée sur une variété reglée, LNM 1266, Springer (1987).

[H] R. Hartshorne: Algebraic Geometry, Springer (1977)

[N] A. Nobile: On families of singular plane projective curves, Ann. Mat. Pura Appl., IV, 138 (1984).

[PS] R. Piene - G. Sacchiero: Duality for rational normal scrolls, Commun. Alg. 12(9) (1984), 1041-1066

[S1] G. Sacchiero, Fibrati normali di curve razionali dello spazio proiettivo, Ann. Univ. Ferrara, Sez. VII, XXXVI, 1980, 33-40.

[S2] G. Sacchiero, On the varieties parametrizing rational space curves with fixed normal bundle, Manuscripta math. 97, (1982), 217-228

[S3] G. Sacchiero, in preparation.

[Se] F. Severi: Sulla classificazione delle curve algebriche e sul teorema d'esistenza di Riemann, Rend. R. Accad. Naz. Lincei, 241 (1915), 877-888.

Address of the authors:
Dipartimento di Scienze Matematiche
Università di Trieste
Piazzale Europa 1
34127 Trieste (ITALY).

Geometry
of
Complete Cuspidal Plane Cubics

J. M. MIRET and S. XAMBÓ DESCAMPS

Dept. Àlgebra i Geometria, Univ. Barcelona
Gran Via 585, 08007-Barcelona, Spain

Abstract. We show how to compute *all* fundamental numbers for plane cuspidal cubics. This updates and extends the work of Schubert on this subject. In our approach we need a far more precise description of the first order degenerations (13 in all) than that given by Schubert and this is obtained by proving a number of key geometric relations that are satisfied by cuspidal cubics. Moreover, our procedure does not require using coincidence formulas to derive the basic degeneration relations.

Introduction

The enumerative theory of cuspidal cubics was first considered by Maillard (doctoral thesis, 1871) and Zeuthen [1872]. Subsequently they were extensively studied by Schubert. For an exposition of his (and others) results, see Schubert [1879], § 23, pp. 106-143. Schubert also considers cuspidal cubics in $\mathbf{P}^3$, but here for simplicity we will study only cuspidal cubics in a fixed projective plane $\mathbf{P}^2$ over an algebraically closed ground field $\mathbf{k}$. In case the characteristic p of $\mathbf{k}$ is positive we will assume that $p \neq 2, 3$.

Let S be the space of plane non degenerate cuspidal cubics, so that S is an orbit under the action of the group $\mathrm{G} = \mathrm{PGL}(\mathbf{P}^2)$ on the space of plane cubics. Each cuspidal cubic determines a triangle, called *singular triangle* (*Singularitätendreieck*, Schubert [1879], p. 106), whose verteces c, v, y are, respectively, the cusp, the inflexion and the intersection point of the cuspidal and inflexional tangents. The sides of this triangle, denoted q, w, z are, respectively, the cuspidal tangent, the inflexional tangent and the line cv (see Fig. 1 at the end).

The conditions that were first considered in the enumerative theory of cuspidal cubics were the *characteristic conditions* μ, ν (i.e., going through a point and being tangent to a line, respectively). Schubert also considers conditions imposing that a given vertex (side) of the singular triangle lies on a line (goes through a point), and denotes any of these six conditions with the same symbol used to denote the corresponding element

The authors were partially supported by the CAYCIT and DGICYT of the singular triangle. Altogether we have eight conditions, which will be called *fundamental conditions* for the cuspidal cubics.

By transversality of general translates (Kleiman [1974]), the cubics satisfying seven (possibly repeated) fundamental conditions whose data are in general position are finite in number and at least in characteristic zero they count with multiplicity 1. In characteristic $p > 0$ each solution may have to be weighted with a multiplicity that is a power of p. The numbers so obtained are called *fundamental numbers* for the cuspidal cubics. The fundamental numbers involving only μ and ν are the *characteristic numbers.*

It turns out that there are 620 *non-zero* fundamental numbers for the cuspidal cubics (discounting those that may be obtained by duality), and of these Schubert gives explicit tables for 391 (*loc. cit.*, pp. 140–142). Of the remaining 229, a few (actually 21) can be deduced from related entries in tables he gives for space cuspidal cubics. As we explain below, Schubert's work is also incomplete on other (more fundamental) counts. The general problem of verifying and understanding all the geometric numbers computed by 19th century geometers, which is the main motivation of this and related works, was stated by Hilbert [1902] as Problem 15 of his list.

Schubert's calculations rely on the method of degenerations, which in turn requires to know, if we want to compute all fundamental numbers,

i) *that the space S^* of complete cubics* (see Section 1) *is smooth in codimension one,*

ii) *how many boundary components* (called *degenerations*) *there are in S^** (see Section 2),

iii) *how to solve a number of related enumerative problems on each of the degenerations* (see Sections 4–7 and 9), and

iv) *to express, on S^*, the fundamental conditions in terms of the degenerations* (*degeneration relations*, see Section 10) *and to establish that a number may be computed by substituting one of its conditions by its expression in terms of the degenerations.*

For a given subset of fundamental numbers much less may be needed. Thus, in order to compute the 8 characteristic numbers, it is enough to know a single degeneration (degeneration σ, whose points consist of a conic and one of its tangent lines), but for this one it is nevertheless still necessary to take care of the points i)-iv) to verify them. This was done recently, in different ways, by Sacchiero [1984] and by Kleiman – Speiser [1986].

Question i) is not considered by Schubert. As far as ii) goes, Schubert constructs, in addition to σ, 12 degenerations, by means of the so called homolography process, but he does not provide any formal verifications, nor does he prove that they are all possible degenerations. These questions were clarified in Miret – Xambó [1987] (see Section 2 below).

Question iii) is rather involved. Since the building elements of some of the degenerations exceed in number what would be allowed by their dimension, they cannot be independent and so *there must exist relations among those elements.* Schubert gives lists of such relations, expressed in enumerative terms (tables of "Stammzahlen", loc. cit., pp. 120-127), and asserts that they were obtained by an indirect process ("a posteriori erschlossen", ibid., p. 119). Now in Miret – Xambó [1987] the Stammzahlen that are needed to describe the degenerations were studied and were showed to be related to basic projective geometry properties of the cuspidal cubics. In this paper we continue the study of this topic and give a detailed *geometric description* of all the degenerations.

Another difference with Schubert arises in the treatment of question iv). Schubert derived degeneration relations by means of coincidence formulas (loc. cit., p. 107 and ff.). This procedure leads, however, to computations of multiplicities that seem very difficult to handle, and which have been verified, as far as the authors know, only in very special cases, like some that arise in the verification of the characteristic numbers. Instead, one may work on the idea, already used by Schubert to cross-check his results, that most geometric numbers can be computed in several different ways. When used systematically, this observation allows to establish, if we already have assembled suitable enumerative information on the various degenerations, the required degeneration relations by simple algebra. This version of the method of degenerations is explained in Section 8.

The organization of this paper is as follows. Section 1 is devoted to the determination of the Picard group of S. At the end we define the space of complete cuspidal cubics. In Section 2 we briefly recall the description of the 13 first order degenerations of the cuspidal cubics. Then in Section 3 we prove a few geometric properties of cuspidal cubics that supplement and refine those given in Miret-Xambó [1987]. In Sections 4–7 we carry out systematic enumerative computations on the various degenerations (Stammzahlen) based on the properties inherited by the degenerations from corresponding properties of the cuspidal cubics. Then in Section 8 we outline, as we said above, a setup for the method of degenerations. In Section 9 we include a number of tables of degeneration numbers; they are obtained from the elementary numbers by direct arithmetic calculation. In Section 10 we determine the degeneration relations for the cuspidal cubics, that is, the expressions of the first order conditions in terms of the degenerations and of the condition that the cusp of the cubic be on a line. Section 11 contains examples that show how to put together the information gathered before to effectively compute the fundamental numbers of cuspidal cubics. Finally in Section 12 we give the tables of all the fundamental numbers.

Acknowledgements. The second named author wants to thank Steven Kleiman for his suggesting that the method of degenerations be explained in the context of a non-trivial example, rather than in an abstract form, and Robert Speiser for fruitful discussions about issues related to coincidence formulas.

1. Spaces of cuspidal cubics

1.1. Let $\mathbf{P}^2$ be the *complex* projective plane. The homogeneous coordinates of $\mathbf{P}^2$ will be denoted (x_0, x_1, x_2). The point $P_0 = (1,0,0)$ will be called the origin of coordinates. The space parametrizing plane cubics is isomorphic to $\mathbf{P}^9$ and we will identify these spaces. We shall let S denote the 7 dimensional locally closed subset whose points represent non-degenerate cuspidal cubics. Thus S is an orbit of the natural action of the group $G = PGL(\mathbf{P}^2)$ on $\mathbf{P}^9$. In particular S is a smooth variety.

1.2. If X is a point or a line, we shall set S_X to denote the subvariety of S whose points are cuspidal cubics with its cusp on X. Similarly, if P is a point and L is a line, $P \in L$, then $S_{P,L}$ will denote the cycle of cuspidal cubics that have the cusp at P with cuspidal tangent L. The cycle $S_{P,L}$ is irreducible, because it is an open set of a linear space. From this it follows that the cycle S_X is also irreducible. The class of S_L in Pic(S) will be denoted c and the class of the cycle of cuspidal cubics whose cuspidal tangent goes through a point will be denoted q.

1.3. Theorem. *Pic(S)* $= \mathbf{Z} \oplus \mathbf{Z}/(5)$. *The free generator of this group is c and the generator corresponding to* $\mathbf{Z}/(5)$ *is the projection of q.*

Proof: Let L be a given line, and let U be the open set of S whose points are cuspidal cubics with the cusp not on L. Thus $S - U = S_L$ and hence we have an exact sequence

$$A^0(S_L) \to A^1(S) \to A^1(U) \to 0.$$

From this we see that $\mathrm{Pic}(S) = A^1(S)$ is generated by c and $A^1(U)$. Now we have an isomorphism $U \simeq \mathbf{A}^2 \times S_{P_0}$, induced by translations in $\mathbf{A}^2 \simeq \mathbf{P}^2 - L$, and so $A^1(U) \simeq A^1(S_{P_0})$.

To study the last group, let T denote the space of cubics that have a double point, and let T_P denote the 6 dimension linear space of cubics that have a double point at P. Thus cubics in T_{P_0} have equations of the form

$$x_0 f_2 + f_3 = 0, \tag{1}$$

where f_i, $i = 2, 3$, is a homogeneous polynomial of degree i in the variables x_1, x_2. It is clear that $\overline{S} \subset T$, where $\overline{S}$ is the closure of S in T. Now $\overline{S}_{P_0}$ is a quadratic cone of rank 3 in T_{P_0}, for it is clear that (1) has a double tangent at P_0 if and only if $\mathrm{Disc}(f_2) = 0$. Moreover, if $\tilde{F}$ is the quintic hypersurface of T_{P_0} given by the equation $\mathrm{Res}(f_2, f_3) = 0$, and $F = \tilde{F} \cap \overline{S}_{P_0}$, then points in F_{red} represent degenerate cuspidal cubics and conversely. Indeed, if in (1) $f_2 = w^2$, where w is a linear form in x_1, x_2, then the cubic $x_0 w^2 + f_3 = 0$ is a degenerate cuspidal cubic if and only if w divides f_3.

We will show that $[F] = 2[F_{\text{red}}]$, and that F_{red} is irreducible. If this is so, from the exact sequence

$$A^0(F_{\text{red}}) \to A^1(\overline{S}_{P_0}) \to A^1(S_{P_0}) \to 0$$

and the fact, also proved below, that

$$A^1(\overline{S}_{P_0}) \simeq \mathbf{Z},$$

generated by a ruling of the cone, we deduce that $A^1(S_{P_0}) = \mathbf{Z}/(5)$, because a quintic hypersurface section is rationally equivalent to 10 rulings and so F_{red} is equivalent to 5 rulings. Now observe that the rulings of the cone are the subspaces of cuspidal cubics that have a given cuspidal tangent, and that one of these rulings generates, by translations, the cycle of cuspidal cubics whose cuspidal tangent goes through a fixed point.

To prove that $[F] = 2[F_{\text{red}}]$, consider an affine space $\mathbf{A}^5$ and define a map

$$f: \mathbf{A}^5 \longrightarrow \overline{S}_{P_0}$$

by transforming (s, b_0, b_1, b_2, b_3) into the cubic

$$x_0(x_1 + sx_2)^2 = b_0x_1^3 + b_1x_1^2x_2 + b_2x_1x_2^2 + b_3x_2^3.$$

This induces an isomorphism of $\mathbf{A}^5$ with $\overline{S}_{P_0} - R$, where R is the ruling of $\overline{S}_{P_0}$, given by the cuspidal cubics whose cuspidal tangent is the line $\{x_2 = 0\}$. The pull-back under f of the subscheme F is the subscheme given by the equation $\operatorname{Res}((x_1 + sx_2)^2, f_3) = 0$. Now using Fulton [1984], Example A.2.1, p. 410, it is easy to see that

$$\operatorname{Res}((x_1 + sx_2)^2, f_3) = \operatorname{Res}(x_1 + sx_2, f_3)^2$$

and so on the open set $\overline{S}_{P_0} - R$ we see that F is divisible by 2, and that the restriction of $\frac{1}{2}F$ to each ruling is a hyperplane of the ruling. Hence the equality $[F] = 2[F_{\text{red}}]$ is correct on the complementary set of any ruling, and therefore it holds globally.

To end the proof we have to see that a rank three projective quadratic cone K satisfies $A^1(K) = \mathbf{Z}$, generated by a ruling. To see this notice that in order to compute $A^1(K)$ we may throw away the vertex of the cone, because its codimension is 2. Having done that, K is a fibre bundle over a smooth conic C with fibre $\mathbf{A}^1$. Hence $A^1(K)$ is isomorphic to $A^1(C)$. But $A^1(C) \simeq \mathbf{Z}$, generated by the class of a point of C, and from this the claim follows. ◇

1.4. Corollary. *The Picard group of the space of non degenerate nodal cubics is generated by the class of the cycle of nodal curves with node on a fixed line.*

PROOF: Let T_L be the cycle of nodal curves that have its node on a line L. This cycle is irreducible and we have an exact sequence

$$A^0(T_L) \to A^1(T) \to A^1(V) \to 0, \quad V = T - T_L.$$

So $\mathrm{Pic}(T) = A^1(T)$ is generated by the class $b = [T_L]$ and by $A^1(V)$. Now $V \simeq \mathbf{A}^2 \times T_P$, so $A^1(V) = A^1(T_P)$. Now cubics that have a double point at P form a 6 dimensional linear space, which is nothing but $\overline{T}_P$. In this space we have the quadratic cone $D = \overline{S}_P$ and the hypersurface E whose points consist of cubics that split in a conic and a line, and, up to subvarieties of codimension 2 or higher, $\overline{T}_P - T_P = D \cup E$. Thus we have an exact sequence

$$A^0(D \cup E) \to A^1(\overline{T}_P) \to A^1(T_P) \to 0$$

So it is clear that $A^1(T_P) = \mathbf{Z}/(m)$, where $m = \gcd(d, e)$, d and e the degrees of D and E in $\overline{T}_P$, respectively. Now D has degree 2, as we noticed above, and E is a Segre variety, which has degree 5. So we conclude that $A^1(T_P) = 0$ and so our statement follows. ◇

1.5. Complete cuspidal cubics. We will use the letters b, c, v, y, z, q, w also to denote the maps that transform a given cubic C in S into, respectively, the dual cubic C^*, the cusp, the inflexion point, the intersection of the cuspidal and inflexional tangents, the line joining the inflexion and the cusp. the cuspidal tangent, and the inflexion tangent.

Set

$$\mathbf{P} = \mathbf{P}^{9^*} \times (\mathbf{P}^2)^3 \times (\mathbf{P}^{2^*})^3$$

and consider the map

$$h: S \to \mathbf{P}, \quad h = (b, c, v, y, z, q, w).$$

Let S^* be the closure of the graph of h in $Z = \overline{S} \times \mathbf{P}$. The space S^* will be referred to as the space of *complete cuspidal cubics*. The points in $S^* - S$ will be called *degenerate* cuspidal cubics, where the inclusion of S in S^* is given by $\mathrm{id} \times h$. Since the composition of h with the projection of $\mathbf{P}$ onto its first factor is $b: S \to \mathbf{P}^{9^*}$, it is natural to define $b: S^* \to \mathbf{P}^{9^*}$ as the restriction to S^* of the projection onto $\mathbf{P}^{9^*}$. Given a point C' of S^*, we shall say that $b(C')$ is the *tangential cubic* associated to the complete cubic C'. In the same way we can define morphisms c, v, y, z, q and w from S^* to the corresponding factors of Z. Given $C' \in S^*$, $c(C')$ will be called *the cusp* of C' and similarly with the other maps.

For a non-degenerate cuspidal cubic, the triangle whose vertexes are c, v, and y, and whose sides are z, q, w, is called *singular triangle*. The same notion can now be defined for degenerate cuspidal cubics in S^*. In other words, given a degenerate complete cuspidal cubic C', the six-tuple

$$(c(C'), v(C'), y(C'), z(C'), q(C'), w(C'))$$

will be called singular triangle of C', the first three elements being the vertices and the last three the sides. The cubic is degenerate if and only if its singular triangle is a degenerate triangle.

The projection of a point $C' \in S^*$ to $\overline{S}$ will be referred to as the *point cubic* associated to C'.

1.6. Theorem (see Miret-Xambó [1987]). *The variety S^* of complete cuspidal cubics is non-singular in codimension 1.*

In next section we give a description of the boundary components of S^*.

1.7. Conventions. Henceforth we will say that a point P is general with respect to a cuspidal cubic C if it does not lie on C nor on any side of the singular triangle. A point P of C will be said to be general if it is different from the cusp and the inflexion.

Given four colinear points A, B, C, D we shall write $\rho(A, B, C, D)$ to denote their cross ratio.

We also recall here that given a cuspidal cubic of the form $x_0x_2^2 = x_1^3$ then the dual cubic has equation $27u_0u_2^2 + 4u_1^3 = 0$.

2. Degenerations

The boundary $S^* - S$ has 13 irreducible components D_i, all of dimension 6 (see Miret-Xambó [1987]). The brief descriptions given below are intended to outline the structure of the general points of D_i, $i = 0, \ldots, 12$ (see the drawings at the end). In each case we indicate what the corresponding point and line cycles are, as well as the sides and verteces of the singular triangle. The degenerations $D_1, \ldots, D_{12}$ can be obtained by applying a homolography to a non-degenerate cuspidal cubic with suitable choices of its center P and axis L. This means that points on D_i, $i = 1, \ldots, 12$, are the limit cycles for $t = 0$ or $t = \infty$ of the cycles obtained transforming the given cuspidal cubic by a homology of modulus t with center at P and axis L.

In what follows instead of saying "the pencil of lines through point P is a component of the dual cubic" we will say that "P is a focus of the cubic". Thus, if three points are declared as foci, this means that the dual cubic decomposes into the three pencils of lines through the given points.

2.1. D_0. General points in D_0 consist of a smooth conic K together with a distinguished tangent line L of K. The three sides of the singular triangle of such a pair coincide with L, while the three verteces coincide with the contact point, say P. The tangential cubic consists of the dual conic K^* and the pencil of lines through P.

2.2. D_1 and D_{12}. Points in D_{12} consist of a triple line L with three distinct foci on it. The sides of the singular triangle coincide with L and its three vertices are three distinct points on L disjoint from the foci. The degeneration D_1 is dual of D_{12}.

2.3. D_2 and D_{11}. Points in D_{11} consist of a triple line L with three distinct foci on it. The vertices c and y fall together on a focus, and the vertex v is a point on L which is not a focus. The sides w and z coincide with L and q is a line through $c = y$ different from L. The degenerations D_2 is dual of D_{11}.

2.4. D_3 and D_{10}. Points in D_{10} consist of a triple line L with three distinct foci on it. The sides q and w coincide with L and z is a line different from L that does not go through a focus. The verteces c and v fall together on the intersection of z and L and y is a point on L different from $c = v$ and which is not a focus. The degeneration D_3 is dual of D_{10}.

2.5. D_4 and D_9. Points on D_9 consist of a triple line L with a simple focus and a double focus. The sides q and z coincide with L, while w is a line through the double focus distinct from L. The verteces $v = y$ fall on the double focus and c is a point on L different from the foci. The degeneration D_4 is dual of D_9.

2.6. D_5 and D_8. Points in D_8 consist of a triple line L with a simple focus and a double focus. The side z coincides with L, while q and w are lines different from L that go through the simple and the double focus, respectively. The intersection of q and w is the vertex y, while c falls on the simple focus and v on the double focus.

2.7. D_6 and D_7. Degenerations of type D_7 consist of a double line L and a simple line L', with a simple focus Q on L and a double focus R that falls on $L \cap L'$. The three sides of the singular triangle coincide with L, while the verteces are three distinct points of L disjoint from the foci. The degeneration D_6 is dual of D_7.

It is to be remarked that the elements with which a degeneration is built up need not be independent. Take, for instance, D_{12}. We have six points on a line. Such configurations fill a space $\overline{D}_{12}$ of dimension 8. Since D_{12} has dimension 6 we see that D_{12} is a codimension 2 subvariety of $\overline{D}_{12}$. Similarly we can define varieties $\overline{D}_{11}$, $\overline{D}_{10}$ and $\overline{D}_7$ of dimensions 7, 8 and 8 that contain the degenerations D_{11}, D_{10} and D_7 as subvarieties of codimensions 1, 2 and 2, respectively. Thus $\overline{D}_{11}$ may be described as the variety whose points are ordered pairs of lines with three distinguished points on the first, and $\overline{D}_{10}$ and $\overline{D}_7$ as varieties whose points are ordered pairs of lines with four distinguished points on the first line. Of course, similar remarks can be made for the dual degenerations D_1, D_2, D_3 and D_5.

The enumerative geometry of D_7, D_{10}, D_{11} and D_{12} will be studied in Sections 4, 5, 6 and 7, respectively.

3. Projective properties of cuspidal cubics

3.1. Proposition. *Let C be a non-degenerate cuspidal cubic and P a general point with respect to C. Let L_1, L_2, L_3 be the tangent lines to C through P and set $\rho_i = \rho(Pc, Pv, Py, L_i)$. Then*

$$\frac{1}{\rho_1} + \frac{1}{\rho_2} + \frac{1}{\rho_3} = 3$$
$$\rho_1\rho_2\rho_3 = 1.$$

Conversely, given non-zero scalars ρ_i, $i = 1,2,3$, satisfying the two equations above, three distinct concurrent lines L_1, L_2, L_3, say at the point P, and a triangle c, v, y with no vertex on the lines such that $\rho_i = \rho(Pc, Pv, Py, L_i)$, then there exists a cuspidal cubic C with singular triangle c, v, y which is tangent to the lines L_i, $(i = 1,2,3)$. (The proof actually shows that C is unique.)

Proof: Take the singular triangle of C as the reference triangle and take a general point of C as the unit point. Let $P = (a, b, 1)$. The projection of y from P on the line $z = cv$ is $y' = (a, 0, 1)$. Let $M = (m, 0, 1)$ be the point where a tangent to C through P meets the line $z = cv$. Then imposing that the line PM satisfies the dual equation we find that m has to satisfy the relation

$$m^3 + (27b^3 - 3a)m^2 + 3a^2m - a^3 = 0.$$

Let m_i, $i = 1,2,3$, be the roots of this equation and M_i the corresponding points. One computes that $\rho(c, v, y', M_i) = m_i/a$ and from this the first part of the proposition follows easily.

To see the converse, take $(c, y, v; P)$ as a reference. With respect to this reference the line L_i has coordinates $(1, \rho_i - 1, -\rho_i)$. We know that the cuspidal cubics with singular triangle c, v, y are of the form $\alpha x_1^3 = x_0 x_2^2$, $\alpha \neq 0$. Using the line equation of this cubic we see that it is tangent to the line L_i if and only if

$$\rho_i^3 + (\frac{27}{4}\alpha - 3)\rho_i^2 + 3\rho_i - 1 = 0.$$

Thus if the ρ_i satisfy the conditions in the first part of the statement, then in order that the cubic be tangent to the three lines it is necessary and sufficient that $\frac{27}{4}\alpha - 3 = -(\rho_1 + \rho_2 + \rho_3)$. Since this equation has a unique solution with respect to α, which is non-zero, this ends the proof. ◇

The preceeding result still holds if P is a point on C not on the singular triangle, taking the tangent to C at P twice. In this case, however, we have a more precise statement:

3.2. Proposition. *Given a point P of C, let L be the tangent to C at P and L' the tangent to C through P other than L. Then the cross-ratio of any four of the lines Pc, Pv, Py, L, L' is independent of P. In fact we have that*

$$\begin{aligned} \rho(Pc, Pv, Py, L) &= -2 \\ \rho(Pc, Pv, Py, L') &= \tfrac{1}{4} \\ \rho(Pc, Pv, L, L') &= -\tfrac{1}{8} \\ \rho(Pc, Py, L, L') &= \tfrac{1}{4} \\ \rho(Pv, Py, L, L') &= -\tfrac{1}{2}. \end{aligned}$$

Notice that any two of these relations imply the other three.

Conversely, given a triangle c, v, y and two lines L and L' meeting at a point P not on the sides of the triangle and in such a way that two (and hence all) of the equations above are satisfied, then there exists a cuspidal cubic C with singular triangle c, v, y that is tangent to L at P and also tangent to L' (necessarily at a point different from P).

Proof: A straightforward computation as in the proof of 3.1. ◇

3.3. Proposition. *Given a point P of the cuspidal tangent q of a non-degenerate cuspidal cubic C, different from c, then the pair of lines q, Pv is harmonic with respect to the pair of tangents to C through P other than q. Conversely, given a harmonic tetrad of concurrent lines q, L, L' and L'' (say at P), and points c on q and v on L, both different from P, there exists a cuspidal cubic C with cusp at c and inflexion at v such that the tangent lines to C from P are q, L' and L''.*

Proof: Taking (c, v, y) as reference triangle and the unit point on C then the equation of C has the form $x_1^3 = x_0 x_2^2$ and the point P is of the form $(a, 1, 0)$. Let u, u' be the tangents to C, other than q, through P. Let $Q = (m, 0, 1)$ and $Q' = (m', 0, 1)$ be the intersections of u and u' with the line cv. It suffices to show that the pairs of points (c, v) and (Q, Q') are harmonic. Imposing that the lines $u = PQ$ and $u' = PQ'$ are tangent to C (using the dual equation) we find that $m + m' = 0$, and this ends the first part of the proof. The converse part can be seen in the same way as the converse part of 3.1. ◇

3.4. Proposition.

- **(a)** *Given a point P of the line z of a non-degenerate cuspidal cubic C, different from c and v, then the cross ratio of the lines z, Py and any pair of tangents to C from P is a primitive cube root of unity.*
- **(b)** *The line z and the three tangents to C from P form an equianharmonic tetrad, that is, its cross-ratio is a primitive cube root of -1.*
- **(c)** *The line Py and the three tangents to C from P form also an equianharmonic tetrad.*
- **(e)** *Conversely, given a triple of concurrent lines $\{L, L', L''\}$, say at a point P, and a pair of points c, y not on those lines, there is a cuspidal cubic with singular triangle c, y, v, where v is a point on the line cP, and which is tangent to the lines L, L' and L'' if either the cross ratio of Pc, Py and any pair of L's is a primitive cube root of unity or the tetrads Pc, L, L', L'' and Py, L, L', L'' are equianharmonic.*

Proof: Take the same reference as in the proof of **3.1** Let $P = (a, 0, 1)$. Then the line joining P and the point $M = (m, 1, 0)$ on the line q is given by the equation $-x_0 + mx_1 + ax_2 = 0$. Imposing that it satisfies the dual equation we get the relation

$$4m^3 = 27a^2,$$

whose solutions are of the form $m_i = \xi^k m_0$, $k = 0, 1, 2$, where ξ is a primitive cube root of unity and $m_0/3$ is a fixed cube root of $a^2/4$. Computation shows that

$\rho(c, y, M_i, M_j) = \xi^{j-i}$, which proves part **(a)**. Similarly, $\rho(c, M_0, M_1, M_2) = \xi + 1$, which proves **(b)**. The proof of **(c)** is similar. The converse part can be seen in the same way as the converse part of 3.1. ◇

We also collect here a three lemmas about cross ratios because we do not know a reference for them. The proofs are obtained by straightforward analytic computations.

3.5. Lemma. *Given three non-concurrent lines L_1, L_2, L_3, a point P not lying on any of them and a scalar $k \neq 1$, there exists a unique line L through P such that $\rho(P, L \cap L_1, L \cap L_2, L \cap L_3) = k$.*

3.6. Lemma. *Given a four lines L_1, L_2, L_3, L_4 such that no three of them are concurrent, a point P not lying on any of them and a scalar $k \neq 1$, there exist exactly two lines L through P such that the $\rho(L \cap L_1, L \cap L_2, L \cap L_3, L \cap L_4) = k$.*

3.7. Lemma. *Given five lines $L_1, \ldots, L_5$ in general position and two scalars k_1 and k_2 different from 1, there exists a unique line L such that*

$$\rho(L \cap L_1, L \cap L_2, L \cap L_3, L \cap L_4) = k_1$$
$$\rho(L \cap L_1, L \cap L_2, L \cap L_3, L \cap L_5) = k_2.$$

We also need a few cycle identities for ordered and unordered triples of collinear points. First recall that for flags "point-line" in the projective plane, $\{p, g\}$, we have the relation $gp = g^2 + p^2$, where g is the condition that the line goes through a point and p the condition that the point be on a line. Now consider configurations $(L; c, v, y)$ consisting of a line L and three distinguished points c, v, y on L. The variety V parametrizing such configurations is smooth and complete. Moreover, it follows easily from the relation just recalled that on V we have the following relations:

3.8. Lemma.

$$L^2 + c^2 = Lc,$$
$$L^2 + v^2 = Lv,$$
$$L^2 + y^2 = Ly.$$

Now consider configurations consisting of a line L together with a zero cycle Z of degree r on L. The points in the support of Z will be called foci of the configuration. The variety V' of such configurations is smooth and complete. In fact, V' can be defined as the projective bundle associated to the vector bundle $S^r(E^*)$, where E is the tautological rang 2 bundle on $\check{\mathbf{P}}^2$. Given j lines in general position, and a point $(L; Z)$ of V', write $Z = Z' + Z''$, where the support of Z' lies on the union of the lines and the support of Z'' is disjoint from them. We shall write Q_j to denote the subvariety of V' whose points $(L; Z)$ satisfy that on each of the lines there is at least a point of Z (hence of Z') and that $\deg(Z'') \leq r - j$. It is not hard to see that Q_j is irreducible of codimension j. Now let Σ be the set of $\binom{j}{2}$ points of intersection of the j lines. For each $P \in \Sigma$, let Q_j^P denote the subvariety of V' whose points $(L; Z)$ satisfy that $P \le Z'$, that on each of the j lines there is at least a point of Z, and that $\deg Z'' \le r - j + 1$. It is also easy to see that Q_j^P is an irreducible subvariety of codimension j. For each pair of points $P, Q \in \Sigma$, $P \neq Q$, let $Q_j^{P,Q}$ denote the subvariety of V' whose points $(L; Z)$ satisfy that $P + Q \le Z'$, that on each of the j lines there is at least a point of Z, and that $\deg Z'' \le r - j + 2$. It is also easy to see that $Q_j^{P,Q}$ is an irreducible subvariety of codimension j.

3.9. Lemma.

$$Q^j = Q_j + \sum_P Q_j^P + \sum_{P,Q} Q_j^{P,Q}$$

Proof: That the left hand sides are equal to the right hand sides up to multiplicities follows from simple combinatoric arguments. That the multiplicities are equal to 1 in all cases can be seen by the principle of general translates (see Kleiman [1974] and Laksov-Speiser [1987]). ◇

With the same notations, let Q and P denote the conditions that a configuration has, respectively, a focus on a given line and a focus at a given point. If the number of foci is 2 or 3, from the preceeding lemma we conclude:

3.10. Lemma.

$$\begin{aligned} [Q^2] &= [Q_2] + [P] \\ [Q^3] &= 3[PQ] \\ [Q^4] &= 3[P^2] \\ [Q^5] &= 0 \end{aligned} \qquad \text{resp.} \qquad \begin{aligned} [Q^2] &= [Q_2] + [P] \\ [Q^3] &= [Q_3] + 3[PQ] \\ [Q^4] &= 6[PQ_2] + 3[P^2] \\ [Q^5] &= 15[P^2Q] \end{aligned}$$

◇

4. Stammzahlen for D_7

We shall use the notations introduced in **2.7**.

4.1. Proposition. *The singular triangle c, v, y of a degeneration of type D_7 may be any triple of distinct collinear points. The simple focus Q and the double focus R are collinear with c, v, y and are uniquely determined by the relations $\rho(c, v, y, Q) = 1/4$ and $\rho(c, v, y, R) = -2$. The simple line may be any line through R.*

Proof: It is a direct consequence of **3.2** and the way the degeneration is obtained by a homolography. ◇

Let $\overline{D}_7'$ be the variety of ordered 5-tuples of distinct collinear points c, v, y, Q, R. Let D_7' be the subvariety of $\overline{D}_7'$ given by the relations in **4.1**. Let $\overline{\pi}\colon \overline{D}_7 \to \overline{D}_7'$ be the map that forgets the simple line L' and $\pi\colon D_7 \to D_7'$ the restriction of $\overline{\pi}$ to D_7. Next lemma shows that the computation of the Stammzahlen for D_7 is equivalent to the computation of Stammzahlen for D_7'.

4.2. Lemma. *Let N be a fundamental number for D_7.*

(a) *If the exponent of L' in N is 0 or at least 3, then $N = 0$.*

(b) *If the condition L' appears just once in N, then $N = N'$, where N' is the number on D_7' obtained dropping the condition L' from N.*

(c) *If the condition L' appears just twice in N, say $N = L'^2 x$, then $N = R'x'$, where the product x' on D_7' corresponds to the product x on D_7 (that is, $x = \pi^*(x')$) and where R' is the condition on D_7' that the double focus be on a line.*

Proof: Follows easily using the projection formula and we omit it. ◇

4.3. Theorem. *The number $L^{i_1} Q^{i_2} R^{i_3} c^{i_4} v^{i_5} y^{i_6}$, $i_1 + \ldots + i_6 = 5$, is equal to 1 if one exponent is 2 and the others are at most 1 or if $i_1 = 0$ and the other exponents are at most 2; is equal to 2 if $i_1 = 1$ and the remaning are at most 1; otherwise is 0.*

Proof: If $i_1 = 2$, then the line is fixed and so by **4.1** the number must be one if the remaining exponents are at most 1 and 0 otherwise. The similar reasoning works if $i_1 = 1$ and some other exponent is 2 or if two exponents are 2. If $i_1 = 0$ and there is a single square, then the conclusion follows from **3.5** and **4.1**. If $i_1 = 1$ and the remaining exponents are at most one, then the value is 2 by **3.6** and **4.1**. If $i_1 = 0$ and the others are at most 1 (and hence all equal to 1), then we can apply **3.7**. ◇

4.4. Remark The expression of $[D_7]$ in the Chow ring of $\overline{D}_7$ is as follows:

$$[D_7] = L^2 - 2Lc - 2Lv - 2Ly - 2LQ - 2LR \\ + cv + cy + cQ + cR + vy + vQ + vR + yQ + yR + QR.$$

The proof of this relation and of the similar relations for D_{10}, D_{11} and D_{12} (see 5.4, 6.4 and 7.4) are similar and we will give details only for the case of D_{12}. The method of proof consists in writing the corresponding D_k as a linear combination of a basis of the corresponding Chow group, with undetermined coefficients, and then to establish enough linear relations among the coefficients by multiplying with suitable monomials in the fundamental conditions, using the tables of Stammzahlen in each case. One reason for bothering only about D_{12} is that in this case the expression is actually used to complete the computation of the Stammzahlen , while in the remaining three cases we do not need the expression for such a purpose.

5. Stammzahlen for D_{10}

5.1. Proposition. *The three foci of a degeneration of type D_{10} may be any unordered triple of collinear points. For each such triple there are two possible pairs $\{c, y\}$ and z is any line through c. More precisely,*

(a) *The cross ratio of c, y and any two foci is a primitive cube root of unity.*

(b) *The point c and the three foci form an equianharmonic tetrad.*

(c) *The point y and the three foci form also an equianharmonic tetrad.*

Proof: It is a direct consequence of **3.4** and the definition of D_{10} by the homolography process. ◇

Let $\overline{D}'_{10}$ be the variety whose points are unordered triples Q_1, Q_2, Q_3 of colinear points (that will be called foci) together with two distinguished points $c = v$ and y of the line defined by the foci. Let D'_{10} be the subvariety of $\overline{D}'_{10}$ given by the relations in **5.1**. Let $\overline{\pi}\colon \overline{D}_{10} \to \overline{D}'_{10}$ be the map that forgets the line z and $\pi\colon D_{10} \to D'_{10}$ the restriction of $\overline{\pi}$ to D_{10}. Next lemma shows that the computation of the Stammzahlen for D_{10} is equivalent to the computation of Stammzahlen for D'_{10}.

5.2. Lemma. *Let N be a fundamental number for D_{10}.*

(a) *If the exponent of z in N is 0 or at least 3, then $N = 0$.*

(b) *If the condition z appears just once in N, then $N = N'$, where N' is the number on D'_{10} obtained dropping the condition z from N.*

(c) *If the condition z appears just twice in N, say $N = z^2x$, then $N = c'x'$, where the product x' on D'_{10} corresponds to the product x on D_{10} (that is, $x = \pi^*(x')$) and where c' is the condition on D'_{10} that the cusp be on a line.*

Proof: Projection formula. ◇

5.3. Theorem. *The fundamental numbers of D'_{10} are given in the following table:*

$L^2Q^3 = 2$	$LQ^2c^2 = 2$	$Q^4y = 6 \cdot 2 + 3 \cdot 2$
$L^2Q^2c = 2$	$LQ^2cy = 4 + 1$	$Q^3c^2 = 2 + 3 \cdot 2$
$L^2Q^2y = 2$	$LQ^2y^2 = 2$	$Q^3cy = 2 + 3 \cdot 2$
$L^2Qcy = 1$	$LQc^2y = 1$	$Q^3y^2 = 2 + 3 \cdot 2$
$LQ^4 = 6 \cdot 2$	$LQcy^2 = 1$	$Q^2c^2y = 2 + 1$
$LQ^3c = 4 + 3 \cdot 2$	$Q^5 = 15 \cdot 2$	$Q^2cy^2 = 2 + 1$
$LQ^3y = 4 + 3 \cdot 2$	$Q^4c = 6 \cdot 2 + 3 \cdot 2$	$Qc^2y^2 = 1$

In this table an expression of the form $m \cdot \mathbf{n}$ on the right hand side means that the factor m has a combinatorial origin and that $\mathbf{n}$ is due to the nature of the relations that exist among the elements of the degeneration. On the other hand, the reason why we decompose some of the numbers as the sum of two expressions comes from using lemma 3.10, as will be seen along the proof (cf. 7.4).

Proof: From 5.1 we immediately get the relations

$$L^2Q_3 = 2, L^2Q_2c = 2, L^2Q_2y = 2, L^2Qcy = 1.$$

From 5.1 and 3.5 we get

$$\begin{array}{ll} PQ_2c = 2 & Q_3c^2 = 2 \\ PQ_2y = 2 & Q_3y^2 = 2 \\ PQcy = 2 & Q_2c^2y = 2 \\ & Q_2cy^2 = 2 \end{array}$$

Similarly, from 5.1 and 3.6 we get

$$LQ_3c = 4, LQ_3y = 4, LQ_2cy = 4.$$

Finally from 5.1 and 3.7 we get

$$Q_3cy = 2.$$

Now using 3.10 we see that the proof is reduced to computations. ◇

5.4. Remark The expression of $[D_{10}]$ in terms of the fundamental conditions of $\overline{D}_{10}$ is the following (cf. 4.4):

$$[D_{10}] = 5L^2 - 4Lc - 4Ly + Q^2 - 5QL + 2Qc + 2Qy + 2cy.$$

6. Stammzahlen for D_{11}

6.1. Proposition. *For D_{11} the point $c = y$ and the two foci Q, Q' other than c can be any triple of collinear points and q can be any line through c. The point v is uniquely determined from Q, Q' and c by the relation that the pair (Q, Q') is harmonic with respect to (c, v).*

Proof: This is a direct consequence of **3.1** and the description of D_{11} by homolographies. ◇

Given that the only relation among the elements of the degeneration D_{11} is the one given in **6.1**, we may work, in order to find the Stammzahlen of D_{11}, on the variety D'_{11} whose points parametrize unordered pairs of distinct points $\{Q, Q'\}$ together with two distinguished points c, v on the line QQ' that are harmonic with respect to the pair $\{Q, Q'\}$. In fact, if $\pi: D_{11} \to D'_{11}$ is the map which forgets the line q, then next lemma reduces the computation of the Stammzahlen for D_{11} to the computation of certain numbers on D'_{11}.

6.2. Lemma. *Let N be a fundamental number for D_{11}.*

(a) *If the exponent of q in N is 0 or at least 3, then $N = 0$.*

(b) *If the condition q appears just once in N, then $N = N'$, where N' is the number on D'_{11} obtained dropping the condition q from N.*

(c) *If the condition z appears just twice in N, say $N = q^2x$, then $N = c'x'$, where the product x' on D'_{11} corresponds to the product x on D_{11} (that is, $x = \pi^*(x')$) and where c' is the condition on D'_{11} that the cusp be on a line.*

Proof: Projection formula. ◇

6.3. Theorem. *The fundamental numbers of D'_{11} are given in the following table:*

$L^2Q^2c = 1$	$LQ^2cv = 2+1$	$Q^3c^2 = 3\cdot 1$
$L^2Q^2v = 1$	$LQ^2v^2 = 1$	$Q^3cv = 3\cdot 1$
$L^2Qcv = 1$	$LQc^2v = 1$	$Q^3v^2 = 3\cdot 1$
$LQ^3c = 2+3\cdot 1$	$LQcv^2 = 1$	$Q^2c^2v = 1+1$
$LQ^3v = 2+3\cdot 1$	$Q^4c = 3\cdot 1$	$Q^2cv^2 = 2+1$
$LQ^2c^2 = 1$	$Q^4v = 3\cdot 1$	$Qc^2v^2 = 1$

Proof: If the number contains L^2 then line is fixed. The three remaining conditions fix three points and **6.1** fixes the last one. Hence all numbers containing L^2 are equal to 1.

The same reasoning is valid if the number contains Lc^2, Lv^2, LP, c^2v^2, Pc^2, Pv^2 or P^2.

From 6.1 and 3.5 one sees that

$$PQcv = 1, Q_2c^2v = 1, Q_2cv^2 = 1.$$

From 6.1 and 3.6 we see that $LQ_2cv = 2$.

Using now 3.10 it a simple computation to find the values in the table. ◇

6.4. Remark Let $\overline{D}_{11}$ be the variety parametrizing configurations consisting of an unordered pair Q, Q' of points together with two distinguished points c, y on the line QQ' and a line q through c. Then the expression of D_{11} in terms of the first order fundamental conditions of $\overline{D}_{11}$ (with the obvious notations) is the following (cf. 4.4).

$$[D_{11}] = c + v + Q - 2L.$$

7. Stammzahlen for D_{12}

7.1. Proposition. *Given six distinct collinear points c, v, y and Q_1, Q_2, Q_3, let $\rho_i = \rho(c, v, y, Q_i)$. Then in order that c, v, y is the singular triangle and $\{Q_1, Q_2, Q_3\}$ the foci of a degeneration of type D_{12} it is necessary and sufficient that*

$$\frac{1}{\rho_1} + \frac{1}{\rho_2} + \frac{1}{\rho_3} = 3$$

$$\rho_1\rho_2\rho_3 = 1.$$

Proof: It is a direct consequence of **3.1** and the way the degeneration is obtained by a homolography. ◇

7.2. Theorem. *The fundamental numbers of D_{12} are given by the following table:*

$L^2Q^3c = 4$	$LQ^3cy = 6 + 3 \cdot 2$	$LQcv^2y = 1$	$Q^3c^2y = 4 + 3 \cdot 2$
$L^2Q^3v = 1$	$LQ^3v^2 = 1$	$LQcvy^2 = 1$	$Q^3cv^2 = 3 + 3 \cdot 3$
$L^2Q^3y = 2$	$LQ^3vy = 3 + 3 \cdot 1$	$Q^5c = 15 \cdot 4$	$Q^3cvy = 4 + 3 \cdot 3$
$L^2Q^2cv = 3$	$LQ^3y^2 = 2$	$Q^5v = 15 \cdot 1$	$Q^3cy^2 = 2 + 3 \cdot 2$
$L^2Q^2cy = 2$	$LQ^2c^2v = 3$	$Q^5y = 15 \cdot 2$	$Q^3v^2y = 1 + 3 \cdot 1$
$L^2Q^2vy = 1$	$LQ^2c^2y = 2$	$Q^4c^2 = 6 \cdot 4$	$Q^3vy^2 = 2 + 3 \cdot 1$
$L^2Qcvy = 1$	$LQ^2cv^2 = 3$	$Q^4cv = 6 \cdot 4 + 3 \cdot 3$	$Q^2c^2vy = 4 + 1$
$LQ^4c = 6 \cdot 4$	$LQ^2cvy = 5 + 1$	$Q^4cy = 6 \cdot 4 + 3 \cdot 2$	$Q^2cv^2y = 3 + 1$
$LQ^4v = 6 \cdot 1$	$LQ^2cy^2 = 2$	$Q^4v^2 = 6 \cdot 1$	$Q^2cvy^2 = 2 + 1$
$LQ^4y = 6 \cdot 2$	$LQ^2v^2y = 1$	$Q^4vy = 6 \cdot 2 + 3 \cdot 1$	$Qc^2v^2y = 1$
$LQ^3c^2 = 4$	$LQ^2vy^2 = 1$	$Q^4y^2 = 6 \cdot 2$	$Qc^2vy^2 = 1$
$LQ^3cv = 7 + 3 \cdot 3$	$LQc^2vy = 1$	$Q^3c^2v = 6 + 3 \cdot 3$	$Qcv^2y^2 = 1$

Proof: The numbers that contain L^2 have been determined in Miret-Xambó [1987] (Theorem 4, Table 1).

The computation of the remaining numbers of the table will be based on lemma **7.4**, in which we first compute six auxiliary numbers; on lemma **3.10**, which allows to relate the auxiliary numbers to those we need, and on lemma **7.5**, in which we give an expression of the class $[D_{12}]$ in terms of a basis of the codimension 2 Chow group of $\overline{D}_{12}$.

Given j lines in general position ($j = 2, 3$), we shall write Q_j to denote the condition that there is exactly one focus on each of the j lines. We will also write P to denote the codimension 2 condition that one focus coincides with a given point. With these notations we have:

7.4. Lemma.

(1) $Q_3cv^2 = 3$.
(2) $Q_3cvy = 4$.
(3) $Q_3cy^2 = 2$.
(4) $Q_3v^2y = 1$.
(5) $Q_2c^2vy = 4$.
(6) $QPcvy = 3$.

Proof: The proofs can be done, in more or less straightforward manner, choosing a suitable reference and imposing the conditions **7.1** that a degeneration of type D_{12} must satisfy. We will only give details of **(1)**.

To establish **(1)** the reference we choose is the following. Let L_1, L_2, L_3 be the lines in general position required to define Q^3, M the line required to define the condition c and A the point v^2. Then we take the points $M \cap L_1, L_2 \cap L_3, A$ as the vertices of the reference triangle and $L_1 \cap L_2$ as unit point. Thus we have that

$$L_1: \quad x_1 = x_2,$$

$$L_2: \quad x_0 = x_2,$$

$$L_3: \quad ax_0 = x_2,$$

$$M: \quad x_1 = mx_2,$$

where $a, m \neq 0, 1$.

Let L the axis of the degeneration, so that L goes through A and hence $L: \quad x_1 = \lambda x_0$. Let $Q_i = L \cap L_i$ be the foci of the degeneration. A simple computation shows that

$$Q_1 = (1, \lambda, \lambda), Q_2 = (1, \lambda, 1), Q_3 = (1, \lambda, a).$$

Let $y = (1, \lambda, \mu)$. Then a computation of cross ratios shows that if we put $\rho_i = \rho(c, v, y, Q_i)$ then $\rho_1 = \mu/\lambda$, $\rho_2 = \mu$ and $\rho_3 = \mu/a$. The equations **7.1** are equivalent to the conditions $\lambda = 3\mu - a - 1$ and $\mu^3 = a(3\mu - a - 1)$, and hence there are exactly 3 degenerations of type D_{12} that satisfy the conditions Q_3cv^2. ◇

7.5. Lemma.

$$[D_{12}] = 7L^2 - 3Lc - 6Lv - 7Ly - 6LQ + Qc + 2Qv + 3Qy + Q^2 + 2cv + cy + 4vy.$$

Proof: From the fact that $\overline{D}_{12}$ is a projective bundle over $\check{\mathbf{P}}^2$ it follows that the Chow group $A^2(\overline{D}_{12})$ is freely generated by the degree 2 monomials in $\overline{L}, \overline{c}, \overline{v}, \overline{y}, \overline{Q}$. Hence there exist integers $m_1, \ldots, m_4, n_1, \ldots, n_4, r_1, \ldots, r_4$ and $s_1, \ldots, s_3$ such that

$$(*) \quad [D_{12}] = m_1\overline{c}^2 + m_2\overline{v}^2 + m_3\overline{y}^2 + m_4\overline{L}^2 + n_1\overline{Lc} + n_2\overline{Lv} + n_3\overline{Ly} + n_4\overline{LQ} + r_1\overline{Qc} + r_2\overline{Qv} + r_3\overline{Qy} + r_4\overline{Q}^2 + s_1\overline{cv} + s_2\overline{cy} + s_3\overline{vy}.$$

Now from the values of the three numbers that contain L^2 which are equal to 1 we see that if $\overline{u}$ is any of the first order conditions on $\overline{D}_{12}$ then $\overline{u}|_{D_{12}} = u$. More generally, given a monomial $\overline{x}$ on the first order conditions on $\overline{D}_{12}$, let x denote its restriction to D_{12}, so that x is obtained replacing the first order conditions in $\overline{x}$ by the corresponding conditions on D_{12}. It turns out that $x = \overline{x} \cdot D_{12}$. Using this relation with the 7 numbers that contain L^2 it is easy to find the values of the r_i and s_j, $i = 1, \ldots, 4$, $j = 1, \ldots, 3$.

Notice that from **3.8** we may compute the following values:

(1) $Q^3cv^2 = \mathbf{3} + 3 \cdot \mathbf{3} = 12$.
(2) $Q^3cvy = \mathbf{4} + 3 \cdot \mathbf{3} = 13$.
(3) $Q^3cy^2 = \mathbf{2} + 3 \cdot \mathbf{2} = 8$.
(4) $Q^3v^2y = \mathbf{1} + 3 \cdot \mathbf{1} = 4$.
(5) $Q^2c^2vy = \mathbf{4} + \mathbf{1} = 5$.

Now we have:

$5 = Q^2c^2vy = n_4 + r_2 + r_3 + 6r_4 = n_4 + 11$, so $n_4 = -6$.

$4 = Q^3v^2y = m_1 + n_1 + 6r_1 + s_2$, and so $n_1 = -(m_1 + 3)$.

$8 = Q^3cy^2 = m_2 + n_2 + 6r_2 + s_1 = m_2 + n_2 + 14$, and so $n_2 = -(m_2 + 6)$.

$12 = Q^3cv^2 = m_3 + n_3 + 6r_3 + s_2 = m_3 + n_3 + 19$, and so $n_3 = -(m_3 + 7)$.

$13 = Q^3cvy = m_4 + n_1 + n_2 + n_3 + 6n_4 + 6r_1 + 6r_3 + 15r_4 + s_1 + s_2 + s_3$, and so $m_4 = m_1 + m_2 + m_3 + 7$.

The conclusion follows from the relations **3.6**. ◇

With the expression $(*)$ and the knowledge of the fundamental numbers of $\overline{D}_{12}$ (which can be obtained by combinatorial arguments and so here will be assumed to be known) we can now obtain the values of the table **7.3**. We omit the details. There is, however, one aspect of the table which we want to comment, namely, the boldfaced numbers. We will do this by looking at an example. Take the number Q^4cv. Its value can be obtained as follows:

$$Q^4cv = D_{12}\cdot\overline{Q}^4\overline{cv} = -7\overline{LQ}^4\overline{cvy}+3\overline{Q}^5\overline{cvy}+\overline{Q}^4\overline{c}^2\overline{vy}+4\overline{Q}^4\overline{cv}^2 = -7\cdot6+3\cdot15+6+4\cdot6 = 33.$$

Now by **3.8**

$$Q^4cv = 6PQ_2cv + 3P^2cv = 6PQ_2cv + 3\cdot L^2Q^2cv = 6PQ_2cv + 3\cdot 3,$$

from which it follows that

$$PQ_2cv = 4.$$

This has been taken into account in the form we write the value of Q^4cv in the table decomposed as $6\cdot 4 + 3\cdot 3$. ◇

8. On the method of degenerations

In this section we introduce a version of the method of degenerations, especially as used by Schubert, which does not rely on coincidence formulas. Then in next section we indicate how we have used it to derive the degeneration relations (**9.1**) for the plane cuspidal cubics. To see how conditions arise in practice, and also for additional terminology, see **8.11**.

8.1. Let S be a smooth variety and let $d = \dim S$. Let

(8.1.1) $$X_1, \ldots, X_p, Z_1, \ldots, Z_s \ (p \geq 1, s \geq 0)$$

be subvarieties of S, where the X_i are *hypersurfaces* and the Z_j have at least codimension 2. The varieties **(8.1.1)** will be referred to as *conditions*. The codimension of a condition will also be called *order* of the condition. Conditions of order one are said to be *simple* conditions. We shall assume that the given list of conditions satisfies the conditions **A1** and **A2** below. In this paper we will not use higher order conditions (the Z's); they are included here because they are needed in other cases, like in twisted cubics.

A1. The sum of the codimensions of the Z_j ($j = 1, \ldots, s$) is $d-p$, and the intersection of all the varieties $X_1, \ldots, X_p, Z_1, \ldots, Z_s$ is a finite set.

A2. The intersection of all the varieties

$$X_1, \ldots, X_{i-1}, X_{i+1}, \ldots, X_p, Z_1, \ldots, Z_s$$

is a reduced curve C_i, $(i = 1, \ldots, p)$.

We shall let N denote the number of points in this set, counting multiplicities if they are present and we will write

$$N = X_1 \cdots X_p \cdot Z_1 \cdots Z_s$$

We shall say that N is the number of figures of type S that satisfy the conditions $X_1, \ldots, X_p, Z_1, \ldots, Z_s$.

We shall also assume that we have hypersurfaces $Y_1, \ldots, Y_q$ of S that satisfy the following condition:

A3. The classes $[Y_1], \cdots, [Y_q]$ generate $\mathrm{Pic}(S)_{\mathbf{Q}}$ (as a $\mathbf{Q}$-vector space).

8.2. In order to explain how we will approach the computation of N, let us first remark that if S were *complete*, then we would have

$$N = \deg_S [X_1] \cdots [X_p] \cdot [Z_1] \cdots [Z_s],$$

where $[Z]$ denotes the rational class of the cycle Z, which often is an affordable computation, inasmuch as under the completeness assumption one sometimes knows the rational intersection ring of S. This is the case, for example, if S is a Grassmannian, or a flag manifold, in which case the computation is just "Schubert calculus", but it is not the case for, say, smooth conics and quadrics or plane cuspidal cubics. So to end the description of our setup we need a modified procedure, with respect to the complete case, that is sufficient for the the computation of N.

8.3. To that end we shall assume that there exists a smooth variety S' (not necessarily complete) that satisfies the conditions **D1-D3** below (*axioms for degenerations*). Given any subset A of S, we shall write A' to denote its closure of A in S'.

D1. $S \subseteq S'$ and $D := S' - S = D_1 \cup \ldots \cup D_r$, where $D_1, \ldots, D_r$ are smooth irreducible hypersurfaces of S' and $D_i \cap D_j = \phi$. The varieties D_i will be called *degenerations*.

D2. Let

$$D_i \cdot X'_j = \textstyle\sum_k m_{ijk} X_{ijk},$$
$$D_i \cdot Z'_j = \textstyle\sum_k n_{ijk} Z_{ijk},$$

where the X_{ijk}, Z_{ijk} are the irreducible components of $D_i \cap X'_j$ and $D_i \cap Z'_j$, so that they have the same codimension in D_i as X_j, Z_j in S, respectively, and m_{ijk}, n_{ijk} are the corresponding multiplicities. Then we assume that for any choice of integers $k_1, \ldots, k_p, h_1, \ldots, h_s$, each in its appropriate range, the varieties

$$X_{i1k_1}, \ldots, X_{ipk_p}, Z_{i1h_1}, \ldots, X_{ish_s},$$

have empty intersection, and that omiting any of the X's, say X_{ijk_j}, the remaining have finite intersection. The number of points in this intersection, counting multiplicities if present (computed on D_i), will be denoted by

$$N_{ij}[k_1, \ldots, k_p, h_1, \ldots, h_s] = N_{ij}[k, h].$$

These numbers will be called *elementary numbers* with respect to the problem of computing N.

D3. Let C'_j be the intersection of the varieties $X'_1, \ldots, X'_p, Z'_1, \ldots, Z'_s$, except X'_j; by assumptions **A2** and **C2**, C'_j is a curve. We shall assume that this curve is *complete* and that the inclusion

$$u_j \colon C'_j \to S'$$

is a regular embedding.

8.4. Lemma. *The classes*

$$[D_1], \ldots, [D_r], [Y'_1], \ldots, [Y'_q]$$

generate $Pic(S')_{\mathbf{Q}}$.

Proof: We have an exact sequence (Fulton [1984], Prop. 1.8)

$$\text{(8.4.1)} \qquad \to A^0(D)_{\mathbf{Q}} \to A^1(S')_{\mathbf{Q}} \to A^1(S)_{\mathbf{Q}} \to 0.$$

By **A3**, $A^1(S)_{\mathbf{Q}}$ is generated by $[Y_1], \cdots, [Y_q]$. On the other hand, the classes of the components of D form a free $\mathbf{Q}$-basis of $A^0(D)_{\mathbf{Q}}$. The conclusion follows readily. ◇

8.5. We may inparticular express the classes $[X'_j]$ as rational linear combinations of $[D_1], \ldots, [D_r], [Y'_1], \ldots, [Y'_q]$,

$$\text{(DR)} \qquad [X'_j] = a_{1j}[D_1] + \ldots + a_{rj}[D_r] + b_{1j}[Y'_1] + \ldots + b_{qj}[Y'_q].$$

Any such equation will be called a *degeneration relation* for X'_j. The rational numbers a_{kj}, b_{kj} will be called coefficients of the degeneration relation. A priori they need not be uniquely determined, but in concrete applications they will. Notice that they are uniquely determined if $[D_1], \ldots, [D_r], [Y'_1], \ldots, [Y'_q]$ are $\mathbf{Q}$-linearly independent. Conversely, if the coefficients in a degeneration relation are all non-zero and unique, then $[D_1], \ldots, [D_r], [Y'_1], \ldots, [Y'_q]$ are $\mathbf{Q}$-linearly independent. This is the criterion we shall use to determine $\mathrm{Pic}(S')_{\mathbf{Q}}$ in our examples. We could also proceed observing that the sequence **(8.4.1)** is exact to the left if and only if the map

$$cl_S \colon \mathrm{Pic}(S)_{\mathbf{Q}} \to \mathrm{H}_2(S)_{\mathbf{Q}}$$

is an isomorphism and using the fact that the latter holds, for instance, if S has a cellular decomposition, or even in more general cases (see Rosselló-Xambó [1987]).

8.6. Let

$$d_i \colon D_i \to S'$$

be the inclusions. Then we will write $N_{ij} = \deg(\dot{D}_i \cdot C'_j)$ and we will say that the N_{ij}, $i = 1, \ldots, r$, are the *degeneration numbers* of C_j. Since C'_j is a complete curve, we also have

$$N_{ij} = \deg_{C'_j}[D_i \cdot C'_j] = \deg_{C'_j}(u_j^*[D_i]).$$

8.7. Degeneration lemma.

(a) *$N = \deg_{C'_j}(u_j^*[X'_j])$ for all $j = 1, \ldots, p$.*

(b) *Given a degeneration relation* **DR** *for X'_j, then*

$$N = \textstyle\sum_i a_{ij} N_{ij} + N',$$

for any $i = 1, \ldots, p$, where

$$N' = \textstyle\sum_i b_{ij} \deg_{C'_j}(u_j^*[Y'_j])$$

(so N' does not involve X_j).

(c) *If we let*

$$M_{ij}(k,h) = \Big(\prod_{l \neq j} m_{ilk_l}\Big) \cdot \Big(\prod_l n_{ilh_l}\Big)$$

then we have

$$N_{ij} = \textstyle\sum_{k,h} M_{ij}(k,h) N_{ij}[k,h].$$

Proof:

(a) By definition $N = \deg(X_j \cdot C_j)$, and $N = \deg_{C'_j}(u_j^*(X'_j))$ by **D2**. Now the fact that C'_j is complete implies that $N = \deg_{C'_j}([u_j^* X'_j]) = \deg_{C'_j}(u_j^*[X'_j])$.

(b) It is a direct consequence of **(a)** and the definitions.

(c)
$$\begin{aligned} N_{ij} &= \deg(D_i \cdot C'_j) = \deg d_i^*(C'_j) \\ &= \deg d_i^*(X'_1 \cdots X'_{j-1} \cdot X'_{j+1} \cdots X'_p \cdot Z'_1 \cdots Z'_s), \\ &= \deg d_i^*(X'_1) \cdots d_i^*(X'_{j-1}) \cdot d_i^*(X'_{j+1}) \cdots d_i^*(X'_p) \cdot d_i^*(Z'_1) \cdots d_i^*(Z'_s). \end{aligned}$$

From this, the expression of **D2** and the definitions of $N_{ij}[k,h]$ and $M_{ij}(k,h)$, the stated expression for N_{ij} follows immediately. ◇

8.8. The degeneration lemmma gives a foundation to the "method of degenerations", especially as used by Schubert. The expression of N given in **(b)** breaks up the problem of computing N into *(i)* the determination of the degeneration coefficients, *(ii)* the computation of the degeneration numbers N_{ij} and *(iii)* the computation of the numbers N'. Part **(c)** of the lemma reduces the computation of degeneration numbers into the determination of the varieties X_{ijk} and Z_{ijk}, the multiplicities m_{ijk} and n_{ijk} with which they appear, and the computation of the elementary numbers $N_{ij}[k,h]$. The latter are enumerative problems in a space of dimension $d-1$ and for their determination usually the same method can be applied, so that the whole procedure has a recursive quality. As far as (*iii*) goes, in practice the numbers N' will be easier to compute than the number N itself.

8.9. Part **(a)** of the degeneration lemma *gives p expressions for the number N*. So in particular we have equalities

$$\deg_{C'_j}(u^*_j[X'_j]) = \deg_{C'_{j'}}(u^*_{j'}[X'_{j'}])$$

for any j, j' in $\{1,\ldots,p\}$. Thus if we know degeneration relations **DR** for X'_j and $X'_{j'}$, then we get an equation of the form

$$\textbf{(8.9.1)} \qquad a_{1j}N_{1j} + \ldots a_{rj}N_{rj} + N' = a_{1j'}N_{1j'} + \ldots a_{rj'}N_{rj'} + N''.$$

This yields a necessary condition that the coefficients of the degeneration relations must satisfy. It turns out that in interesting enumerative situations a suitable selection of equations of the form **(8.9.1)** is enough to determine them. If some of the multiplicities m, n that appear in the definition of the degeneration numbers were also unknown, they may as well be left in **(8.9.1)** as integer unknowns.

8.10. Classically degeneration relations were established through the use of "coincidence formulas", which often lead to elusive computations of multiplicities. For example, Schubert's derivation of the 4 degeneration relations for twisted cubics (Schubert [1879], p. 168) has not been made rigorous because of his application of the coincidence formulas (or rather the way he suggests to apply them) leaves undetermined certain fundamental multiplicities. The approach advanced here suffices to determine those degeneration formulas without needing coincidence formulas. Below we will show how to find suitable degeneration relations for the cuspidal cubics.

8.11. Let us discuss how conditions arise. A common way to describe cycles on a variety S which parametrizes a certain kind of figures is by means of geometric relations imposed to the figures ("räumliche Bedingungen" in Schubert's terminology; see Schubert [1879], p. 5). The geometric relations will involve some other kind of figure. When we allow the latter to move we obtain an *algebraic family of cycles* on S. Such algebraic families of cycles are the usual source for supplying conditions in the sense given above.

In order to simplify notations, we shall use the conventions, which go back to Schubert and before, that we explain presently. Suppose S is a smooth variety of dimension d and that X is an algebraic family of cycles on S. Then given an integer n, X^n will mean that we take n (independent) general values of the parameter space of the family and that we consider as conditions the cycles $X_1, \ldots, X_n$ corresponding to those values. Given families

$$X, X', \ldots, Z, Z', \ldots$$

$(X, X', \ldots$ of codimension 1, $Z, Z', \ldots$ of codimension at least 2) and integers

$$n, n', \ldots, m, m', \ldots$$

the expression

$$N = X^n X'^{n'} \cdots Z^m Z'^{m'} \cdots$$

will mean the enumerative problem whose conditions are n general cycles of the family X, n' general cycles of the family X', and so on. In order for the problem to be well posed we need that the sum of the codimensions be equal to d. In the explicit examples the assumptions **A1**, **A2** and **D2** can be ascertained from general principles such as the transversality of the general translates (Kleiman [1974]), or a generalized version in which it is not required that the group acts transitively on S (Casas [1987], Laksov-Speiser [1988]).

In specific examples, the conditions in the list $X, X', \ldots, Z, Z', \ldots$ will be selected so that they express basic geometric relationships that our figures satisfy and will be referred to as *fundamental conditions*. The numbers formed with fundamental conditions will be called *fundamental numbers*. If the only conditions involved are (simple) contact conditions with linear varieties then the numbers are referred to as *characteristic numbers*.

9. Tables of degeneration numbers

In Sections 4-7 we have studied the elementary numbers with respect to the fundamental conditions for cuspidal cubics. With the elementary numbers we can compute the degeneration numbers. In this section we assemble the tables of all degeneration numbers that are needed to compute all fundamental numbers. Each table is labled with a monomial α in the variables c, v, y, q, w, z and the monomials are ordered lexicographically. The numbers to the right of a given D_j are the degeneration numbers of the form $D_j \cdot (X_0^{6-d-i} X_1^i \alpha)$, $i = 0, \ldots, 6-d$, where d is the degree of α, X_0 the condition of going through a point and X_1 of being tangent to a line. Thus there are $7-d$ numbers in each row. A row corresponding to a degeneration is omited if it turns out to be identically 0.

Table 1

D_0	42	87	141	168	141	87	42

Table c

D_0	27	45	54	45	27	12
D_7	0	24	78	78	24	0
D_{12}	0	0	0	36	72	60

Table v

D_0	27	45	54	45	27	12
D_2	45	54	27	0	0	0
D_7	0	24	78	78	24	0
D_{12}	0	0	0	9	18	15

Table y

D_0	27	45	54	45	27	12
D_3	30	36	18	0	0	0
D_7	0	24	78	78	24	0
D_{12}	0	0	0	18	36	30

Table c^2

D_0	5	8	8	5	2
D_7	0	6	21	18	0
D_{12}	0	0	0	12	24

Table cv

D_0	5	8	8	5	2
D_2	18	9	0	0	0
D_5	24	54	36	0	0
D_7	24	60	57	18	0
D_{12}	0	0	27	48	33

Table cy

D_0	5	8	8	5	2
D_3	12	6	0	0	0
D_7	24	60	57	18	0
D_{12}	0	0	18	36	30

Table cz

D_0	7	13	16	13	7
D_1	12	6	0	0	0
D_5	24	54	36	0	0
D_6	0	18	21	6	0
D_7	0	6	21	18	0
D_{10}	0	0	18	30	18
D_{12}	0	0	0	12	24

Table cq

D_0	7	13	16	13	7
D_1	6	3	0	0	0
D_6	0	18	21	6	0
D_7	0	6	21	18	0
D_{11}	0	0	9	18	15
D_{12}	0	0	0	12	24

Table cw

D_0	7	13	16	13	7	D_7	0	6	21	18	0
D_1	24	12	0	0	0	D_9	0	0	36	54	24
D_4	24	54	36	0	0	D_{12}	0	0	0	12	24
D_6	0	18	21	6	0						

Table v^2

D_0	5	8	8	5	2	D_7	0	6	21	18	0
D_2	15	18	9	0	0	D_{12}	0	0	0	3	6

Table vy

D_0	5	8	8	5	2	D_7	24	60	57	18	0
D_2	18	9	0	0	0	D_{12}	0	0	9	18	15
D_3	12	6	0	0	0						

Table y^2

D_0	5	8	8	5	2	D_8	0	0	36	54	24
D_3	18	30	18	0	0	D_{12}	0	0	0	6	12
D_7	0	6	21	18	0						

Table c^2v

D_2	3	0	0	0	D_7	6	15	9	0
D_5	6	15	9	0	D_{12}	0	0	9	15

Table c^2y

D_3	2	0	0	0	D_7	6	15	9	0	D_{12}	0	0	6	10

Table c^2z

D_0	1	2	2	1	D_7	0	1	4	0
D_1	2	0	0	0	D_{10}	0	0	6	8
D_5	6	15	9	0	D_{12}	0	0	0	4
D_6	0	4	1	0					

Table c^2q

D_0	1	2	2	1	D_7	0	1	4	0
D_1	1	0	0	0	D_{11}	0	0	3	3
D_6	0	4	1	0	D_{12}	0	0	0	4

Table c^2w

D_0	1	2	2	1	D_7	0	1	4	0
D_1	4	0	0	0	D_9	0	0	12	18
D_4	6	15	9	0	D_{12}	0	0	0	4
D_6	0	4	1	0					

Table cv^2

D_2	6	3	0	0	D_7	6	15	9	0
D_5	12	24	18	0	D_{12}	0	0	9	12

Table cvy

D_2	3	0	0	0	D_7	24	27	9	0
D_3	2	0	0	0	D_{12}	0	9	18	13
D_5	18	12	0	0					

Table cvz

D_0	1	2	2	1	D_6	0	4	1	0
D_1	2	0	0	0	D_7	6	16	13	0
D_2	6	3	0	0	D_{10}	0	0	6	8
D_5	18	39	27	0	D_{12}	0	0	9	16

Table cvq

D_0	1	2	2	1	D_6	0	4	1	0
D_1	1	0	0	0	D_7	6	16	13	0
D_2	15	9	0	0	D_{11}	0	9	15	9
D_5	6	15	9	0	D_{12}	0	0	9	16

Table cy^2

D_3	10	6	0	0	D_8	0	18	24	12
D_7	6	15	9	0	D_{12}	0	0	6	8

Table cyz

D_0	1	2	2	1	D_6	0	4	1	0
D_1	2	0	0	0	D_7	6	16	13	0
D_3	10	6	0	0	D_{10}	0	9	15	8
D_5	18	12	0	0	D_{12}	0	0	6	12

Table v^2y

D_2	6	3	0	0	D_7	6	15	9	0
D_3	2	0	0	0	D_{12}	0	0	3	4

Table v^2q

D_0	1	2	2	1	D_7	0	1	4	0
D_1	1	0	0	0	D_{11}	0	0	9	12
D_2	9	15	9	0	D_{12}	0	0	0	1
D_6	0	4	1	0					

Table vy^2

D_2	3	0	0	0	D_8	0	9	15	6
D_3	10	6	0	0	D_{12}	0	0	3	5
D_7	6	15	9	0					

Table vyz

D_0	1	2	2	1	D_6	0	4	1	0
D_1	2	0	0	0	D_7	6	16	13	0
D_2	6	3	0	0	D_{10}	0	9	15	8
D_3	10	6	0	0	D_{12}	0	0	3	6
D_5	18	12	0	0					

Table y^2z

D_0	1	2	2	1	D_7	0	1	4	0
D_1	2	0	0	0	D_8	0	0	12	18
D_3	8	15	9	0	D_{10}	0	0	6	8
D_6	0	4	1	0	D_{12}	0	0	0	2

Table $c^2v^2 = c^2vz$

D_2	1	0	0	D_7	1	3	0
D_5	4	9	9	D_{12}	0	0	3

Table c^2vy

D_5	5	3	0	D_7	5	3	0	D_{12}	0	3	5

Table c^2vq

D_2	3	0	0	D_7	1	3	0	D_{12}	0	0	3
D_5	1	3	0	D_{11}	0	3	2				

Table c^2vw

D_2	1	0	0	D_5	4	9	9	D_9	0	3	5
D_4	5	3	0	D_7	1	3	0	D_{12}	0	0	3

Table c^2y^2

D_3	2	0	0	D_8	0	6	4
D_7	1	3	0	D_{12}	0	0	2

Table c^2yz

D_3	2	0	0	D_7	1	3	0	D_{12}	0	0	2
D_5	5	3	0	D_{10}	0	3	3				

Table c^2yw

D_3	2	0	0	D_7	1	3	0	D_9	0	3	5
D_4	5	3	0	D_8	0	6	4	D_{12}	0	0	2

Table c^2zq

D_1	1	0	0	D_6	0	3	1	D_{11}	0	0	1
D_5	1	3	0	D_{10}	0	0	2				

Table c^2zw

D_1	2	0	0	D_5	4	9	9	D_9	0	0	4
D_4	1	3	0	D_6	0	3	1	D_{10}	0	0	2

Table c^2qw

D_1	3	0	0	D_6	0	3	1	D_9	0	0	4
D_4	1	3	0	D_8	0	6	4	D_{11}	0	0	1

Table c^2w^2

D_4	5	3	0	D_9	0	3	5

Table cv^2y

D_2	1	0	0	D_7	5	3	0
D_5	8	6	0	D_{12}	0	3	4

Table cvy^2

D_3	2	0	0	D_7	5	3	0	D_{12}	0	3	3
D_5	4	0	0	D_8	9	9	4				

Table $cvyz$

D_2	1	0	0	D_5	13	9	0	D_{10}	0	3	3
D_3	2	0	0	D_7	6	6	0	D_{12}	0	3	6

Table $cvyq$

D_2	3	0	0	D_7	6	6	0	D_{11}	0	3	2
D_3	2	0	0	D_8	9	9	4	D_{12}	0	3	6
D_5	5	3	0								

Table v^2y^2

D_2	1	0	0	D_7	1	3	0	D_{12}	0	0	1
D_3	2	0	0	D_8	0	3	1				

Table v^2yz

D_2	1	0	0	D_5	8	6	0	D_{10}	0	3	3
D_3	2	0	0	D_7	1	3	0	D_{12}	0	0	1

Table vy^2z

D_2	1	0	0	D_7	1	3	0	D_{10}	0	3	3
D_3	5	3	0	D_8	0	3	5	D_{12}	0	0	1
D_5	4	0	0								

Table c^2v^2y

D_5	3	3	D_7	1	0	D_{12}	0	1

10. Degeneration relations

In next theorem we state the degeneration expressions of the first order conditions for cuspidal cubics and then we indicate how they can be obtained by application of the procedure explained in section 8. Here we see that $\mathrm{Pic}(S)_{\mathbf{Q}}$ is generated by c (see **1.3**) and hence $\mathrm{Pic}(S')_{\mathbf{Q}}$ is generated by c and the 13 degenerations.

10.1. Theorem. *Let $D = D_1 + D_2 + D_3$ and $D' = D_{10} + D_{11} + D_{12}$. Then the expressions on S' of the first order conditions in terms of c and the first order degenerations is as follows:*

1) $5X_0 = 3c + 2D_0 + 3D + 6D_4 + 2D_5 + 3D_6 + 4D_7 + 3D_8 + 9D_9 + 9D'$.

2) $5X_1 = -3c + 8D_0 + 12D + 9D_4 + 3D_5 + 7D_6 + 6D_7 + 2D_8 + 6D_9 + 6D'$.

3) $5v = -4c + 9D_0 + 6D_1 + D_2 + 6D_3 + 2D_4 - D_5 + 6D_6 + 3D_7 + D_8 + 3D_9 + 3D'$.

4) $5y = -c + 6D_0 + 4D_1 + 4D_2 - D_3 + 3D_4 + D_5 + 4D_6 + 2D_7 - D_8 + 2D_9 + 2D'$.

5) $5z = c + 4D_0 + D + 2D_4 - D_5 + D_6 + 3D_7 + D_8 + 3D_9 - 2D_{10} + 3D_{11} + 3D_{12}$

6) $5q = 4c + D_0 - D + 3D_4 + D_5 - D_6 + 2D_7 - D_8 + 2D_9 + 2D_{10} - 3D_{11} + 2D_{12}$.

7) $w = -c + 2D_0 + D + D_6 + D_7 + D'$.

Here is the same information in matrix form:

	D_0	D_1	D_2	D_3	D_4	D_5	D_6	D_7	D_8	D_9	D_{10}	D_{11}	D_{12}	c
$5X_0$	2	3	3	3	6	2	3	4	3	9	9	9	9	3
$5X_1$	8	12	12	12	9	3	7	6	2	6	6	6	6	−3
$5v$	9	6	1	6	2	−1	6	3	1	3	3	3	3	−4
$5y$	6	4	4	−1	3	1	4	2	−1	2	2	2	2	−1
$5z$	4	1	1	1	2	−1	1	3	1	3	−2	3	3	1
$5q$	1	−1	−1	−1	3	1	−1	2	−1	2	2	−3	2	4
w	2	1	1	1	0	0	1	1	0	0	1	1	1	−1

10.1.1. Remark. If we take into account only the degeneration D_0, which is enough to compute the characteristic numbers (see Table **1** in Section 9), then the relations above for X_0 and X_1 become the following:

$$5X_0 = 3c + 2D_0, \quad 5X_1 = -3c + 8D_0.$$

These relations were obtained for the first time, using coincidence formulas, by Zeuthen [1872] and were recently verified by Kleiman-Speiser [1986]. Notice that a priori we know, by **1.3**, that $5X_0$ and $5X_1$ are linear combinations of c and the degenerations with integer coefficients.

Proof: The proof of the seven degeneration relations can be done by a judicious choice of equations of the form **8.9**. To write such equations we need to know enough degeneration numbers. Those that will be used are contained in the tables given in the preceeding section. Since the procedure is straightforward, here we will prove only the first two relations. We shall write a_i and a to denote the coefficients of X_0 with respect to D_i and c and b_i and b for the coefficients of X_1.

We want to determine the values of $a, a_0, \ldots, a_{12}, b, b_0, \ldots, b_{12}$. To this end first notice that $X_0^5c^2 = 2$ and $X_0^4X_1c^2 = 8$. From these relations we obtain, taking into account the degeneration numbers given in Table c^2 and using **8.7 (b)**, the equations $5a_0 = 2$, $5b_0 = 8$, $8a_0 + 6a_7 = 8$. Hence

$$a_0 = 2/5, b_0 = 8/5, a_7 = 4/5.$$

In what follows we briefly point out what relation we take, the equations it leads to and the value of the coefficients they determine.

From $X_0(X_0^2X_1^2c^2) = X_1(X_0^3X_1c^2)$ we get the relation $8a_0 + 21a_7 = 8b_0 + 6b_7$. So

$$b_7 = 6/5.$$

From $X_0(X_0X_1^3c^2) = X_1(X_0^2X_1^2c^2)$ we get the relation $5a_0+15a_7+12a_{12} = 8b_0+21b_7$, and so

$$a_{12} = 9/5.$$

From $X_0(X_1^4c^2) = X_1(X_0X_1^3c^2)$ we get the relation $2a_0 + 24a_{12} = 5b_0 + 18b_7 + 12b_{12}$, and so

$$b_{12} = 6/5.$$

As a corollary we get, using **8.7 (b)**, the following numbers:

$$c^2 = 2,\ 8,\ 20,\ 38,\ 44,\ 32.$$

[By this we mean the numbers $X_0^{5-i}X_1^ic^2$, $i = 0, \dots, 5$].

Using table c and and the numbers for c^2 just obtained we can determine the coefficients a and b. In fact, from the relation $X_0(X_0^4X_1c) = X_1(X_0^5c)$ we get the equation $8a + 45a_0 + 24a_7 = 2b + 27b_0$. Similarly, from the relation $X_0(X_0^3X_1^2c) = X_1(X_0^4X_1c)$ we get the equation $20a + 54a_0 + 78a_7 = 8b + 45b_0 + 24b_7$. Solving for a and b we obtain

$$a = -b = 3/5.$$

From $X_0(X_1^3c^2v) = X_1(X_0X_1^2c^2v)$ we obtain $15a_{12} = 9b_5 + 9b_7 + 9b_{12}$ and so

$$b_5 = 3/5.$$

From $X_0(X_0X_1^2c^2v) = X_1(X_0^2X_1c^2v)$ we obtain $9a_5 + 9a_7 + 9a_{12} = 15b_5 + 15b_7$ which implies that

$$a_5 = 2/5.$$

From $X_0(X_0^2X_1c^2v) = X_1(X_0^3c^2v)$ we obtain $15a_5 + 15a_7 = 3b_2 + 6b_5 + 6b_7$ which implies that

$$b_2 = 12/5.$$

As a corollary we obtain the following numbers:

$$c^2v = 9,\ 18,\ 27,\ 27,\ 18.$$

Using table cv and the numbers for c^2v just obtained we can determine a_2. From the relation $X_0(X_0^3X_1cv) = X_1(X_0^4cv)$ we obtain $60a_7 + 9a_2 + 54a_5 + 8a_0 + 18a = 24b_7 + 18b_2 + 24b_5 + 5b_0 + 9b$ and so

$$a_2 = 3/5.$$

From $X_1(X_0^3c^2y) = X_0(X_0^2X_1c^2y)$ and the table of c^2y we get $6b_7 + 2b_3 = 15a_7$ and hence

$$b_3 = 12/5.$$

From $X_1(X_0^2c^2y^2) = X_0(X_0X_1c^2y^2)$ we obtain $b_7 + 2b_3 = 3a_3 + 6a_8$ and hence

$$a_8 = 3/5.$$

From $X_1(X_0X_1c^2y^2) = X_0(X_1^2c^2y^2)$ we obtain $3b_7 + 6b_8 = 2a_{12} + 4a_8$, and so

$$b_8 = 2/5.$$

Now we have $X_1X_0^2c^2y^2 = 6$ and $X_0^3c^2y^2 = a_7 + 2a_3$.

From the relation $X_0(X_0^2X_1cy^2) = X_1(X_0^3cy^2)$ we obtain $15a_7 + 6a_3 + 18a_8 + 6a = 6b_7 + 10b_3 + b(a_7 + 2a_3)$, so

$$a_3 = 3/5.$$

From $X_0(X_0^2X_1c^2z) = X_1(X_0^3c^2z)$ and $X_0(X_0^2X_1c^2q) = X_1(X_0^3c^2q)$ we obtain

$$\left.\begin{aligned} 4a_6 + a_7 + 15a_5 + 2a_0 &= 2b_1 + 6b_5 + b_0 \\ 4a_6 + a_7 + 2a_0 &= b_1 + b_0 \end{aligned}\right\}$$

which yields

$$b_1 = 12/5, \quad a_6 = 3/5.$$

Now we have $X_0(X_0^3c^2q) = a_0 + a_1$, $X_0^3X_1c^2q = 4$, $X_0^2X_1^2c^2q = 10$.

From $X_1(X_0^4cq) = X_0(X_0^3X_1cq)$ we obtain $7b_0 + 6b_1 + b(a_0 + a_1) = 18a_6 + 6a_7 + 3a_1 + 13a_0 + 4a$, and so

$$a_1 = 3/5.$$

From $X_1(X_0^3X_1cq) = X_0(X_0^2X_1^2cq)$, $X_1(X_0^2X_1c^2q) = X_0(X_0X_1^2c^2q)$, and $X_1(X_0X_1^2c^2q) = X_0(X_1^3c^2q)$, we obtain

$$\left.\begin{aligned} 18b_6 + 6b_7 + 3b_1 + 13b_0 + 4b &= 21a_6 + 21a_7 + 9a_{11} + 16a_0 + 10a \\ 4b_6 + b_7 + 2b_0 &= a_6 + 4a_7 + 3a_{11} + 2a_0 \\ b_6 + 4b_7 + 3b_{11} + 2b_0 &= 3a_{11} + 4a_{12} + a_0 \end{aligned}\right\}$$

Solving for b_6, a_{11} and b_{11} we obtain

$$b_6 = 7/5, \quad a_{11} = 9/5, \quad b_{11} = 6/5.$$

From $X_1(X_0^2X_1c^2z) = X_0(X_0X_1^2c^2z)$ we obtain $4b_6 + b_7 + 15b_5 + 2b_0 = a_6 + 4a_7 + 6a_{10} + 9a_5 + 2a_0$, and so

$$a_{10} = 9/5.$$

From $X_1(X_0X_1^2c^2z) = X_0(X_1^3c^2z)$ we obtain $b_6 + 4b_7 + 6b_{10} + 9b_5 + 2b_0 = 4a_{12} + 8a_{10} + a_0$, and so

$$b_{10} = 6/5.$$

From $X_1(X_0^3c^2w) = X_0(X_0^2X_1c^2z)$ and $X_1(X_0^2c^2qw) = X_0(X_0X_1c^2qw)$ we obtain

$$\left.\begin{aligned} 4b_1 + 6b_4 + b_0 &= 4a_6 + a_7 + 15a_4 + 2a_0 \\ 3b_1 + b_4 &= 3a_6 + 3a_4 + 6a_8 \end{aligned}\right\}$$

Solving for a_4 and b_4 we obtain

$$a_4 = 6/5, b_4 = 9/5.$$

From $X_1(X_0^2X_1c^2w) = X_0(X_0X_1^2c^2w)$ we obtain $4b_6 + b_7 + 15b_4 + 2b_0 = a_6 + 4a_7 + 9a_4 + 12a_9 + 2a_0$, and so

$$a_9 = 9/5.$$

From $(X_0X_1^2c^2w) = X_0(X_1^3c^2w)$ we obtain $b_6 + 4b_7 + 9b_4 + 12b_9 + 2b_0 = 4a_{12} + 18a_9 + a_0$, and so

$$b_9 = 6/5.$$

11. Fundamental numbers

Once we know degeneration relations for the first order conditions and the degeneration numbers, the computation of fundamental numbers is reduced to arithmetic operations (see **8.7 (b)**). This has been applied in the proof of **10.1** to find several fundamental numbers that were needed along the way. Here we include a couple of examples that will further illustrate the use of **8.7**.

11.1. $N' = X_0^3 c^2 v^2$

Since $X_0^2 c^2 v^2$ only contains degenerations of type D_2, D_5 and D_7 (see Table c^2v^2 in section 9), with degeneration numbers 1, 4 and 1, respectively, we have, by **10.1 (1)**, that

$$N' = a_2 + 4a_5 + a_7 = (3 + 8 + 4)/5 = 3.$$

Notice that the term $\frac{3}{5}c$ in the expression of X_0 does not give any contribution to N', because numbers with c^3 are 0 (see **8.8**).

11.2. $N = X_0^4 cv^2$

Since $X_0^3 cv^2$ only contains degenerations of type D_2, D_5 and D_7 (see Table cv^2 in section 9), with degeneration numbers 6, 12 and 6, respectively, we have, by **10.1 (1)**, that

$$N = 6a_2 + 12a_5 + 6a_7 + aN' = (18 + 24 + 24 + 9)/5 = 15.$$

The value of this number that we find in Schubert [1879] (p. 141, line 4) is 17. This looks like a misprint, rather than a mistake, for on p. 138, line −11, we find that the value given to the dual number is 15.

11.3. $M'' = X_0^2 c^2 vyz$

Here it is not hard to see that $X_0 c^2 vyz = X_0 c^2 v^2 y$ and hence this only contains degenerations of type D_5 and D_7 (see Table c^2v^2y in section 9), with degeneration numbers 3 and 1, respectively. Therefore we have, by **10.1 (1)**, that

$$M'' = 3a_5 + a_7 = (6 + 4)/5 = 2.$$

11.4. $M' = X_0^3 cvyz$

Since $X_0^2 cvyz$ only contains degenerations of type D_2, D_3, D_5 and D_7 (see Table $cvyz$ in section 9), with degeneration numbers 1, 2, 13 and 6, respectively, we have, by **10.1 (1)**, that

$$M' = a_2 + 2a_3 + 13a_5 + 6a_7 + aM'' = (3 + 6 + 26 + 24 + 6)/5 = 13.$$

11.5. $M = X_0^3 X_1 vyz$

Here $X_0^3 vyz$ contains degenerations of type D_0, D_1, D_2, D_3, D_5 and D_7 (see Table vyz in section 9), with degeneration numbers 1, 2, 6, 10, 18 and 6, respectively, we have, by **10.1 (2)**, that

$$M = b_0 + 2b_1 + 6b_2 + 10b_3 + 18b_5 + 6b_7 + bM' = (8+24+72+120+54+36-39)/5 = 55.$$

This is one of the numbers that we can not find in Schubert's book.

12. Old and new tables of fundamental numbers of cuspidal cubics

Here we collect the values of all non-zero fundamental numbers (see the Remarks at the end). They have been calculated, as illustrated in the preceeding section, by means of formula **8.7 (b)**, using the degeneration formulas **10.1** (basically **(1)** and **(2)**). Most have been calculated in more than one way. Those not listed in Schubert [1879] (nor anywhere else, as far as we know) are distinguished with a **. A few numbers are marked with *; this means that their value can be deduced from some table of Schubert corresponding to space cuspidal cubics. The arrangement of the tables is as follows. A number like $M = X_0^3 X_1 vyz$ is located at the second place of the row that begins with $vyz =$. The row ends with $= yzq$ because by duality M is equal to $X_0 X_1^3 qzy$. The rows are ordered lexicographically by the leading monomials. To the monomial **1** there corresponds the list of *characteristic numbers*:

$$24, 60, 114, 168, 168, 114, 60, 24.$$

Order 1

$c =$	12	42	96	168	186	132	$72 = w$
$v =$	66	123	177	168	105	51	$18 = q$
$y =$	48	96	150	168	132	78	$36 = z$

Order 2

$c^2 =$	2	8	20	38	44	$32 = w^2$
$cv =$	47	89	128	119	71	$32 = qw$
$cy =$	32	62	92	92	62	$32 = zw$
$^{**}cz =$	22	52	94	112	88	$52 = yw$
$^{*}cq =$	7	25	58	85	79	$52 = vw$
$cw =$	52	106	166	166	106	$52 = cw$
$v^2 =$	20	35	47	38	17	$5 = q^2$
$vy =$	59	89	92	65	35	$14 = zq$
$^{**}vz =$	40	79	121	112	61	$25 = yq$
$vq =$	34	79	139	139	79	$34 = vq$
$y^2 =$	20	44	74	74	44	$20 = z^2$
$yz =$	34	70	112	112	70	$34 = yz$

Order 3

$c^2v =$	9	18	27	27	18	$= qw^2$
$c^2y =$	6	12	18	18	12	$= zw^2$
$c^2z =$	4	10	19	22	16	$= yw^2$
$c^2q =$	1	4	10	13	10	$= vw^2$
$c^2w =$	10	22	37	40	28	$= cw^2$
$cv^2 =$	|15|	27	36	27	9	$= q^2w$
$cvy =$	33	48	45	27	12	$= zqw$
**$cvz =$	19	37	55	49	25	$= yqw$
**$cvq =$	19	49	64	49	28	$= vqw$
*$cvw =$	43	67	73	49	19	$= cqw$
$cy^2 =$	12	30	36	24	12	$= z^2w$
**$cyz =$	22	46	55	40	22	$= yzw$
**$cyq =$	13	34	46	37	22	$= vzw$
**$cyw =$	40	70	73	46	22	$= czw$
$cz^2 = c^2z =$	4	10	19	22	16	$= y^2w$
**$czq =$	7	19	37	43	31	$= vyw$
$cq^2 = c^2q =$	1	4	10	13	10	$= v^2w$
$v^2y =$	15	21	18	9	3	$= zq^2$
$v^2z =$	10	19	28	22	7	$= yq^2$
$v^2q =$	10	22	37	31	10	$= vq^2$
$vy^2 =$	21	30	27	15	6	$= z^2q$
**$vyz =$	31	55	55	31	13	$= yzq$
**$vyq =$	31	61	64	37	16	$= vzq$
$vz^2 = v^2z =$	10	19	28	22	7	$= y^2q$
$y^2z =$	10	22	37	34	16	$= yz^2$

Order 4

$c^2v^2 =$	3	6	9	9	$= q^2w^2$
$c^2vy =$	6	9	9	6	$= zqw^2$
$c^2vz = c^2v^2 =$	3	6	9	9	$= yqw^2$
$c^2vq =$	3	9	9	6	$= vqw^2$
*$c^2vw =$	9	15	18	15	$= cqw^2$
$c^2y^2 =$	2	6	6	4	$= z^2w^2$
$c^2yz =$	4	9	9	6	$= yzw^2$
$c^2yq = c^2y^2 =$	2	6	6	4	$= vzw^2$
**$c^2yw =$	8	15	15	10	$= czw^2$
$c^2zq =$	1	3	6	5	$= vyw^2$
$c^2zw =$	4	9	15	14	$= cyw^2$
$c^2qw =$	3	9	12	9	$= cvw^2$
$c^2w^2 =$	6	9	9	6	$= c^2w^2$
$cv^2y =$	9	12	9	3	$= zq^2w$
$cv^2z = c^2v^2 =$	3	6	9	9	$= yq^2w$
**$cv^2q =$	6	15	18	12	$= vq^2w$
$cv^2w =$	9	12	9	3	$= cq^2w$

$cvy^2 =$	14	15	9	4	$= z^2qw$
$^{**}cvyz =$	13	21	18	9	$= yzqw$
$^{**}cvyq =$	17	24	18	10	$= vzqw$
$^{**}cvyw =$	23	27	18	7	$= czqw$
$cvz^2 = c^2v^2 =$	3	6	9	9	$= y^2qw$
$^{**}cvzq =$	7	18	24	17	$= vyqw$
$^{**}cvzw =$	13	21	24	17	$= cyqw$
$cvq^2 = c^2vq =$	3	9	9	6	$= v^2qw$
$^{**}cvqw =$	21	33	33	21	$= cvqw$
$^{**}cy^2z =$	6	15	15	8	$= yz^2w$
$cy^2q = c^2y^2 =$	2	6	6	4	$= vz^2w$
$cy^2w =$	14	15	9	4	$= cz^2w$
$cyz^2 = c^2yz =$	4	9	9	6	$= y^2zw$
$^{**}cyzq =$	7	18	21	13	$= vyzw$
$^{**}cyzw =$	16	30	30	16	$= cyzw$
$cyq^2 = c^2y^2 =$	2	6	6	4	$= v^2zw$
$cz^2q = c^2qz =$	1	3	6	5	$= vy^2w$
$czq^2 = c^2qz =$	1	3	6	5	$= v^2yw$
$v^2y^2 =$	5	6	3	1	$= z^2q^2$
$v^2yz =$	7	12	9	3	$= yzq^2$
$^{**}v^2yq =$	8	15	12	4	$= vzq^2$
$v^2zq =$	4	9	15	11	$= vyq^2$
$v^2q^2 =$	3	9	9	3	$= v^2q^2$
$^{**}vy^2z =$	9	15	12	5	$= yz^2q$
$vy^2q =$	11	15	9	4	$= vz^2q$
$vyz^2 = v^2yz =$	7	12	9	3	$= y^2zq$
$^{**}vyzq =$	13	27	27	13	$= vyzq$
$y^2z^2 =$	4	9	9	4	$= y^2z^2$

Order 5

$c^2v^2y =$	2	3	3	$= zq^2w^2$
$c^2v^2q =$	1	3	3	$= vq^2w^2$
$c^2v^2w =$	2	3	3	$= cq^2w^2$
$c^2vy^2 =$	3	3	2	$= z^2qw^2$
$c^2vyz = c^2v^2y =$	2	3	3	$= yzqw^2$
$c^2vyq = c^2vy^2 =$	3	3	2	$= vzqw^2$
$^{**}c^2vyw =$	5	6	5	$= czqw^2$
$c^2vzq = c^2v^2q =$	1	3	3	$= vyqw^2$
$c^2vzw = c^2v^2w =$	2	3	3	$= cyqw^2$
$^{*}c^2vqw =$	4	6	5	$= cvqw^2$
$c^2vw^2 = c^2v^2w =$	2	3	3	$= c^2qw^2$
$c^2y^2z =$	1	3	2	$= yz^2w^2$
$c^2y^2w =$	3	3	2	$= cz^2w^2$
$c^2yzq = c^2y^2z =$	1	3	2	$= vyzw^2$
$^{**}c^2yzw =$	3	6	5	$= cyzw^2$

$c^2yqw = c^2y^2q =$	3	3	2	$= cvzw^2$
$c^2yw^2 = c^2y^2w =$	3	3	2	$= c^2zw^2$
$c^2zqw =$	1	3	4	$= cvyw^2$
$cv^2y^2 =$	4	3	1	$= z^2q^2w$
$cv^2yz = c^2v^2y =$	2	3	3	$= yzq^2w$
$^{**}cv^2yq =$	5	6	4	$= vzq^2w$
$cv^2yw = cv^2y^2 =$	4	3	1	$= czq^2w$
$cv^2zq = c^2v^2q =$	1	3	3	$= vyq^2w$
$cv^2zw = c^2v^2w =$	2	3	3	$= cyq^2w$
$cv^2q^2 = c^2v^2q =$	1	3	3	$= v^2q^2w$
$^{*}cv^2qw = cvqw^2 =$	5	6	4	$= cvq^2w$
$^{**}cvy^2z =$	5	6	3	$= yz^2qw$
$cvy^2q = c^2vy^2 =$	3	3	2	$= vz^2qw$
$cvy^2w = cv^2y^2 =$	4	3	1	$= cz^2qw$
$cvyz^2 = c^2v^2y =$	2	3	3	$= y^2zqw$
$^{**}cvyzq =$	6	9	6	$= vyzqw$
$^{**}cvyzw =$	7	9	6	$= cyzqw$
$cvyq^2 = c^2vy^2 =$	3	3	2	$= v^2zqw$
$^{**}cvyqw =$	8	9	6	$= cvzqw$
$cvz^2q = c^2v^2q =$	1	3	3	$= vy^2qw$
$cvz^2w = c^2v^2w =$	2	3	3	$= cy^2qw$
$cvzq^2 = c^2v^2q =$	1	3	3	$= v^2yqw$
$cy^2z^2 = c^2y^2z =$	1	3	2	$= y^2z^2w$
$cy^2zq = c^2y^2z =$	1	3	2	$= vyz^2w$
$^{**}cy^2zw = cyzw^2 =$	5	6	3	$= cyz^2w$
$cyz^2q = c^2y^2z =$	1	3	2	$= vy^2zw$
$cyzq^2 = c^2y^2z =$	1	3	2	$= v^2yzw$
$v^2y^2z =$	2	3	1	$= yz^2q^2$
$v^2y^2q =$	3	3	1	$= vz^2q^2$
$^{**}v^2yzq =$	3	6	4	$= vyzq^2$
$v^2yq^2 = v^2y^2q =$	3	3	1	$= v^2zq^2$
$vy^2z^2 = v^2y^2z =$	2	3	1	$= y^2z^2q$
$^{**}vy^2zq = vyzq^2$	4	6	3	$= vyz^2q$

12.1. Remark. For any condition α in the list $\{c, v, y, z, q, w\}$, it is clear that if a fundamental number N contains α^3, then $N = 0$. We may conveniently phrase this by writing $\alpha^3 = 0$. Similarly, if (α, β) is any pair on the list

$$\{(c,q),(c,z),(v,z),(v,w),(y,w),(y,q)\},$$

then $\alpha^2\beta^2 = 0$, for whenever α and β refer to incident elements of the singular triangle we cannot fix both independently. Finally it is also clear that if (α, β) is a pair of distinct vertices or sides of the singular triangle and γ is the side or vertex defined by the pair, then $\alpha^2\beta^2\gamma = 0$.

12.2. Remark. In the tables above we have used identities of the form $\alpha^2\beta = \alpha\beta^2$, which is valid for any pair (α, β) on the list

$$\{(c,q),(c,z),(v,z),(v,w),(y,w),(y,q)\},$$

inasmuch as they are valid for triangles.

12.3. Remark. We have not listed the table corresponding to order 6. In this case, if the order six monomial involves at least one square and it is not in one of the cases in 12.1, or amenable to such a case by 12.2, then the row corresponding to it is (1,1), for it is not hard to see that such a monomial fixes the singular triangle. On the other hand, the list corresponding to the unique square free monomial $cvyzqw$ is (2,2), for there are 2 triangles satisfying this condition. In any case, the cuspidal cubics having a given triangle as a singular triangle form a pencil and so there is a unique cubic in it going through a point or (by duality) tangent to a line (cf. Schubert [1879], Remark on top of p. 143).

12.4. Remark. For reasons of dimensions, it is clear that all monomials of degree 7 not involving X_0 and X_1 are 0.

12.5. Remark. It turns out that the fundamental numbers which do not satisfy one of the vanishing conditions given in the preceeding remarks are automatically non-zero.

REFERENCES

Casas, E. [1987], *A transversality theorem and an enumerative calculus for proper solutions*, Preprint, 1987.

Fulton, W. [1984], *Intersection Theory*, Ergebnisse NF 2, Springer–Verlag, 1984.

Kleiman, S. [1974], *The transversality of a general translate*, Compositio Math. 38 (1974), 287-297.

Kleiman, S.; Speiser, R. [1986], *Enumerative geometry of cuspidal plane cubics*, Proceedings Vancouver Conference in Algebraic Geometry 1984 (eds. Carrell, Geramita and Russell), CMS-AMS Conf. Proc. Vol 6, 1986.

Laksov, D.; Speiser, R. [1987], *Transversality criteria in any characteristic*, Preprint, 1987.

Maillard, S. [1871], *Récherche des charactéristiques des systèmes élémentaires de courbes planes du troisième ordre*, Thesis, Paris, publ. by Cusset (1871).

Miret, J. M.; Xambó, S. [1987], *On Schubert's degenerations of cuspidal plane cubics*, Preprint Univ. of Barcelona, 1987.

Rosselló, F.; Xambó, S. [1987], *Computing Chow groups*, in: Algebraic Geometry Sundance 1986, LN in Math. 1311, 220-234.

Sacchiero, G. [1984], *Numeri caratteristici delle cubiche piane cuspidale*, Preprint Univ. di Roma II (1984).

Schubert, H. C. H. [1879], *Kalkül der abzählenden der Geometrie*, Teubner, Leipzig, 1879 (reprinted by Springer-Verlag, 1979).

Zeuthen, H. [1872], *Détermination des charactéristiques des systèmes elémentaires des cubiques*, CR. Acad. Sc. Paris 74, 521-526.

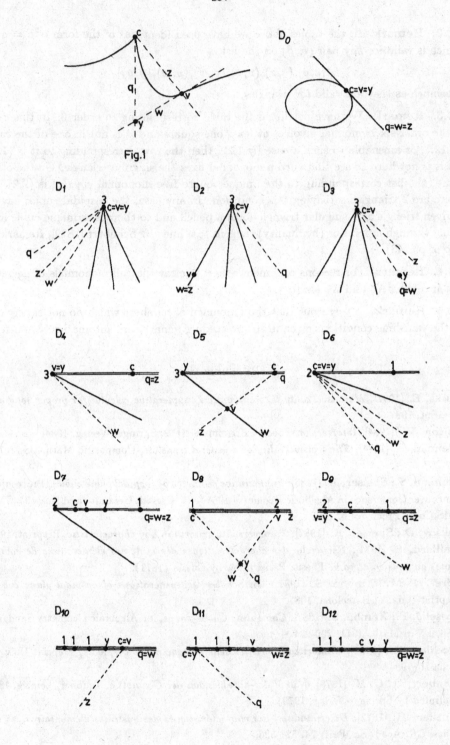

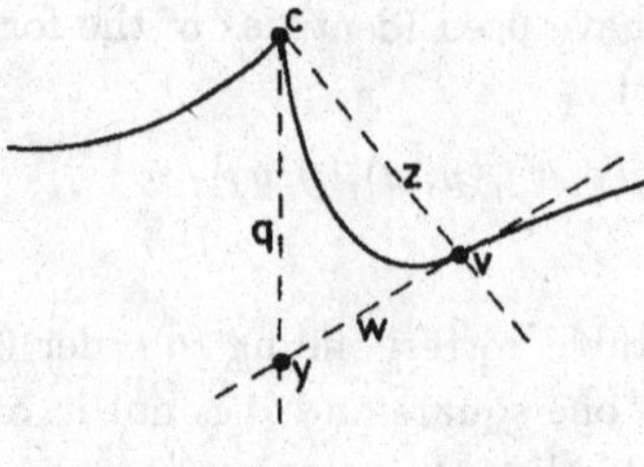
c
z
q
v
w
y

Fig.1

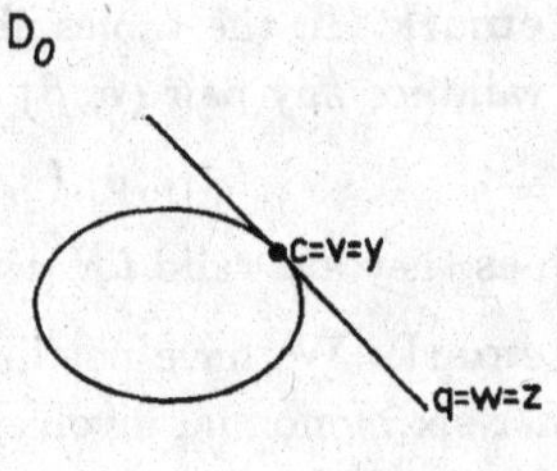
D0
c=v=y
q=w=z

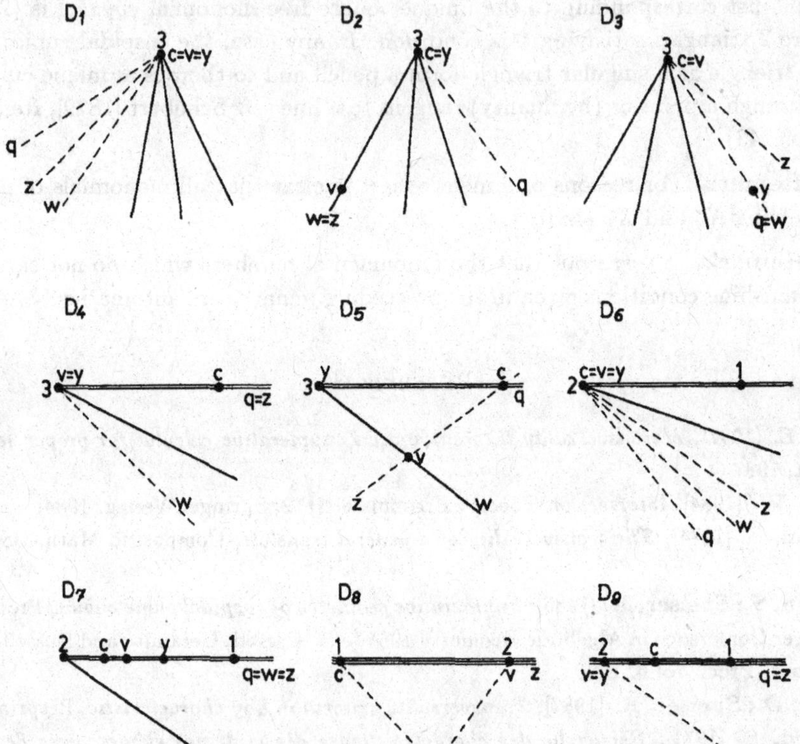
D1
3
c=v=y
q
z
w
D2
3
c=y
v
w=z
q
D3
3
c=v
z
y
q=w
D4
v=y
c
3
q=z
w
D5
y
c
3
q
y
z
w
D6
c=v=y
1
2
z
w
q
D7
2
c
v
y
1
q=w=z
D8
1
2
c
v
z
y
w
q
D9
2
c
1
v=y
q=z
w
D10
1 1 1
y
c=v
q=w
z
D11
1 1 1
v
c=y
w=z
q
D12
1 1 1
c v y
q=w=z

UNOBSTRUCTED ARITHMETICALLY BUCHSBAUM CURVES

ROSA M. MIRÓ–ROIG

FACULTAD DE MATEMÁTICAS. UNIVERSIDAD DE BARCELONA
GRAN VIA 585. 08007 BARCELONA. SPAIN.

INTRODUCTION

This paper contains the material of a talk that the author gave at the conference "Projective curves and Algebraic Geometry" at Cognola (Trento, 1988). The author is very grateful to the organizers for their generous hospitality.

Let X be a curve in $\mathbb{P}^3$. We say that X is unobstructed if the corresponding point of the Hilbert scheme is smooth; otherwise X is called obstructed. A geometric characterization of unobstructedness is not known even for smooth space curves, but several examples of obstructed smooth curves in $\mathbb{P}^3$ are known (see for instance |Mu|, |S|, |EF|, |K1|, |E|).

In the past few years, the subject of arithmetically Buchsbaum curves, as a natural extension of arithmetically Cohen–Macaulay curves, has recieved much attention. In |Eℓ|, Ellingsrud proved that arithmetically Cohen–Macaulay curves are unobstructed. However, this is not true for arithmetically Buchsbaum curves (cf. |EF|) and, in |EF 1|, Ellia–Fiorentini considered the following problem:

PROBLEM 1. To characterize unobstructed arithmetically Buchsbaum curves.

In particular,

PROBLEM 2. Is any arithmetically Buchsbaum curve of maximal rank unobstructed?

The goal of this work is to give sufficient conditions on the numerical character of an arithmetically Buchsbaum curve of maximal rank in order to assure that it is unobstructed (Cf. Theorem 2.1 and Theorem 2.2).

The first section is primarily a review of the results about arithmetically Buchsbaum curves needed later. The heart of this paper is § 2 where we prove the main results.

§ 1. PRELIMINARIES

Let k be an algebraically closed field of characteristic zero, $S = k\ |X_0, X_1, X_2, X_3|$, $\mathfrak{m} = (X_0, X_1, X_2, X_3)$ and $\mathbb{P}^3 = Proj(S)$. By a curve we mean a closed one–dimensional subscheme of $\mathbb{P}^3$ which is locally Cohen–Macaulay and equidimensional. To a curve $C \subset \mathbb{P}^3$

we associate the Hartshorne–Rao module $M(C) = \oplus_{t \in Z} H^1(\mathbb{P}^3, \mathcal{I}_C(t))$, which is a graded S–module of finite lenght. A curve $C \subset \mathbb{P}^3$ is called arithmetically Buchsbaum (briefly a. B.) if the maximal ideal $\mathfrak{m}$ of S annihilates $M(C)$. A curve $C \subset \mathbb{P}^3$ is said to have maximal rank if the restriction map $H^0(\mathbb{P}^3, \mathcal{O}_{\mathbb{P}^3}(t)) \to H^0(C, \mathcal{O}_C(t))$ is of maximal rank for all integer t. Given a curve C in $\mathbb{P}^3$, we let $d =$ degree of C, $p_a :=$ arithmetic genus of C, $s := \min\{t / H^0(\mathcal{I}_C(t)) \neq 0\}$, $e : \max\{t/H^1 \mathcal{O}_C(t) \neq 0\}$, $c = \max\{t/H^1 \mathcal{I}_C(t) \neq 0\}$ ($c := -\infty$, if C is arithmetically Cohen–Macaulay) and $\sigma := \min\{t/H^0 \mathcal{I}_{C \cap H}(t) \neq 0\}$ $H \subset \mathbb{P}^3$ general plane.

Let $(n_1, \ldots, n_r)$ be a sequence of non-negative integers, where $n_1 \neq 0$ and $n_r \neq 0$. Then $L_{n_1, \ldots, n_r}$ is the even liaison class associated to a finite dimensional graded S–module which is annihilated by $\mathfrak{m}$ and whose homogeneous components are vector spaces of dimension $n_1, \ldots, n_r$ respectively. If M is a such module then we say that the diameter of M is r and we write $diam\, M = r$. The Buchsbaum type of $L_{n_1, \ldots, n_r}$ is the integer $N = n_1 + \ldots + n_r$.

Theorem 1.1. (|A 1|, |GM|). Let $C \in L_{n_1, \ldots, n_r}$, $s = s(C)$. Let $H \subset \mathbb{P}^3$ be a general plane.

(a) $s - 1 \leq \sigma \leq s$

(b) $s \geq 2N$

(c) $M(C)_t = 0$ for $t \leq s - 3$

(d) $M(C)_{S-2} \neq 0$ if and only if $\sigma = s - 1$ (and then $h^0 \mathcal{I}_{C \cap H}(s-1) = n_1$). □

If $C \in L_{n_1, \ldots, n_r}$ then the left most non-zero component of $M(C)$ occurs in degree greater or equal to $2N - 2$. We denote by $L^h_{n_1, \ldots, n_r}$ the set of curves $C \in L_{n_1, \ldots, n_r}$ whose first non-zero component occurs in degree $2N - 2 + h$.

Theorem 1.2. (|GM|, |EF|). If C is an a. B. curve of maximal rank, then $diam\,(C) \leq 2$. □

Remark 1.2.1. Let $C \subset \mathbb{P}^3$ be an a. B. curve of maximal rank. If $diam\,(C) = 0$, then C is arithmetically Cohen–Macaulay and so C is unobstructed. Therefore, we will resctrict our attention to the cases $diam\,(C) = 1$ and $diam\,(C) = 2$.

Definition 1.3. Let $X \subset \mathbb{P}^3$ be a curve, and let H be a general plane. The sheaf $\mathcal{O}_{X \cap H}$ has a resolution:

$$0 \to \oplus_0^{\sigma-1} \mathcal{O}_{\mathbb{P}^1}(-n_i) \to \oplus_0^{\sigma-1} \mathcal{O}_{\mathbb{P}^1}(-i) \to \mathcal{O}_{X \cap H} \to 0$$

where $n_0 \geq n_1 \geq \ldots \geq n_{\sigma-1} \geq \sigma$ is a sequence of integers which does not depend on H, and σ is the smallest degree of a curve in H containing the points. If X is integral, this sequence is without gaps. The σ-ple $N = (n_0, \ldots, n_{\sigma-1})$ is said to be the **numerical character** of $X \cap H$ and of X and we denote $A_i(X) = \#\{n_j = i\}$.

In |BM|, |BM 1|, Bolondi and Migliore give the following classification for smooth a. B. curves of maxiamal rank:

Theorem 1.4. Let $N = (n_0, \ldots, n_{\sigma-1})$ be a sequence of integers such that

$$[*] \quad \begin{cases} n_0 \geq n_1 \geq \ldots \geq n_{\sigma-1} \geq \sigma \\ N \text{ is without gaps} \\ \sigma \geq 2n - 1 \\ A_\sigma \geq n - 1 \\ A_{\sigma+1} \geq n, \quad A_{\sigma+1} = n \Rightarrow A_t = 0 \ \forall t > \sigma + 1 \end{cases}$$

Then, there exists a smooth maximal rank curve $Y \in L_n$ whose numerical character is N.

Conversely, let $Y \in L_n$ be a smooth maximal rank curve. Then its numerical character is a sequence of integers without gaps satisfying [*]. □

Theorem 1.5. Let $N = (n_0, \ldots, n_{\sigma-1})$ be a sequence of integers such that

$$[\bullet] \quad \begin{cases} n_0 \geq n_1 \geq \ldots \geq n_{\sigma-1} \geq \sigma \\ N \text{ is without gaps} \\ \sigma \geq 2m + 2n + h - 1, \ h \geq 1 \\ A_\sigma = m - 1 \\ A_{\sigma+1} \geq m + n + 1 \\ A_{\sigma+2} \geq n, \quad A_{\sigma+2} = n \Rightarrow A_t = 0 \ \forall t > \sigma + 2 \end{cases}$$

Then, there exists a smooth maximal rank curve $Y \in L^h_{m\,n}$ whose numerical character is N.

Conversely, let $Y \in L^h_{m\,n}$ be a smooth maximal rank curve. Then its numerical character is a sequence of integers without gaps satisfying [•]. □

From now on we will say that the numerical character $N = (n_0, \ldots, n_{\sigma-1})$ of an a. B. curve of maximal rank satisfies [*] (respectively [•]) if satisfies conditions [*] (Respectively [•]) of Theorem 1.4 (Respectively, Theorem 1.5).

We will use the following results:

Proposition 1.6. Let $C \subset \mathbb{P}^3$ be a curve. Then $H^0 N_C$ is naturally isomorphic to the Zariski tangent space of Hilb $\mathbb{P}^3$ at the point corresponding to C. Moreover, if $H^1 N_C = 0$ then C is unobstructed and the irreducible component of Hilb $\mathbb{P}^3$ passing through C has dimension 4 $deg(C)$. □

Proposition 1.7. (Cf. |K; Corollary 2.3.6|). Let $X \subset \mathbb{P}^3$ be an unobstructed curve. Assume that $H^1I_X(t-4) = H^1I_X(t) = H^1I_X(q-4) = H^1I_X(q) = 0$ and let $Y \subset \mathbb{P}^3$ be a curve linked to X by means of two surfaces of degree t and q. Then Y is unobstructed. □

Proposition 1.8. (Cf. |EH|). Let $C \subset \mathbb{P}^3$ be an a. B. curve of maximal rank. If $e \leq s-2$ then $H^1N_C = 0$. □

I am very grateful to J. O. Kleppe for pointing to me the following result:

Proposition 1.9. (Cf. |K 2|). Let $C \subset \mathbb{P}^3$ be a curve. Assume that $diam\, C = 1$ and $e = c < s$. Then $H^1N_C = 0$. □

§ 2. UNOBSTRUCTED ARITHMETICALLY BUCHSBAUM CURVES.

In this section, we state the main results of this paper. Concretely, we give sufficient conditions on the numerical character of an a. B. curve of maximal rank in order to assure that it is unobstructed.

Although the proof is essentially the same, we analyze separately, the case of diameter one and the case of diamenter two.

Theorem 2.1. Let $Y \in L_n$ be an irreducible a. B. curve of maximal rank with numerical character $N = (n_0, \ldots, n_{\sigma-1})$ satisfying [*]. If $n_0 < \sigma+3$ or $n_0 \geq \sigma+3$ and $A_{\sigma+3} = 1$, then Y is unobstructed.

Proof: We distinguish several cases:

Case 1, $n_0 = \sigma+1$. In this case, $e \leq \sigma-2 \leq s-2$ and Y is unobstructed (Proposition 1.8).

Case 2, $n_0 = \sigma+2$. In this case, $e = c = -1 < s$ and Y is unobstructed (Proposition 1.9).

Case 3, $n_0 = \sigma+3$. In this case, we work by induction on σ. First of all note that $n_0 \geq \sigma+3$ implies $\sigma \geq 2n+2$. If $\sigma = 2n+2$, then $N = (2n+5,\ 2n+4,\ 2n+3^{n+1},\ 2n+2^{n-1})$. In this case, we link Y to an a. B. curve Z by means of two surfaces S_{n_0-2} and $\sum_v$ of degrees n_0-2 and $v >> 0$, respectively; and Z to Y' by means of S_{n_0-2} and a surface $\sum_{v-3}$ of degree $v-3$. We get a maximal rank curve Y' in L_n with $\sigma' = \sigma(Y') = \sigma-3$, $N' = (n'_0, \ldots, n'_{\sigma-4})$ where $n'_i = n_{i+3}-3$ and $n'_0 = \sigma-2 = \sigma'+1$. Thus, by case 1, Y' is unobstructed. Since $H^1I_{Y'}(t) = 0$ for $t = n_0-2,\ v-3,\ n_0-6,\ v-7$, Z is unobstructed (Proposition 1.7). Since $H^1I_Z(t) = 0$ for $t = n_0-2,\ v,\ n_0-6,\ v-4$, Y is unobstructed (Proposition 1.7).

Now let σ be greater than $2n+2$. We distinguish two cases:

a) $n_0 = \sigma + 3$, $A_{\sigma+3} = 1$. In this case, we link Y to an a. B. curve Z by means of two surfaces, $S_{\sigma+2}$ and $\sum_v$ of degrees $\sigma + 2$ and $v >> 0$, respectively, and Z to Y' by means of $S_{\sigma+2}$ and surface $\sum_{v-2}$ of degree $v - 2$. Using the exact sequences:

$$0 \to I_{S_{\sigma+2}\cap\Sigma_v} \to I_Z \to \omega_Y(2-\sigma-v) \to 0$$

and

$$0 \to I_{S_{\sigma+2}\cap\Sigma_{v-2}} \to I_Z \to \omega_{Y'}(4-\sigma-v) \to 0$$

we get that Y' is a maximal rank curve in L_n, $\sigma' = \sigma(Y') = \sigma - 2$, $c' = c(Y') = \sigma - 3 = \sigma' - 1$ and $e' \leq c' = \sigma' - 1 < \sigma'$. Thus Y' is unobstructed. Since $H^1 I_{Y'}(t) = 0$ for $t = \sigma + 2,\ v - 2,\ \sigma - 2,\ v - 6$, Z is unobstructed. Moreover, $H^1 I_Z(t) = 0$ for $t = \sigma + 2,\ v,\ \sigma - 2,\ v - 4$, so Y is unobstructed.

b) $n_0 > \sigma + 3$. In this case, we link Y to an a. B. curve Z by means of surfaces S_{n_0} and $\sum_v$ of degrees n_0 and $v >> 0$ and Z to Y' by means of S_{n_0} and a surface $\sum_{v-1}$ of degree $v - 1$. We get a maximal rank curve $Y' \in L_n$ with $\sigma' = \sigma - 1 < \sigma$, $N' = (n'_0, \ldots, n'_{\sigma-2})$ where $n'_i = n_{i+1} - 1$ and $n'_0 \geq \sigma' + 3$. By hypothesis of induction Y' is unobstructed. Since $H^1 I_{Y'}(t) = 0$ for $t = n_0,\ n_0 - 4,\ v - 1,\ v - 5$, Z is unobstructed. Moreover, $H^1 I_Z(t) = 0$ for $t = n_0,\ n_0 - 4,\ v,\ v - 4$; so Y is unobstructed, which proves what we want. □

Theorem 2.2. Let $Y \in L_{mn}$ be an irreducible a. B. curve of maximal rank with numerical character $N = (n_0, \ldots, n_{\sigma-1})$ satisfying [•]. If $n_0 < \sigma + 3$ or $n_0 > \sigma + 3$ and $A_{\sigma+3} = A_{\sigma+4} = 1$, then Y is unobstructed.

Proof: We distinguish two cases:

Case 1, $n_0 = \sigma + 2$. In this case $e \leq \sigma - 1 \leq s - 2$ and Y is unobstructed (Proposition 1.8).

Case 2, $n_0 > \sigma + 3$ **and** $A_{\sigma+3} = A_{\sigma+4}. = 1$. In this case, we work by induction on σ. First of all note that $n_0 > \sigma + 3$ implies, $\sigma \geq 2m + 2n + 3$. If $\sigma = 2m + 2n + 3$, then $N = (2m + 2n + 7, 2m + 2n + 6, 2m + 2n + 5^{n+1}, 2m + 2n + 4^{m+n+1}, 2m + 2n + 3^{m-1})$. In this case, we link Y to an a. B. curve Z by means of two surfaces S_{n_0-2} and $\sum_v$ of degrees $n_0 - 2$ and $v >> 0$, respectively; and Z to Y' by means of S_{n_0-2} and a surface $\sum_{v-3}$ of degree $v - 3$. We get an a. B. curve of maximal rank $Y' \in L_{mn}$ with numerical character $N' = (2m + 2n + 2^n, 2n + 2m + 1^{m+n+1}, 2m + 2n^{m-1})$. In particular, $n'_0 = \sigma' + 2$. So Y' is unobstructed. Since $H^1 I_{Y'}(t) = 0$ for $t = n_0 - 2, v - 3, n_0 - 6, v - 7$, Z is unobstructed (Proposition 1.7). Since $H^1 I_Z(t) = 0$ for $t = n_0 - 2,\ v,\ n_0 - 6\, v - 4$ Y is unobstructed.

Now let σ be greater than $2m + 2n + 3$. We distinguish two cases:

a) $n_0 = \sigma + 4$, $A_{\sigma+4} = A_{\sigma+3} = 1$. As before, we link Y to an a. B. curve Z by means of two surfaces S_{n_0-2} and Σ_v of degrees $n_0 - 2$ and $v >> 0$, respectively, and Z to Y' by means of S_{n_0-2} and a surface Σ_{v-3} of degree $v - 3$. We get an a. B. curve of maximal rank $Y' \in L_{mn}$ with $\sigma' = \sigma - 3$ and $n'_0 = \sigma + 2$. Thus Y' is unobstructed and the same argument as before shows that Y is unobstructed.

b) $n_0 > \sigma + 4$, $A_{\sigma+3} = A_{\sigma+4} = 1$. In this case, we link Y to an a. B. curve Z by means of two surfaces S_{n_0} and $\sum_v$ of degrees n_0 and $v >> 0$, respectively; and Z to Y' by means of S_{n_0} and a surface $\sum_{v-1}$ of degrees $v-1$. We get an a. B. curve of maximal rank $Y' \in L_{m\,n}$ with $\sigma' = \sigma - 1 < \sigma$, $N' = (n'_0, \ldots, n'_{\sigma-2})$ where $n'_i = n_{i+1} - 1$, $n'_0 \geq \sigma' + 4$ and $A_{\sigma'+3} = A_{\sigma'+4} = 1$. By hypothesis of induction Y' is unobstructed. Since $H^1 I_{Y'}(t) = 0$ for $t = n_0,\ n_0 - 4,\ v-1,\ v-5$, Z is unobstructed. Moreover $H^1 I_{Y'}(t) = 0$ for $t = n_0,\ n_0 - 4,\ v,\ v-4$, so Y is unobstructed, which proves what we want. □

Problem 2.3. Is this fact true for all a. B. curves of maximal rank? More generally, is this fact true for other maximal rank space curves? That is to say, which are the maximal rank curves of $\mathbb{P}^3$ which are unobstructed?

REFERENCES

A M. Amasaki. On the structure of Arithmetically Buchsbaum curves in $\mathbb{P}^3$. Publ. RIMS 20 (1984) 793–837.

BM G. Bolondi–J. Migliore. Classification of maximal rank curves in the liaison class L_n. Math. Ann. 277 (1987) 585–603.

BM 1 G. Bolondi–J. Migliore. Buchsbaum liaison classes. Preprint, 1987.

C M. C. Chang. Buchsbaum subvarieties of codimension 2 in $\mathbb{P}^n$. Preprint, 1987.

El G. Ellingsrud. Sur le schéme de Hilbert des variétés de codimension 2 dans $\mathbb{P}^e$ à Cône de Cohen–Macaulay. Ann. Scient. Éc. N. Sup. 2 (1975) 423–432.

E Ph. Ellia. D'autres composantes non réduites de Hilb $\mathbb{P}^3$. Math. Ann. 277 (1987) 433–446.

EF Ph. Ellia–M. Fiorentini. Défaut de postulation et singularités du Schéme de Hilbert. Annali Univ. di Ferrara 30 (1984) 185–198.

EF 1 Ph. Ellia–M. Fiorentini. Courbes arithmetiquement Buchsbaum de l'espace projectif. Preprint, 1987.

EH Ph. Ellia–A. Hirschowitz. In preparation.

GM A. Geramita–J. Migliore. On the ideal of an Arithmetically Buchsbaum curve. To appear in J. Pure and Appl. Alg.

Mu D. Mumford. Futher pathologies in algebraic geometry. Amer. J. Math. 89 (1962) 642–648.

K J. Kleppe. The Hilbert–flag scheme, its properties and its connection with the Hilbert scheme. Thèse, Oslo 1982.

K 1 J. Kleppe. Non reduced components of the Hilbert scheme of smooth space curves. Preprint, 1985.

K 2 J. Kleppe. To appear in the Proceedings of Cognola.

S E. Sernesi. Un esempio di curva ostruita in $\mathbb{P}^3$. Sem. di variabili Complesse, Bologna 1981, 223–231.

On the Néron-Severi groups of the surfaces of special divisors.

Gian Pietro Pirola[1]
Universitá di Pavia, Dipartimento di Matematica,
Strada Nuova 65, 27100 Pavia, Italia.

Let C be a complete smooth curve of genus g defined over $\mathbb{C}$. Let $W^r_d(C)$ be the variety that parametrizes the line bundles on C of degree $d<g$ and dimension $r>0$:

$$W^r_d = W^r_d(C) = \{L \in \mathrm{Pic}^d(C) : h^0(L) \geq r+1\}$$

if C is general (cf. [1]) W^r_d has dimension $\rho(g,d,r)=g-(r+1)(g-d+r)$, ($W^r_d$ is empty if $\rho<0$), W^r_d is smooth outside of W^{r+1}_d and, moreover, it is irreducible (cf. [3]) if $\rho>0$.

On $W^r_d - W^{r+1}_d$ there is defined (cf. [1]) a tautological vector bundle S of rank r+1, whose fibers have an identification:

$$S_L = H^0(C,L), \text{ where } L \in W^r_d - W^{r+1}_d.$$

Ciro Ciliberto asked about the Néron-Severi group of W^r_d when $\rho=2$ and C is a general curve (W^r_d is then a smooth complex surface). The natural conjecture was that this group should be generated by the class of the restriction of the theta divisor (that we shall denote by Θ), of the Jacobian of C, $J(C) \cong \mathrm{Pic}^d(C)$, and by $c_1(S)=c_1(\det(S))$. When $\rho \geq 3$ this result is proved by the Lefschetz-type theorem of [3] combined with the fact that the Néron-Severi group of J(C) is generated by the class of Θ (cf. [2]).
However if r=1 this turns out to be false. In fact we shall prove:

Proposition

Let g=2k, k≥3. Then the rank of the Néron-Severi group of $W^1_{k+2} = W^1_{k+2}(C)$ of a general curve C of genus 2k is bigger than:

$$N(k) = \frac{(2k)!}{(k+1)!k!}$$

(N(k) is the well-known Castelnuovo number).

Proof

First we notice that

$$\rho(2k,k+1,1)=0 .$$

Then, if C is general (cf. [1]) W^1_{k+1} is a set of N(k) distinct points. Taking $L \in W^1_{k+1}$ we get an embedding $C \to W^1_{k+2}$ by the prescription

$$\text{(a)} \qquad C \ni P \to L(P) = L \otimes \mathcal{O}(P) .$$

---

[1] This research has been done within the framework of the MPI 40% project "Algebraic Geometry".

We will denote by C_i, $i=1,\dots,N(K)$, the $N(k)$ images of the embeddings (a) of C defined by the $L_i \in W^1_{k+1}$, and, by abuse of notation, the corresponding classes in $NS(W^1_{k+2})$, the Néron-Severi group of W^1_{k+2}.
We will prove the following :

Lemma
1) $C_i \cdot C_h = 0$ if $i \neq h$,
2) $C_i^2 < 0$,
<u>where</u> $(\cdot)$ <u>is the intersection form on</u> $NS(W^1_{k+2})$.

If the Lemma is proved, the proposition follows at once. In fact Θ is ample so $\Theta^2 > 0$, whereas $(\cdot)$ is negative definite on the submodule generated by the C_i.

Proof of the Lemma
1) Suppose $M \in C_i \cap C_h$. Then there should exist two points P and Q that belong to C such that

$$M \cong L_i + P \cong L_h + Q .$$

Notice that $P \neq Q$ because $L_i \ncong L_h$. But then the line bundle M should have dimension ≥ 2, that is $h^0(C,M) \geq 3$, and then W^2_{k+2} should be non-empty and C not general because $\rho(2k,k+2,2) < 0$.

2) We remark that $\Theta \cdot (C_i - C_h) = 0$, so applying the index theorem to $C_i - C_h$ we obtain $C_i^2 \leq 0$, and that $C_i^2 = 0$ would imply that C_i and C_h are numerically equivalent.
For the first few cases we can prove directly that for some i and h (and then for all $i \neq h$) this is not the case.

$g=6$: $W = W^1_5$ is the singular locus of the theta divisor of $J(C)$, so there is an involution $x: W \rightarrow W$ induced by $D \rightarrow K_C - D$. The fixed points of x are the theta-characteristics of dimension ≥ 1 so if C is general there are none. For any $i = 1,\dots,5 = N(3)$ we have new embeddings of C in W by composing (a) with x. We denote by C'_i their classes in $NS(W)$:

$$C'_i = x(C_i) .$$

The curves $L_i + p$ and $K_C - L_h - p$, $p \in C$, intersect if and only if

$$h^0(C, K - L_i - L_h) > 0 .$$

But clearly

$$h^0(C, K - L_i - L_h) = \delta_{ih} .$$

Then we obtain:

$$C_i \cdot C'_i = 0 ,$$
$$C_h \cdot C'_i = 2 .$$

So C_i and C_h are not numerically equivalent if $i \neq h$.

$g=8$: In this case let C be the normalization of a *general* plane curve of degree 7 with 7 nodes. C is not general in the sense of the moduli because it has a g^2_7 and $\rho(8,7,2)=-1$. Anyway observe, by a count of parameters, that W^2_6 is empty, in particular C is neither trigonal nor bielliptic and then by Mumford's refinement of Martens' theorem (cf [1]) $W=W^1_6$ has dimension two. We can locate 7 points of W^1_5 by considering the pencils from the seven nodes of the g^2_7, and the remaining ones by taking the nodes of the adjoint linear system $g'^2_7=K_C-g^2_7$ (notice that $N(4)=14$). Let $\{C_i\}_{i=1,\dots,7}$ and $\{C'_i\}_{i=1,\dots,7}$ be the corresponding classes. As before there are new embeddings of C in W by considering the curves

$$g^2_7-P\,,\ P\in C \qquad \text{and} \qquad g'^2_7-P\,,\ P\in C\,,$$

respectively. If we denote with Y and Y' their classes we obtain as before

$$\text{(b)} \qquad \begin{cases} C_i\cdot Y=0 \\ C'_i\cdot Y=2\,. \end{cases}$$

We notice that, by a straightforward analysis of the Petri map, the points of all our embedded curves are smooth points of W, so the usual intersection makes sense. Numerical equivalence is clearly a closed condition, so (b) and a monodromy argument prove that C_i and C_h are not numerically equivalent, if $i\neq h$ and C is general in the sense of moduli.

A similar proof works for $g=10$ and it should be possible to extend the same kind of argument to all even genera $g=2k$ by letting C be the normalization of a general plane curve of degree $k+3$ and geometric genus $2k$; however some tecnical problems should be overcome, so we prefer to proceed differently.

Proof for $k\neq 4,5$.

Now let C be a general curve of genus 2k. The adjunction formula shows that

$$C_i^{\,2}=2g-2-K\cdot C_i\,,$$

where K denotes the class of the canonical bundle of W^1_{k+2}. From [1] (see also [4] and [5] (2,8), page 81) it follows that the normal bundle N to $W^r_d-W^{r+1}_d$ in $\mathrm{Pic}^d(C)$ is isomorphic to $S^*\otimes Q$, where S^* is the dual of S and Q is the adjoint bundle, that is there is an identification

$$Q_L=H^1(C,L)\,,$$

where $L\in W^r_d-W^{r+1}_d$. On $W^r_d-W^{r+1}_d$ there is an exact sequence of vector bundles (cf. [1], page 176)

$$0\to S\to E\to F\to Q\to 0$$

and moreover (cf. [1], chapter 7, section 4),

$$c_1(E)=-\Theta \text{ and } c_1(F)=0\,.$$

Then

$$c_1(Q)=c_1(S)+\Theta$$

and

$$c_1(S^*\otimes Q)=-(g-d+r)c_1(S)+(r+1)c_1(Q)=(d+1-g)c_1(S)+(r+1)\Theta .$$

Finally, from the fact that the tangent bundle to $Pic^d(C)$ is trivial it follows that

$$K \equiv c_1(N) = 2\Theta-(k-3)c_1(S) .$$

On the other hand Poincaré's formula gives:

$$\Theta \cdot C_i = g ,$$

so we obtain

$$C_i^2+2\Theta \cdot C_i -(k-3)c_1(S)\cdot C_i = 2g-2 ,$$

and then

$$C_i^2-(k-3)c_1(S)\cdot C_i=2g-2-2g=-2 .$$

Now, if $C_i^2=0$ we obtain

$$(k-3)c_1(S)\cdot C_i=2 ,$$

which is impossible if $k\neq 4,5$ because $c_1(S)\cdot C_i$ is an integer Q.E.D..

Remark

When k=2, g=4, W_3^1 is a set of two points and W_4^1 is just the two-fold symmetric product C(2) of the curve C. In this case, as Ciro Ciliberto pointed out to us, the classes of the two embeddings C_i are numerically equivalent, and in fact $C_i^2=0$.

REFERENCES

[1] E. Arbarello, M. Cornalba, P. Griffiths and J. Harris, **Geometry of Algebraic Curves,** vol. I, Springer Verlag, New York - Berlin - Heidelberg - Tokyo (1985).

[2] C. Ciliberto, J. Harris, M. Teixidor, *On the endomorphisms of* $J(W^r_d(C))$ *when* $\rho=1$ *and* C *has general moduli,* preprint 1988.

[3] W. Fulton, R. Lazarsfeld, *On the connectedness of degeneracy loci and special divisors*, Acta Math. **146** (1981), 251-275.

[4] J. Harris, L. Tu, *Chern numbers of kernel and cokernel bundles,* Inventiones Math. **75** (1984), 467-475.

[5] G. Pirola, *Chern character of degeneracy loci and curves of special divisors*, Annali di Matematica Pura e Applicata **142** (1985), 77-90.

DEFORMATIONS OF MAPS

Ziv Ran*
Department of Mathematics & Computer Science
University of California
Riverside, CA 92521

During the 1970's, Horikawa [3] developed a powerful analytic deformation theory for holomorphic maps of compact complex manifolds, extending earlier work by Kodaira [4] in the case of embeddings. The aim of the work reported here is to extend some of Kodaira and Horikawa's results to the case of maps of *singular* compact complex spaces where, among other things, the deformations in question will, in general, be nonlocally trivial. Full details will be given elsewhere.

We will work in the category of compact complex spaces and holomorphic maps. By introducing polarizations, it is presumably possible to work out a projective algebraic analogue, valid for *separable* morphisms, but as it stands the theory is not applicable to inseparable morphisms.

1. Introduction

In Sections 1, 3 X and Y are assumed reduced. We begin with a formal definition.

Definition 1.1. *Let $f: X \to Y$ be a morphism and $(S, 0)$ a pointed analytic space. A deformation of f parametrized by $(S, 0)$ is a commutative diagram*

$$\begin{array}{ccccc} X & & \longrightarrow & \tilde{X} & \\ & {}^{f}\searrow & & | & \searrow^{\tilde{f}} \\ | & & Y & \longrightarrow & & \tilde{Y} \\ \downarrow & \swarrow & & {}^{g}\downarrow & \swarrow_{h} & \\ 0 & & \longrightarrow & S & \end{array} \qquad (1)$$

where g, h are flat and $X = g^{-1}(0), Y = h^{-1}(0)$. The functor of equivalence classes of deformations of f is denoted $\mathrm{Def}(X, f, Y)$ *or when f is an inclusion, by* $\mathrm{Def}(X, Y)$.

Some special cases of this are the following.

1. When $X = \emptyset$, it reduces to the usual deformation functor of Y.
2. When f is an inclusion, the subfunctor of $\mathrm{Def}(X, Y)$ corresponding to diagrams (1) in which $\tilde{Y}$ is the trivial deformation $Y \times S$ is the quotient of a germ of the Douady

---

* A. P. Sloan fellow; partially supported by NSF.

space (or, in the algebraic case, the Hilbert scheme) by a germ of the automorphism group of Y.

3. When X and Y are smooth, $\mathrm{Def}(X,f,Y)$ essentially coincides with Horikawa's deformation functor.

Remark 1.2. An elementary approach to $\mathrm{Def}(X,f,Y)$ is the following. Assume for simplicity f is an embedding and let $\tilde{Y}$ be the miniversal deformation of Y, which exists as an analytic space by the theorem of Douady-Grauert. Then if Y has no automorphisms, $\mathrm{Def}(X,f,Y)$ coincides with the germ at $[X]$ of the Douady space of compact analytic subspaces of $\tilde{Y}$, and may be studied as such. The problem with this approach is that it leads to an obstruction group that is "too large," i.e., fails to vanish even in excellent circumstances, such as those of Theorem 3.2 below; e.g. when X is 1-dimensional and $H^1(\mathcal{O}_X) \neq 0$, the obstruction group never vanishes. Thus to make good on this approach one would have to analyze the obstructions themselves. This, in fact, is what Horikawa does in the smooth case. Here, however, we are going to take another tack and set up our deformation problem differently, so as to yield a smaller obstruction group.

Nonetheless, the foregoing considerations do show, for an arbitrary $f: X \to Y$, at least if X, Y have no infinitesimal automorphisms, that $\mathrm{Def}(X,f,Y)$ is representable by an analytic space, namely a Douady space (the general case reduces to the embedding case by considering the graph). This was observed by Horikawa for X, Y smooth, but his argument works in the singular case as well.

Our basic idea for studying $\mathrm{Def}(X,f,Y)$ goes as follows. Recall that first-order deformations $\tilde{X}$ of X and $\tilde{Y}$ of Y are classified respectively by the sheaves $\Omega_{\tilde{X}} \otimes \mathcal{O}_X$ and $\Omega_{\tilde{Y}} \otimes \mathcal{O}_Y$, which are extensions

$$0 \to \mathcal{O}_X \to \Omega_{\tilde{X}} \otimes \mathcal{O}_X \to \Omega_X \to 0 \tag{2}$$

$$0 \to \mathcal{O}_Y \to \Omega_{\tilde{Y}} \otimes \mathcal{O}_Y \to \Omega_Y \to 0, \tag{3}$$

and this yields an identification of the tangent spaces to $\mathrm{Def}(X)$ and $\mathrm{Def}(Y)$ with T^1_X and T^1_Y, respectively, where we use the standard notation, for any space Z,

$$T^i_Z = \mathrm{Ext}_{\mathcal{O}_Z}(\Omega_Z, \mathcal{O}_Z).$$

Now if $f: X \to Y$ deforms along with X and Y, then in addition to (2) and (3) we also get a commutative diagram

$$\begin{array}{ccccc} f^*\mathcal{O}_Y & \to & f^*(\Omega_{\tilde{Y}} \otimes \mathcal{O}_Y) & \to & f^*\Omega_Y \\ \delta_0 \downarrow & & \downarrow & & \downarrow \delta_1 \\ \mathcal{O}_X & \to & \Omega_{\tilde{X}} \otimes \mathcal{O}_X & \to & \Omega_X \end{array} \tag{4}$$

where δ_0 and δ_1 are the canonical maps. It is then fairly clear that first-order deformations of $f : X \to Y$ are classified by the data (2), (3), (4). To make use of this observation, we would need to define and study a group which might and will be called $\mathrm{Ext}'(\delta_1, \delta_0)$, whose elements correspond to such data. This is a piece of homological algebra, possibly of independent interest, to which we turn next. [After this work was done, I learned that in the *affine algebraic* case (deformations of ring-homomorphisms) some similar constructions had been considered by M. Gerstenhaber and S. D. Schack (*TAMS* **279**(1983), 1–50). The global case appears nevertheless to be new.]

2. Ext of homomorphisms

We begin with some notations. Let $f : X \to Y$ be a morphism of ringed spaces. If A, B are, respectively, $\mathcal{O}_X$ and $\mathcal{O}_Y$-Modules, put

$$\mathrm{Hom}_f(B, A) = \mathrm{Hom}_{\mathcal{O}_X}(f^*B, A) = \mathrm{Hom}_{\mathcal{O}_Y}(B, f_*A);$$

its elements are called f-linear homomorphisms. We denote by $\mathrm{Ext}^i_f(B, A)$ the derived functors of $\mathrm{Hom}_f(B, A)$, in either variable, and note that we have 2 Grothendieck spectral sequences

$$E_2^{p,q} = \mathrm{Ext}^p_X(L^q f^*B, A) \Rightarrow \mathrm{Ext}^i_f(B, A) \tag{5}$$

$$E_2^{p,q} = \mathrm{Ext}^p_Y(B, R^q f_*A) \Rightarrow \mathrm{Ext}^i_f(B, A) \tag{6}$$

Now for f-linear homomorphisms $\delta_j \in \mathrm{Hom}_f(B_j, A_j)$, $j = 0, 1$, it is possible to define functorial groups $\mathrm{Ext}^i(\delta_1, \delta_0)$, having the following properties.

(2.0) $\mathrm{Ext}^0(\delta_1, \delta_0) = \mathrm{Hom}(\delta_1, \delta_0)$ is the set of pairs (α, β) where $\alpha : A_1 \to A_0$, $\beta : B_1 \to B_0$ and the diagram

$$\begin{array}{ccc} f^*B_0 & \xleftarrow{f^*\beta} & f^*B_1 \\ \delta_0 \downarrow & & \downarrow \delta_1 \\ A_0 & \xleftarrow{\alpha} & A_1 \end{array}$$

commutes.

(2.1) $\mathrm{Ext}^1(\delta_1, \delta_0)$ is the set of pairs of extensions

$$0 \to A_0 \to A_2 \to A_1 \to 0$$

$$0 \to B_0 \to B_2 \to B_1 \to 0$$

plus commutative diagrams

$$\begin{array}{ccccc} f^*B_0 & \to & f^*B_2 & \to & f^*B_1 \\ \delta_0\downarrow & & \downarrow & & \downarrow\delta_1 \\ A_0 & \to & A_2 & \to & A_1 . \end{array}$$

(2.2) There is an exact sequence

$$0 \to \operatorname{Hom}(\delta_1, \delta_0) \to \operatorname{Hom}(A_1, A_0) \oplus \operatorname{Hom}(B_1, B_0) \to \operatorname{Hom}_f(B_1, A_0)$$

$$\xrightarrow{\partial} \operatorname{Ext}^1(\delta_1, \delta_0) \to \operatorname{Ext}^1(A_1, A_0) \oplus \operatorname{Ext}^1(B_1, B_0) \to \operatorname{Ext}^1(B_1, A_0) \ldots$$

where the coboundary map ∂ is given by $\partial(\epsilon) =$ trivial module extensions, plus the diagram

$$\begin{array}{ccccc} f^*B_0 & \to & f^*B_0 \oplus f^*B_1 & \to & f^*B_1 \\ \delta_0\downarrow & & \downarrow\delta & & \downarrow\delta_1 \\ A_0 & \to & A_2 & \to & A_1 \end{array} \qquad \delta = \begin{pmatrix} \delta_1 & 0 \\ \epsilon & \delta_2 \end{pmatrix}.$$

(2.3) If f is an inclusion and $\delta_1 : B_1 \to f_*A_1$ is surjective with kernel K, then we have an exact sequence

$$0 \to \operatorname{Hom}(\delta_1, \delta_0) \to \operatorname{Hom}(B_1, B_0) \to \operatorname{Hom}(K, B_0) \to \operatorname{Ext}^1(\delta_1, \delta_0) \ldots$$

To define these Ext groups and establish their properties, two approaches are available. My original, brute-force approach was to do homological algebra in the category of f-linear Module homomorphisms. Here, however, we will sketch another, slicker approach, based in part on a suggestion of D. Buchsbaum, which realizes $\operatorname{Ext}^i(\delta_1, \delta_0)$ as ordinary Ext groups of Modules, albeit over a sheaf of noncommutative rings.

To begin with, we associate with our morphism $f : X \to Y$ a *Grothendieck topology* (cf. [1]) $T = T(f)$ as follows. The open sets of T are pairs (U, V) where $U \subset X, V \subset Y$ are open and $f(U) \subset V$; a covering of (U, V) is a collection of open sets $\{(U_\gamma, V_\gamma) : \gamma \in \Gamma\}$ such that the U_γ cover U and the V_γ cover V. Now on T define a structure sheaf $\mathcal{O}_T$ of noncommutative rings by

$$\mathcal{O}_T(U, V) = \left\{ \begin{pmatrix} a & 0 \\ b & c \end{pmatrix} : a \in \mathcal{O}_Y(V), b, c \in \mathcal{O}_X(U) \right\}$$

with multiplication

$$\begin{pmatrix} a & 0 \\ b & c \end{pmatrix} \begin{pmatrix} a' & 0 \\ b' & c' \end{pmatrix} = \begin{pmatrix} aa' & 0 \\ f^*(a')b + b'c & cc' \end{pmatrix}$$

(for $f =$ identity, this was suggested by Buchsbaum). Then there are mutually inverse equivalences of categories

$$\{f\text{-linear Module homo.}\} \underset{\tau}{\overset{\sigma}{\rightleftarrows}} \{\text{left } \mathcal{O}_T\text{-modules}\}.$$

σ associates to a triple $(A, B, \delta : f^*B \to A)$ the abelian sheaf $B \oplus A$ on T (which means the obvious thing), with module structure

$$\begin{pmatrix} a & 0 \\ b & c \end{pmatrix}\begin{pmatrix} y \\ x \end{pmatrix} = \begin{pmatrix} ay \\ b\delta(y) + cx \end{pmatrix}.$$

The inverse τ associates to a left $\mathcal{O}_T$ Module E, $A = \begin{pmatrix} 0 & 0 \\ 0 & 1 \end{pmatrix} \cdot E$, $B = \begin{pmatrix} 1 & 0 \\ 0 & 0 \end{pmatrix} \cdot E$ (which are in fact $\mathcal{O}_X$ and $\mathcal{O}_Y$-Modules respectively), and the map δ given by multiplication by $\begin{pmatrix} 0 & 0 \\ 1 & 0 \end{pmatrix}$. Using this equivalence of categories, we may simply define

$$\mathrm{Ext}^i(\delta_1, \delta_0) = \mathrm{Ext}^i_{\mathcal{O}_T}(\sigma(\delta_1), \sigma(\delta_0))$$

and the necessary properties may be verified easily.

3. Deformation theory

We now apply the homological considerations of §2 to deformation theory. The basic result suggested earlier is the following.

Proposition 3.1. *The first-order deformations of a morphism $f : X \to Y$ are classified by*

$$\mathrm{Ext}^1(\delta_1, \delta_0)$$

where $\delta_1 : f^*\Omega_Y \to \Omega_X, \delta_0 : f^*\mathcal{O}_Y \to \mathcal{O}_X$ are the natural maps; obstructions lie in $\mathrm{Ext}^2(\delta_1, \delta_0)$. *In particular, if* $\mathrm{Ext}^2(\delta_1, \delta_0) = 0$, *then* $\mathrm{Def}(X, f, Y)$ *is unobstructed.*

From now on we will denote $\mathrm{Ext}^i(\delta_1, \delta_0)$ by T^i_f. We will now give 2 typical applications of the general machinery to *stability theorems*; these are statements that under suitable hypotheses, a given morphism $f : X \to Y$ extends along with an arbitrary (small) deformation of X or Y, i.e., that one or the other morphism of functors $\mathrm{Def}(X, f, Y) \to \mathrm{Def}(X), \mathrm{Def}(X, f, Y) \to \mathrm{Def}(Y)$ is smooth.

Our first stability theorem generalizes a theorem proven by Kodaira [4] in the smooth case. We begin with a definition. An embedding $X \subset Y$ is said to be *very regular* if X is locally defined by a sequence of functions with independent differentials; or equivalently, if it is a regular embedding and

$$T^1_Y \otimes \mathcal{O}_X \simeq T^1_X.$$

Thus X being very regular means it is "as close as possible" to being smooth, given that it is regularly embedded in Y; in particular, if $X \subset Y$ is very regular and Y is smooth, then X is smooth.

Theorem 3.2. *Let $X \subset Y$ be a very regular embedding with normal bundle N, and assume that*

$$T_X^2 = T_Y^2 = H^1(N) = 0.$$

Then $\mathrm{Def}(X,Y)$ *is unobstructed and the natural morphism* $\mathrm{Def}(X,Y) \to \mathrm{Def}(Y)$ *is smooth. In particular, if Y is smoothable, then so is X.*

Proof (sketch): There is no loss of generality in assuming X is purely of codimension > 1 in Y. Property (2.2) above yields an exact sequence

$$T_f^1 \xrightarrow[\alpha]{} T_X^1 \oplus T_Y^1 \xrightarrow[\beta]{} \mathrm{Ext}_f^1(\Omega_Y, \mathcal{O}_X) \to T_f^2 \to T_X^2 \oplus T_Y^2 = 0$$

Claim: β is surjective.

Proof: We use Property (2.3), which shows that $\operatorname{coker}\beta$ sits in $\mathrm{Ext}^1(K, \mathcal{O}_X)$, $K = \ker(\delta_1)$. It is not hard to check in our case that $\mathcal{H}om(K, \mathcal{O}_X) = N, \mathcal{E}xt^1(K, \mathcal{O}_X) = 0$ (this uses $\operatorname{codim} X > 1$), hence $\mathrm{Ext}^1(K, \mathcal{O}_X) = 0$.

The claim yields that $T_f^2 = 0$ and α is surjective. ∎

Remark. It is also possible in this case to give a criterion for smoothness of $\mathrm{Def}(X,Y) \to \mathrm{Def}(Y)$; namely replace the hypothesis $H^1(N) = 0$ by $T_Y^1 \to \mathrm{Ext}^1(\Omega_Y, \mathcal{O}_X) \to 0$.

Our next stability theorem essentially generalizes to the singular case a result of Horikawa [3-III].

Theorem 3.3. *Let $f : X \to Y$ be a morphism with*

$$f_*\mathcal{O}_X = \mathcal{O}_Y, \qquad R^i f_*\mathcal{O}_X = 0, \qquad i = 1, 2.$$

Then $\mathrm{Def}(X, f, Y) \to \mathrm{Def}(X)$ *is smooth.*

Remark. Horikawa, in the smooth case, does not assume $R^2 f_*\mathcal{O}_X = 0$.

Proof: Using our hypotheses and the spectral sequence (6), we conclude that $T_Y^i \xrightarrow{\sim} \mathrm{Ext}_f^i(\Omega_Y, \mathcal{O}_X)$, $i = 1, 2$, hence in the usual exact sequence

$$T_f^1 \xrightarrow[\alpha]{} T_X^1 \oplus T_Y^1 \to \mathrm{Ext}_f^1(\Omega_Y, \mathcal{O}_X) \to T_f^2 \xrightarrow[\beta]{} T_X^2 \oplus T_Y^2 \to \mathrm{Ext}_f^2(\Omega_Y, \mathcal{O}_X)$$

we have that

(a) α is surjective;

(b) β is injective.

Now (a) means that $\mathrm{Def}(X, f, Y) \to \mathrm{Def}(X)$ is surjective on first-order deformations; but as obstructions to extending an n-th order infinitesimal deformation to an $(n+1)$-st order one lie in T^2, (b) now means that $\mathrm{Def}(X, f, Y) \to \mathrm{Def}(X)$ is surjective on all infinitesimal deformations, hence is smooth. ∎

Note that the hypotheses of the theorem apply e.g. whenever f is a resolution of a rational singularity. In particular, the following case is slightly amusing.

Example 3.4. Let X be the blowup of $\mathbb{P}^2$ at 6 points lying on a conic, and $C \subset X$ the proper transform of the conic, which is a (-2)-curve. Let $f : X \to Y$ be the blowing down of C, and note that Y is just a nodal cubic surface in $\mathbb{P}^3$. Deforming X to the blowup X' of $\mathbb{P}^2$ at 6 general points, f deforms to an *isomorphism* of X' with a nonsingular cubic surface. Thus f is stable, while the inclusion $C \subset X$ is not. This is, of course, a well-known phenomenon, first discovered by Atiyah.

Our next application asserts the *rigidity of targets of finite flat morphisms with fixed source.*

Theorem 3.5. *Let $f : X \to Y$ be a finite flat morphism. Then any deformation of f inducing a trivial deformation of X also induces a trivial deformation of Y.*

The proof will be omitted, being analogous to that of of Theorem 3.3 (use the fact that $\mathcal{O}_Y$ is a direct summand of $f_*\mathcal{O}_X$). We note that in the same situation we can make the stronger statement that $\mathrm{Def}(X, f, Y) \to \mathrm{Def}(X)$ is *injective*, provided we have surjectivity of

$$\mathrm{Hom}(\Omega_X, \mathcal{O}_X) \oplus \mathrm{Hom}(\Omega_Y, \mathcal{O}_Y) \to \mathrm{Hom}_f(\Omega_Y, \mathcal{O}_X) = \mathrm{Hom}(f^*\Omega_Y, \mathcal{O}_X),$$

and it is not too hard to think up a variety of reasonably natural conditions under which the latter can be assured (e.g. suitable negativity of $\Omega_Y^\vee$).

As a final application, we reprove a (global version of) a result of Kollàr (cf. [2,§6]).

Theorem 3.6 (Kollàr). *Let $f : X \to Y$ be a finite morphism etale in codimension 2, and assume X is locally S_3 (e.g. Cohen-Macaulay). Then any deformation of Y is induced by a unique deformation of f.*

Proof: Consider the exact sequence

$$0 \to K \to f^*\Omega_Y \xrightarrow{df} \Omega_X \to C \to 0$$

(this defines K, C). As K, C are supported in codimension 3 and X is S_3, we have

$$\mathcal{E}xt^i(K, \mathcal{O}_X) = \mathcal{E}xt^i(C, \mathcal{O}_X) = 0, \quad i \leq 2,$$

and it follows that

$$\mathrm{Ext}^i(\Omega_X, \mathcal{O}_X) \to \mathrm{Ext}^i(f^*\Omega_Y, \mathcal{O}_X) = \mathrm{Ext}^i_f(\Omega_Y, \mathcal{O}_X)$$

is an isomorphism for $i \leq 1$ and injective for $i = 2$. Given this, we can argue as in the proof of Theorem 3.3. ∎

This theorem implies that if Y is a variety with canonical singularities, then deformations of Y lift to deformations of its canonical index-1 cover (cf. [2]).

Acknowledgment. I am grateful to D. Buchsbaum for a helpful suggestion concerning the homological material of §2, and to H. Clemens for his encouragement and for sending a copy of [2].

REFERENCES

1. Artin, M., *Grothendieck Topologies*, Harvard notes.

2. Clemens, H., J. Kollàr, and S. Mori, "Higher-dimensional complex geometry, preprint.

3. Horikawa, E., "Deformations of holomorphic maps, I," *J. Math Soc. Japan* **25**(1973), 372–396; II (*ibid* **26**(1974), 647–667); III (*Math. Ann.* **222**(1976), 275–282).

4. Kodaira, K., "On stability of compact submanifolds of complex manifolds," *Amer. J. Math.* **85**(1963), 79–84.

GREEN'S CONJECTURE FOR GENERAL P-GONAL CURVES OF LARGE GENUS

Frank-Olaf Schreyer

Fachbereich Mathematik, Universität Kaiserslautern

Erwin-Schrödinger-Str., D-6750 Kaiserslautern, F.R.Germany

Introduction.

Let $C \subseteq \mathbb{P}^{g-1}$ be a (nonhyperelliptic) smooth canonically embedded curve of genus g defined over $\mathbb{C}$, and let S_C denote its homogeneous coordinate ring. For the minimal free resolution

$$0 \leftarrow S_C \leftarrow F_0 \leftarrow F_1 \leftarrow F_2 \leftarrow \ldots$$

as an (graded) $S = \mathbb{C}[x_0,\ldots,x_{g-1}]$ - module with $F_i = \bigoplus_j S(-j)^{\beta_{ij}}$, ie. $\beta_{ij} = \dim \operatorname{Tor}_i^S(S_C,\mathbb{C})_j$ are the "graded betti numbers", we have

Conjecture (M. Green, [G])

$\beta_{i,i+2} \neq 0 \iff$ *there exists a linear series of divisors of degree* $d \leq g-1$ *and dimension* $r \geq 1$ *on* C *with* $d-2r \leq i$.

One implication " $\Leftarrow$ " was proved by M. Green and R. Lazarsfeld [G]. For the converse we do not know much: The case $i = 1$ is Petri's classical theorem [P]. The case $i = 2$ was proved for $g = 7,8$ in [S1], for $g \geq 11$ by C. Voisin [V] and for arbitrary genus in [S2]. However for a general curve of genus $g = 2^n - 1$ for $n \geq 3$ the analogous conjecture over a field of characteristic 2 does not hold, cf. [S1], [S2].
In [GL] M. Green and R. Lazarsfeld state more general conjectures, which they prove in some special cases.

This paper contains some more evidence for the Conjecture.

Theorem

$\beta_{i,i+2} = 0$ for $i \le p-3$ for a general p-gonal curve C of genus $g > (p-1)(p-2)$, ie. Green's Conjecture is valid for C.

By the semicontinuity of the graded betti numbers and the irreducibility of the moduli-space $\mathcal{M}_{g,p}$ of p-gonal curves (cf. [F]) it suffices to prove this for a single p-gonal curve of genus g. I check this for a curve of class (p,q) on $\mathbb{P}^1 \times \mathbb{P}^1$ with δ nodes in general position for $g = (p-1)(q-1) - \delta$ and $0 \le \delta \le p-2 \le q-2$ using the methods of [S1].

Corollary

$\beta_{i,i+2} = 0$ for a general curve of genus $g > (i+2)(i+1)$.

Note that Green conjectures $\beta_{i,i+2} = 0$ for a general curve of genus $g > 2i+2$.

Proof of the Theorem.

Write $g = (p-1)(q-1) - \delta$ for $0 \le \delta \le p-2$, hence $p \le q$. Choose δ points $p_1, \ldots, p_\delta$ in general position on $\mathbb{P}^1 \times \mathbb{P}^1$ and consider the blow-up $\sigma: Y \longrightarrow \mathbb{P}^1 \times \mathbb{P}^1$ of these points. Let $E = E_1 + \ldots + E_\delta$ denote the exceptional divisor of σ and let (p,q) denote the divisor class $\sigma^*(p,q)$ by abuse of notation.

Proposition 1.

A general element $C \in |(p,q)-2E|$ is a smooth connected curve of genus g.

Proof. $|(2,2)-2E_i-2E_j|$ for $i \ne j$ is base-point free. So $|C|$ is base-point free linear series since $\delta \le p-2$ and $p \le q$, and a general member C is smooth by Bertini's Theorem. To see that C is connected check $H^1(Y,\mathcal{O}(-C)) = 0$. The genus of C is g by the adjunction formula. □

Since Y is rational the canonical series on C is cut out by the adjoint series on Y:

$$H^0(C,\omega_C) \cong H^0(Y,\mathcal{O}(p-2,q-2)-E))$$

The canonical image of C is contained in the image of $Y \subseteq \mathbb{P}^{g-1}$.

We will study the syzygies of C and Y with the methods of [S1],§2-5. Consider the variety

$$X = \bigcup_{D\in|(0,1)|} \bar{D} \subseteq \mathbb{P}^{g-1} ,$$

where $\bar{D}$ denotes the linear span of D in $\mathbb{P}^{g-1}$.

Proposition 2.

X is a (p-1)-dimensional rational normal scroll of type

$$S(\underbrace{q-2,\dots,q-2}_{p-1-\delta},\underbrace{q-3,\dots,q-3}_{\delta})$$

Proof. Scrolls are classified by partitions (cf. [S1],§1 or [ACGH]). The partition in question is associated to the filtration

$$0 \subseteq H^0(Y,\mathcal{O}(p-2,0)-E)) \subseteq H^0(Y,\mathcal{O}(p-2,1)-E)) \subseteq \dots$$

$$\dots \subseteq H^0(Y,\mathcal{O}(p-2,q-3)-E)) \subseteq H^0(Y,\mathcal{O}(p-2,q-2)-E))$$

by [S1],§2. Since the points $p_1,\dots,p_\delta$ are in general position and $\delta \le p-2$ we have $h^0(Y,\mathcal{O}(p-2,b)-E) = (p-1)(b+1) - \delta$, and we obtain the predicted type. □

Let H and R denote the generators of Pic(X) as in [S1]. The ideal of Y in X is determinantal by [S1],§5:

Proposition 3.

The ideal sheaf $\mathcal{J}$ of $Y \subseteq X$ is generated by the maximal minors of a map φ: $\mathcal{F} \longrightarrow \mathcal{G}$ with $\mathcal{F} = \oplus_i\ \mathcal{O}_X(-H+a_iR)$ and $\mathcal{G} = \mathcal{O}_X \oplus \mathcal{O}_X$

where $(a_1,\dots,a_{p-2}) = (\underbrace{q-2,\dots,q-2}_{p-2-\delta},\underbrace{q-3,\dots,q-3}_{\delta})$.

Proof. Consider $\pi\colon X \to \mathbb{P}^1$. By [S1],§5 we have

$$\mathcal{F}(H) \cong \pi^*\pi_*\mathcal{O}_Y((p-3,q-2)-E) \text{ and } \mathcal{G}^* \cong \pi^*\pi_*\mathcal{O}_Y(1,0)$$

and the map φ is the induced by composition

$$\mathcal{F}(H) \otimes \mathcal{G}^* \to \pi^*\pi_*\mathcal{O}_Y((p-2,q-2)-E) \to \mathcal{O}_X(H).$$

The splitting type of $\mathcal{F}$ can be computed from the filtration

$$0 \subseteq H^0(Y,\mathcal{O}((p-3,0)-E)) \subseteq H^0(Y,\mathcal{O}((p-3,1)-E)) \subseteq \dots$$
$$\dots \subseteq H^0(Y,\mathcal{O}((p-3,q-3)-E)) \subseteq H^0(Y,\mathcal{O}((p-3,q-2)-E)),$$

hence has the desired type. To see that the minors of φ generate $\mathcal{J}$ we can apply [S1],(5.7): Y is contained in the determinatal locus of φ and has the same degree and dimension. So they coincide. □

Corollary 4.

The Eagon-Northcott complex

$$0 \leftarrow \mathcal{O}_Y \leftarrow \mathcal{O}_X \leftarrow \Lambda^2\mathcal{F} \leftarrow \dots \leftarrow D_{p-4}(\mathcal{G}^*)\otimes\Lambda^{p-2}\mathcal{F} \leftarrow 0$$

resolves $\mathcal{O}_Y$ *as an* $\mathcal{O}_X$*-module*. □

Each term in this complex is a direct sum of linebundles $\mathcal{O}_X(-mH+bR)$ with $b \geq 2(q-2)$ which can be resolved as an $\mathcal{O} = \mathcal{O}_{\mathbb{P}^{g-1}}$-module by an Eagon-Northcott type complexe $\mathcal{C}^b(-m)$ associated to the $2\times(g-p+1)$ matrix whose minors define X (cf. [S1],§1,§3). The resulting iterated mapping cone is a not necessarily minimal resolution of $\mathcal{O}_Y$ as an $\mathcal{O}$-module, which is exact on global sections for each twist. Hence we can use this complex to get some bounds on the graded betti numbers $\beta_{ij}(Y)$ of Y.

Proposition 5.

(a) $Y \subseteq \mathbb{P}^{g-1}$ *is arithmetically Cohen-Macaulay and*

$\beta_{ij}(Y) = 0$ *for* $i \geq 1$ *and* $j \notin \{i+1,i+2\}$.

(b) $\beta_{i\,i+2}(Y) = 0$ *for* $i \leq 2q-3$.

(c) $\beta_{i\,i+1}(Y) = \beta_{i\,i+1}(X) = i\binom{g-p+1}{i+1}$ *for* $i \geq g-q+1$.

In particular $\beta_{i\,i+1}(Y) = 0$ *for* $i \geq g-p+1$.

Proof. For (a) just observe that the iterated mapping has the right length and only terms outside the given range. For (b) and (c) we look at the iterated mapping cone more closely. Let

$$\psi\colon F(-1) \longrightarrow G \text{ with } F \cong \mathcal{O}^f \text{ and } G \cong \mathcal{O}^2,$$

where $f = g-p+1$ is the degree of X, be the map whose 2x2 minors define X. Set $a = \sum a_i$. The cruical terms of the mapping cone are indicated in the diagram on the next page.

(b) follows from $a_i+a_j+1 \geq 2q-3$. For (c) we note

$$a + p - 2 = (p-2-\delta)(q-2)+\delta(q-3) + p - 2$$

$$= pq - 2q - p - \delta + 2 = g - q + 1.$$

Hence the complex is minimal in the range $i \geq g-q+1$ and the result follows. □

Finally to construct syzygies of C we consider the short exact sequence

$$0 \longleftarrow \mathcal{O}_C \longleftarrow \mathcal{O}_Y \longleftarrow \mathcal{O}_Y(-C) \longleftarrow 0$$

Since $\mathcal{O}_Y(-C) \cong \omega_Y(-H) \cong \mathcal{E}xt^{g-3}_{\mathcal{O}}(\mathcal{O}_Y, \mathcal{O}(-g-1))$ the dual of the resolution of $\mathcal{O}_Y$ gives a resolution of $\mathcal{O}_Y(-C)$ and once more taking a mapping cone yields a not necessarily minimal resolution of $\mathcal{O}_C$.

Proposition 6.

$\beta_{i\,i+1}(C) = \beta_{i\,i+1}(Y)$ *for* $i \geq g-2q+1$.

Proof. By Proposition 5 (b) the syzygies of $\mathcal{O}_Y(-C)$ contribute nothing to the syzygies of $\mathcal{O}_C$ in this range since

$$g-2 - (2q-3) = g-2q+1. \ \square$$

$$\mathcal{O}_X \leftarrow \underset{i<j}{\oplus}\, \mathcal{O}_X(-2H+(a_i+a_j)R) \leftarrow \ldots \leftarrow \mathcal{O}_X(-(p-2)H+aR)^{\oplus(p-3)} \leftarrow 0$$

$$\mathcal{O} \leftarrow \underset{i<j}{\oplus}\, S_{a_i+a_j}G(-2) \leftarrow \ldots \leftarrow S_a G(-p+2)^{\oplus(p-3)}$$

$$\overset{2}{\wedge} F(-2) \leftarrow \underset{i<j}{\oplus}\, S_{a_i+a_j-1}G(-3)\otimes F \leftarrow \ldots \leftarrow S_{a-1}G(-p+1)\otimes F^{\oplus(p-3)}$$

$$\underset{i<j}{\oplus}\, \overset{a_i+a_j}{\wedge} F(-a_i-a_j-2)$$

$$\underset{i<j}{\oplus}\, \overset{a_i+a_j+2}{\wedge} F(-a_i-a_j)$$

$$\overset{a}{\wedge} F(-a-p+2)^{\oplus(p-3)}$$

$$\overset{a+2}{\wedge} F(-a-p)^{\oplus(p-3)}$$

$$D_{f-2}G^*\otimes\overset{f}{\wedge} F(-f) \leftarrow \ldots \qquad \ldots \leftarrow D_{f-a-2}G^*\otimes\overset{f}{\wedge} F(-f-p+2)^{\oplus(p-3)}$$

$$0 \qquad\qquad\qquad 0$$

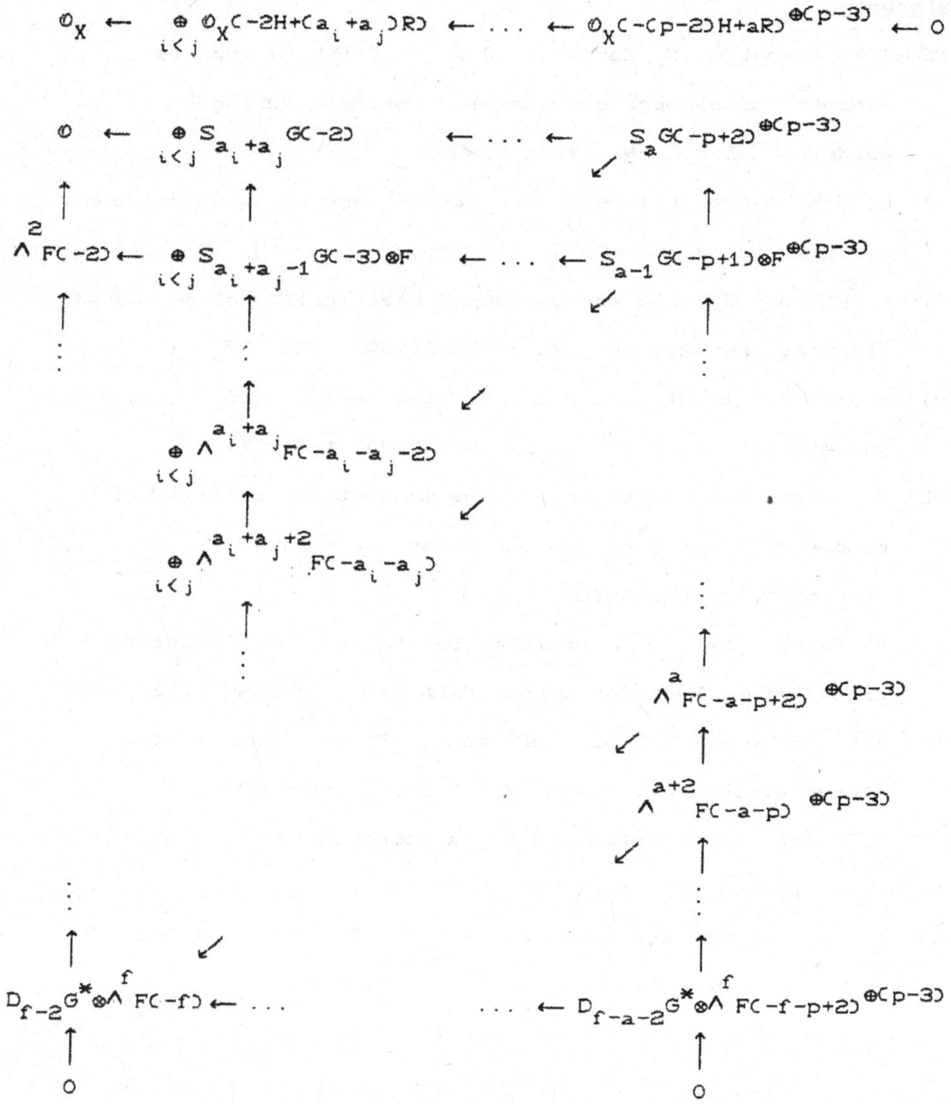

Proof of the Theorem. The resolution of $\mathcal{O}_C$ is selfdual, in particular $\beta_{i,i+2}(C) = \beta_{g-2-i,g+1-i}(C)$. So the result follows from Proposition 5 (c) and 6 because

$$g-2-i \geq g-p \iff i \leq p-2.$$

Actually we have proved a bit more. □

References

[ACGH] E. Abarello, M. Cornalba, P.A. Griffith, J. Harris: Geometry of algebraic curves I , Springer Verlag Heidelberg-Berlin-New York 1985.

[E] L. Ein: A remark on syzygies of the generic canonical curve, J. Diff. Geom. 26 (1987), 361-365.

[F] W. Fulton: Hurwitz schemes and irreducibility of moduli of algebraic curves, Ann. Math. 90 (1969), 542-575.

[G] M. Green: Koszul cohomology and the geometry of projective varieties I, J. Diff. Geom. 19 (1984), 125-171.

[GL] M. Green, R. Lazarsfeld: On the projective normality of complete linear series on an algebraic curve, Invent. math. 83 (1986), 73-90.

[P] K. Petri: Über die invariante Darstellung algebraischer Funktionen einer Veränderlichen, Math. Ann. 88 (1923), 242-289.

[S1] F.O. Schreyer: Syzygies of canonical curves and special linear series, Math. Ann. 275 (1986), 105-137.

[S2] F.O. Schreyer: A standard basis approach to syzygies of canonical curves, preprint.

[V] C. Voisin: Courbes tétragonales et cohomologie de Koszul, J. reine angew. Math. 387 (1988), 111-121.

On rank-1 degenerations of abelian varieties.

by Gerard van der Geer

The subject of this note is a variation on a theme of [vGvdG] related to the Schottky problem. There we introduced for a principally polarized abelian variety $X = (A,\Theta)$ of dimension g a subspace $\Gamma_{00} \subset \Gamma(A,O(2\Theta))$ by

$$\Gamma_{00} = \{ \; s \in \Gamma(A,O(2\Theta)) \; ; \; m_0(s) \geq 4 \; \},$$

where m_0 stands for the multiplicity of vanishing at zero. The codimension of Γ_{00} in $\Gamma(A,O(2\Theta))$ equals $\frac{1}{2}g\,(g+1)+1$ if (A,Θ) is indecomposable (i.e. if Θ is irreducible). The space Γ_{00} is related to the tangent space of the moduli space of principally polarized abelian varieties (with (2,4) level structure) at the point corresponding to X.

Let $V(\Gamma_{00})$ be the set of common zeroes of the elements of Γ_{00}. Welters proved that for $g \neq 4$ one has $V(\Gamma_{00}) = C - C \subset Jac(C)$ for a jacobian variety $x = Jac(C)$, thereby confirming our conjecture made in [vGvdG]. For $g=4$ he proved $V(\Gamma_{00}) = C - C$ plus two points. We conjectured in [vGvdG] that $V(\Gamma_{00}) = \{0\}$ if X is not a jacobian of a curve.

Recently, Donagi put forward a bold conjecture whose validity would imply our conjectures and more. His conjecture describes the so-called Schottky Locus in the moduli space RA_g of principally polarized abelian varieties with a non-trivial point of order 2. By going to the boundary components of lowest codimension of RA_g and by computing the tangent cones to the conjectured components one finds (a version of) our conjectures of [vGvdG]. One can however go deeper into the boundary. We do this in one case by computing the limit of Γ_{00} as X goes to a so-called rank-1 degeneration . The limit of our conjectures gives rise to a new conjecture on abelian varieties. For example we find an analogue of the Novikov conjecture involving values of (derivatives of) theta functions at other points than just the origin. Our ground field is $\mathbb{C}$.

1. Rank-1 degenerations of abelian varieties.

(1.1) Definition. A rank-1 degeneration X of principally polarized abelian varieties is a pair $(\underline{G},D)$, where $\underline{G}$ is a g-dimensional complete variety and D is an ample divisor on $\underline{G}$ such that the following hold :

i) there exists a principally polarized abelian variety $Y = (B, \Xi)$ and a semi-abelian variety G which is an extension

$1 \to G_m \to G \to B \to 0.$

ii) If $\mathbb{P}(G)$ is the associated $\mathbb{P}^1$-bundle of G and $\mathbb{P}(G) - G = G_0 \cup G_\infty$ then $\underline{G}$ is obtained by glueing the two sections G_0 and G_∞ by a shift over an element $b \in B$.

iii) D is the divisor of any non-zero section of a line bundle L on $\underline{G}$ whose pull-back to $\mathbb{P}(G)$ is $L = O_{\mathbb{P}(G)}(G_\infty + \pi^{-1}(\Xi)) = O_{\mathbb{P}(G)}(G_0 + \pi^{-1}(\Xi_b))$. Here π denotes the map $G \to B$.

Two such objects X and X' are called isomorphic if there exists an isomorphism of complete varieties $\underline{G} \to \underline{G}'$ sending D to D' which induces an isomorphism of the underlying principally polarized abelian varieties Y and Y'. One sees easily that the isomorphism classes of rank-1 degenerations of dimension g are in 1-1 correspondence with the pairs consisting of an isomorphism class of a principally polarized abelian variety $Y = (B,\Xi)$ of dimension $g-1$ plus a point of $b \in B/\mathrm{Aut}(Y)$. The moduli space $\underline{A}_g$ of principally polarized abelian varieties of dimension g and rank-1 degenerations of dimension g is the blow-up of a partial Satake-compactification $A_g \cup A_{g-1}$. Mumford studied such objects in [M].

We can also define the <u>theta group</u> $T(X,M)$ for a pair (X,M), where X is a rank-1 degeneration and M is a line bundle on X. We fix an origin of X and require that M is symmmetric. First note that the translation by an element g of G extends to a morphism t_g of $\mathbb{P}(G)$ and $\underline{G}$ inducing a translation by $\pi(g) \in B$ on G_0 and G_∞. One can define a group scheme $T(X,M)$ whose points are :

$T(X,M)(R) = \{ (x,\phi) : x \in G(R),\ \phi$ is an isomorphism of $t_x^* M$ with M over $R\}$

and with group law $(x,\phi) \cdot (y,\psi) = (x + y, t_y^*\phi \cdot \psi)$. There is an exact sequence

$$1 \to \mathbb{C}^* \to T(X,M) \to G[2] \to 0,$$

where $G[2]$ is the kernel of multiplication by 2. The group $G[2]$ fits into the exact sequence

$$0 \to G_m[2] \to G[2] \to B[2] \to 0$$

and we have a commutative diagram of exact sequences

$$\begin{array}{ccccccc}
 & & 1 & & 1 & & \\
 & & \downarrow & & \downarrow & & \\
 & & G_m[2] & \to & G_m[2] & & \\
 & & \downarrow & & \downarrow & & \\
1 \to & G_m & \to T(X,M) & \to & G[2] & \to 0 & \\
 & \downarrow & \downarrow & & \downarrow & & \\
1 \to & G_m & \to T(B,M|B) & \to & B[2] & \to 0 & \\
 & & \downarrow & & \downarrow & & \\
 & & 0 & & 0. & &
\end{array}$$

The image of $G_m[2]$ belongs to the center of $T(X,M)$.

Let $M = L^{\otimes 2}$. The elements of $T(X,M)$ act on $\Gamma(P(G),L^{\otimes 2})$ and on $\Gamma(\underline{G},M)$ by

$$s \to \phi(t_x^* s).$$

The theta group $T(B,M|B)$ -- which is isomorphic to a Heisenberg group -- acts on the (+ 1 and − 1) eigenspaces of the kernel $G_m[2]$ of $T(X,M) \to T(B,M|B)$.

We introduce the following notation. By the Heisenberg group H_g we mean the group which is the extension

$$1 \to k^* \to H_g \to (\mathbb{Z}/2)^g \times ((\mathbb{Z}/2)^\vee)^g \to 0$$

such that multiplication is given by $(r,a,b)\cdot(r',a',b') = (rr'a(b'),a+a',b+b')$. The 2^g dimensional irreducible representation of the Heisenberg group H_g, where G_m acts by multiplication of scalars is denoted by U_g.

Let V be the space

$$V = \Gamma(\underline{G},M) = \{\ s \in \Gamma(\mathbb{P}(G),L^{\otimes 2}) : i_0(s) = i_\infty(t_b(s))\ \}.$$

(1.2) Lemma. The vector space $V = \Gamma(\underline{G},M)$ has dimension 2^g. Let $\bar{\alpha} \in G_m[2]$ be the non-trivial element of the kernel $G[2] \to B[2]$. Then V splits as a direct sum $V = V^1 \oplus V^2$ of eigen spaces of dimension 2^{g-1} under the action of a lift $\alpha \in T(X,M)$ of $\bar{\alpha}$. Both V^1 and V^2 are representations of $T(B,M|B) = H_{g-1}$, each isomorphic to U_{g-1}.

<u>Proof</u>. Consider the homomorphism

$$r : \Gamma(\mathbb{P}(G),L) \to H^0(B,O(2\Xi)) \oplus H^0(B,O(2\Xi)),\quad s \to (i_0(s),i_\infty(t_b s)).$$

Using the Leray spectral sequence the kernel of r can be identified with $H^0(\mathbb{P}(G),\pi^*(O(2\Xi))) \cong H^0(B,O(2\Xi))$. The map r is surjective. The action of $(\alpha,1)$ on $\Gamma(\mathbb{P}(G),L)$ is by ± 1 on the kernel of r, while it is by the opposite sign on the image of r. The space V is the direct sum of the kernel of r and the diagonal of $H^0(B,O(2\Xi)) \oplus H^0(B,O(2\Xi))$. ◊

We can study the Kummer map defined by (a basis of the space of the) sections of M. Indeed, they define a map

$$\phi : \underline{G} \to \mathbb{P}^N = \mathbb{P}(V) \qquad (N = 2^g - 1)$$

which factors through the canonical involution

$\underline{j} : \underline{G} \to \underline{G}$ which extends $j : G \to G,\ g \to -g$.

(1.3) Proposition. The morphism ϕ has the following properties: i) it is of degree 2 if and only if X is indecomposable; ii) ϕ restricted to $\mathrm{Sing}(\underline{G}) = B$ is the Kummer map of B to $\mathbb{P}(U_{g-1})$.

Proof. Left to the reader. ◊

Note that X is indecomposable if and only if Y is indecomposable and $b \neq 0$.

In order to do this in a canonical way we choose an isomorphism of T(X,M) with a rank-1 degeneration of a Heisenberg group H_g. This is a group scheme obtained as follows. Take H_g and take a non-zero element $\overline{\alpha} \in H_g - G_m$. Lift it to an element α of H_g. Let $Z(\alpha)$ be the centralizer of α. Then $Z(\alpha)/\{\alpha,1\}$ is isomorphic to H_{g-1}. Note that α belongs to the center of $Z(\alpha)$. A theta structure on a rank-1 degeneration is

i) a choice of a non-zero $\overline{\alpha} \in H_g/G_m$

ii) a decomposition $U_g = U^1 \oplus U^2$ of U_g in the two eigen spaces of a lift α of $\overline{\alpha}$

iii) an isomorphism of T(X,M) with $Z(\alpha)$ which is the identity on G_m and which sends $\{\alpha,1\}$ to the kernel of $G[2] \to B[2]$.

Now choose a theta structure on X. Then we can identify the two eigenspaces of $\overline{\alpha}$ with U_{g-1} in a canonical way up to scalars. However, there is no way of identifying V with U_g in a canonical way up to scalars. (But this will not affect our computation of the limit of Γ_{00}.)

Example. Let g=2. The image of X is then a quartic surface in $\mathbb{P}^3$. This is a limit of classical Kummer surfaces. It has a double line and 8 isolated singular points. It is studied by Klein under the name : "Plückersche Komplexfläche", see [K].

2. The analytic set up.

Instead of this algebraic formulation one can treat these things analytically. Let $A = \mathbb{C}^g/\Lambda$, where $\Lambda = \mathbb{Z}^g + \tau\,\mathbb{Z}^g$ with τ in the Siegel upper half space H_g be a principally polarized complex abelian variety. We define

$$\theta_\sigma = \Sigma_{n\in\mathbb{Z}} \exp(2\pi i({}^t(n+\sigma)\tau(n+\sigma) + 2{}^t(n+\sigma)z),$$

the classical 2nd order theta function associated to $\sigma \in (\frac{1}{2}\mathbb{Z}/\mathbb{Z})^g$. They define a basis of $O(2\Theta)$. We write

$$\tau = \begin{pmatrix} \tau' & \omega \\ \omega & \tau'' \end{pmatrix} \quad \text{with } \tau' = it \ \ t \to \infty,\ \tau'' \in H_{g-1},\ \omega \in \mathbb{C}^g$$

and consider the limit of these θ_σ as $t \to \infty$. We write $n = n'n''$, $\sigma = \sigma'\sigma''$ with $n \in \mathbb{Z}^{g-1}$, $\sigma' \in \frac{1}{2}\mathbb{Z}^{g-1}/\mathbb{Z}^{g-1}$ in the formula for θ_σ. If we let $t \to \infty$ and take the limit then if $\sigma' = 0$ only terms with $n' = 0$ survive giving together as limit $\theta_{\sigma'}$. If $\sigma'' = \frac{1}{2}$ no terms survive and the limit is zero. The terms that vanish are divisible by $\exp(-\frac{\pi i}{2}\tau'')$. We therefore renormalize and replace z by

$$z - (\tfrac{\tau''}{2}, 0, \ldots, 0).$$

This is allowed since we do not fix an origin and consider D or Θ up to translation. We now get the series

$$\Sigma_n \exp(2\pi i[(n'+\sigma')^2\tau' - (n'+\sigma')\tau' + \ldots]).$$

If $\sigma' = 0$ then in the limit terms with $n' = 0$ survive giving $\theta_{\sigma''}(\tau'', z'')$ while those with $n' = 1$ also survive and give $\exp(4\pi i z') \cdot \theta_{\sigma''}(\tau'', z''+b)$. If $\sigma' = \frac{1}{2}$ we find similarly in the limit $\exp(2\pi i z') \exp(-\frac{\pi i}{2}\tau'')\, \theta_{\sigma''}(\tau'', z'' + \frac{b}{2})$.

We put

$$u = \exp(2\pi i z')$$

(the natural coordinate on $\mathbb{C}^*$) and consider the functions

$$\rho_\sigma = \theta_\sigma(\tau,z) + u^2\, \theta_\sigma(\tau, z+b)$$
$$u\, \theta_\sigma(\tau, z+\tfrac{b}{2}),$$

where now $\tau \in H_{g-1}$, $z \in \mathbb{C}^{g-1}$, $\sigma \in (\frac{1}{2}\mathbb{Z}/\mathbb{Z})^{g-1}$. The ρ_σ form a basis of V^+, while the functions $u\,\theta_\sigma(\tau, z+\frac{b}{2})$ define a basis of V^-. In fact, α acts by sending u to $-u$. We now renormalize again by multiplying by u^{-1} and consider instead the functions

$$\eta_\sigma = u^{-1}\, \theta_\sigma(\tau, z-\tfrac{b}{2}) + u\theta_\sigma(\tau, z+\tfrac{b}{2})$$
$$\theta_\sigma = \theta_\sigma(\tau, z),$$

where σ runs through $(\frac{1}{2}\mathbb{Z}/\mathbb{Z})^{g-1}$. Using these functions we get a morphism

$$\underline{G} \to \mathbb{P}^N$$

and this morphism factors through $(u,z) \to (u^{-1}, -z)$. The Kummer variety of Y is the locus of non-normal singularities. Moreover the points of order two on G map to isolated singular points.

3. The limit of Γ_{00}.

Recall that Γ_{00} is the subspace of sections of a principally polarized abelian variety vanishing with multiplicity at least 4 at the origin, see [vGvdG]. Here we shall always assume that the polarized abelian variety is indecomposable, i.e. the theta divisor is irreducible. To find the corresponding notion for rank-1 degenerations note that Γ_{00} is the space of sections

$$\Sigma_\sigma\, a_\sigma\theta_\sigma$$

such that

i) $\Sigma_\sigma\, a_\sigma\theta_\sigma(\tau,0) = 0$

ii) $\Sigma_\sigma\, a_\sigma\partial_i\,\partial_j\theta_\sigma(\tau,0) = 0$ for all $1 \le i,j \le g$.

In the limit the matrix

$$(\theta_\sigma \quad \partial_1\partial_1\theta_\sigma \quad \ldots \quad \partial_g\partial_g\theta_\sigma)(\tau,0)_\sigma$$

becomes (after replacing τ'' by τ, σ'' by σ)

$$\begin{matrix} \eta_\sigma(\tau,0) & \theta_\sigma(\tau,\frac{b}{2}) & 2\partial_1\theta_\sigma(\tau,\frac{b}{2}) \ldots & 2\partial_{g-1}\theta_\sigma(\tau,\frac{b}{2}) & 2\partial_1\partial_1\theta_\sigma(\tau,\frac{b}{2}) \ldots & 2\partial_{g-1}\partial_{g-1}\theta_\sigma(\tau,\frac{b}{2}) \\ \theta_\sigma(\tau,0) & 0 & 0 & 0 & 2\partial_1\partial_1\theta_\sigma(\tau,0) \ldots & 2\partial_{g-1}\partial_{g-1}\theta_\sigma(\tau,0) \end{matrix}$$

(3.1) Lemma. If $b \neq 0$ and Y is indecomposable (i.e. if X is indecomposable) then the rank of this matrix is $\frac{1}{2}g(g+1)+1$.

Let X be a rank-1 degeneration. We denote by F_0 the completion in $\underline{G}$ of the the fibre $\pi^{-1}(0)$ of the map $G \to B$.

(3.2) Definition-Lemma. The subspace $\Gamma_{00}(X)$ of $\Gamma(X,M)$ is the space of sections which vanish with multiplicity at least 4 on F_0. The codimension of Γ_{00} in $\Gamma(X,M)$ is $\frac{1}{2}g(g+1)+1$.

<u>Proof</u>. An element $\Sigma_\sigma\, a_\sigma\theta_\sigma + \Sigma_\sigma\, b_\sigma\eta_\sigma$ belongs to Γ_{00} if and only if

i) $\Sigma\, a_\sigma\theta_\sigma(\tau,\frac{b}{2}) = 0$,

ii) $\Sigma\, a_\sigma\partial_j\theta_\sigma(\tau,\frac{b}{2}) = 0$, for $j = 1, .., g-1$,

iii) $\Sigma\, b_\sigma\theta_\sigma(\tau,0) = 0$,

iv) $\Sigma\, 2a_\sigma\partial_i\partial_j\theta_\sigma(\tau,\frac{b}{2}) + \Sigma\, b_\sigma\partial_i\partial_j\theta_\sigma(\tau,0) = 0$, for $1 \le i,j \le g-1$.

This means that $(\ldots,a_\sigma,\ldots, b_\sigma,\ldots)$ belongs to the kernel of the linear map given by the matrix of (3.1). Therefore Lemma (3.1) implies the claim about the dimension.◊

(3.3) Remarks. 1) Note that all sections of $\Gamma(X,M)$ vanish with even multiplicity on F_0.

2) The conditions i) and ii) mean that the hyperplane $\Sigma_\sigma a_\sigma\theta_\sigma = 0$ contains the tangent space to the Kummer variety of B in $\phi_B(\frac{b}{2})$.

3) We have $\Sigma_\sigma a_\sigma\eta_\sigma \in \Gamma_{00}$ if and only if $\Sigma_\sigma a_\sigma\theta_\sigma$ vanishes with multiplicity ≥ 3 in $\frac{b}{2}$.

4) An element $\Sigma\, b_\sigma \eta_\sigma$ belongs to $\Gamma_{00}(X)$ if and only if $\Sigma\, b_\sigma\, \theta_\sigma$ belongs to $\Gamma_{00}(B)$.

From this last remark it follows that the zero locus of $\Gamma_{00}(X)\cap G$ is contained in $\pi^{-1}(V(\Gamma_{00}(B)))$.

We can now formulate our conjectures.

Let C' be a complete non-singular irreducible curve of genus $g-1$ and let p, q be two different points on C'. The curve $C = C'/(p{\approx}q)$ of genus g is obtained by identifying p and q. We consider the generalized jacobian $J_0(C)$ of C. Its points are the divisor classes of divisors D on C' of degree zero relatively prime to p and q modulo the equivalence

$D \approx D'$ if and only if there exists a function f on C' with

i) $(f) = D - D'$,

ii) $\mathrm{ord}_p(1-f) > 0$ and $\mathrm{ord}_q(1-f) > 0$.

Another description is the set of isomorphism classes of line bundles on C which when pulled back to C' are of degree zero. The identification is made by associating to a divisor D the line bundle O(D) on C'. This is a line bundle on C' with a section (the function 1). By identifying the fibres of O(D) over p and q using the non-zero section 1 we find a line bundle on C. Two equivalent divisors give rise to isomorphic line bundles. The generalized jacobian has a compactification Jac(C)which is a rank-1 degeneration. Its points are the isomorphism classes of torsion-free O_C - modules of rank 1 and of degree 0.

Using the first description we have a natural map

$C'' \times C'' \to \mathrm{Jac}(C),\quad (a,b) \to \mathrm{cl}[a-b],$

where $C'' = C' - \{p,q\}$. Alternatively, we can consider the line bundles on C of degree zero having a section with at most a single zero and a single pole. The locus of such line bundles forms a surface. Let F be the closure in $\mathrm{Jac}(C)$ of this surface.

(3.4) Conjecture. 1) If the rank-1 degeneration X is the jacobian of a stable curve $C = C'/(p{\approx}q)$ and the genus $g(C) \neq 4$ then $V(\Gamma_{00}) = F$. For $g = 4$ we have $V(\Gamma_{00}) = F \cup \{a,-a\}$, where a is the difference of the two g^1_3 's on C.

2) If X is not a jacobian then $V(\Gamma_{00}) = \{F_0\}$.

Part 1) of this conjecture can for example be checked for hyperelliptic jacobians. Maybe Welters' results can be carried over to this case.

As in the case of usual abelian varieties one can consider the condition that $V(\Gamma_{00}) \neq F_0$ infinitesimally and deduce a differential equation for the second order theta functions. In order to have that $V(\Gamma_{00}) \neq F_0$ we must have that $V(\Gamma_{00}(B)) \neq (0)$. Infinitesimally this is expressed by the fact that the theta functions θ_σ of B satisfy a differential equation of the form

$$(D_1^4 + \text{lower order terms}\,)\,\theta_\sigma\,)(\tau,0\,) = 0$$

with $D_1 \neq 0$.

(3.5) Proposition. Suppose that B is a jacobian. Then $V(\Gamma_{00}(X))$ contains a scheme of the form $\mathrm{Spec}(\mathbb{C}[\varepsilon]/\varepsilon^4)$ not contained in F_0 if and only if there exists a translation invariant differential operator $D_1 \neq 0$ on B and a non-zero constant c such that a differential equation of the following form is satisfied

$$(D_1^4 + cD_1^3 + \text{lower order terms}\,)\,\theta_\sigma\,)(\tau,\tfrac{b}{2}) = 0.$$

In view of the above conjecture we find :

(3.6) Conjecture. Let $Y = (B,\Theta)$ be the jacobian of a smooth curve of genus $g-1$ with $g \neq 4$ and let b be a point of B. Then $b \in C - C$ if and only if the 2nd order theta functions θ_σ of Y satisfy a differential equation of the form

$$(D_1^4 + cD_1^3 + \text{lower order terms}\,)\,\theta_\sigma\,)(\tau,\tfrac{b}{2}) = 0$$

with $D_1 \neq 0$ and $c \neq 0$.

References.

[D] Donagi, R. : The Schottky Problem. Montecatini Lectures, U.M.I. 1987. In : The Theory of Moduli. (Ed. E. Sernesi).Springer Lecture Notes in Math. **1337** (1988), pp 84-137.

[vGvdG] Van Geemen, B., Van der Geer, G.: Kummer varieties and the moduli spaces of abelian varieties. Am. J. **108**, (1986),615-642.

[K] Klein : Über die Plückersche Komplexfläche. Math. Annalen **7** (1874) .

[M] Mumford,D. : On the Kodaira dimension of The Siegel Modular Variety. In : Algebraic Geometry, Open Problems. (Eds. C. Ciliberto, F. Ghione, F. Orecchia) Lecture Notes in Math. **997**,pp. 348-375.

[W] Welters,G.: The surface C–C on Jacobi varieties and 2nd order theta functions. Acta Math. **157** (1986),pp. 1-22.

Author's Address: Mathematisch Instituut
Universiteit van Amsterdam
Roetersstraat 15
1018 WB Amsterdam
The Netherlands.

Sur une conjecture de Griffiths et Harris

Claire Voisin
Département de Mathématiques, Bâtiment 425
Université de Paris–Sud
91405 ORSAY Cedex

0.– Dans [4], Griffiths et Harris proposaient l'éventail suivant de conjectures concernant les courbes contenues dans une hypersurface générale de $\mathbb{P}^4$ de degré $d \geq 6$ (i.e. de fibré canonique ample).

i) on a : d divise $\deg(C)$

ii) l'image de l'application d'Abel–Jacobi : $\varphi_X : \mathrm{Hom}^2(X)/\mathrm{Rat}^2(X) \longrightarrow J^2(X)$ est réduite à zéro

iii) le groupe $\mathrm{Hom}^2(X)/\mathrm{Alg}^2(X)$ est trivial

iv) le groupe $\mathrm{Alg}^2(X)/\mathrm{Rat}^2(X)$ est trivial

v) si C est lisse, C est intersection complète $X \cap \Sigma$, de X et d'une surface Σ de $\mathbb{P}^4$.

Mark Green a expliqué dans son exposé les progrès récents concernant ii); cette note se propose de montrer que v) est faux, ainsi d'ailleurs que l'énoncé v′) suivant, qui est plus faible;

v′) la suite exacte normale de $C \subset X \subset \mathbb{P}^4$ est scindée.

Je remercie le C.I.R.M. et l'Université de Trento pour l'excellent accueil qui nous a été fait lors de ce congrès, ainsi que C. Ciliberto et E. Ballico pour m'avoir autorisée à inclure ces remarques dans leurs "proceedings".

1.– **Contre–exemple à v)**

1.1.– On supposera $d > 2$, le cas $d = 2$ étant trivial, puisque toute quadrique contient une droite.

Soit $X \subset \mathbb{P}^n$, $n \geq 4$, une hypersurface lisse de degré d, et soit $C \subset X$ une courbe lisse; supposons qu'il existe une surface $\Sigma \subset \mathbb{P}^n$ telle que C soit l'intersection complète de X et de Σ. Comme C est lisse, Σ est lisse le long de C, de sorte que $\mathrm{Sing}\,\Sigma$ est constitué de points isolés non situés sur C. Soit $\tau : \tilde{\Sigma} \longrightarrow \Sigma$ une désingularisation de Σ. On a une inclusion naturelle $C \subset \tilde{\Sigma}$, et C est un membre du système linéaire $|\tau^* \mathcal{O}_\Sigma(d)|$ sur $\tilde{\Sigma}$. La classe de C dans $H^2(\tilde{\Sigma}, \mathbb{Z})$ est donc divisible par d, ce qui entraîne :

– d divise $(C^2)_\Sigma$

– d divise $(K_\Sigma \,.\, C)_\Sigma$

la formule d'adjontion donne alors :

1.2.– d divise $\deg(K_C)$.

1.3.– Considérons maintenant la courbe à point double ordinaire D constituée de deux sections planes lisses $P_1 \cap X = C_1$, $P_2 \cap X = C_2$ de X, se rencontrant transversalement en un point p. Une telle courbe existe car $n \geq 4$.

On a, pour $i = 1,2$:

a) d divise $\deg(C_i)$

b) d divise $\deg(K_{C_i})$.

D'après b) on a alors : $\deg(K_D) = \deg(K_{C_1}) + \deg(K_{C_2}) + 2 \equiv 2$ (modulo d). Soit alors S une surface lisse intersection complète $X \cap X_1 \cap .. \cap X_{n-3}$, contenant D, et soit $D' \subset S$ un membre lisse du système linéaire $|\, m\,H + D|$ sur S; (D' existe pour m suffisamment grand).

On a :

$$\begin{aligned}\deg(K_{D'}) &= D'^2 + K_S \,.\, D' = (D+mH)^2 + (K_S \,.\, D+mH)\\ &= \deg(K_D) + 2m \deg(D) + m^2 H^2 + m\, K_S \,.\, H\end{aligned}$$

les deux derniers termes sont divisibles par d; d'après a) $\deg(D)$ l'est également, d'où : $\deg(K_{D'}) \equiv 2$ (modulo d).

Comme $d > 2$, D' ne satisfait pas la condition 1.2, et fournit un contrexemple à v).

2.– Contre–exemple à v')

2.1.– On supposera désormais (pour simplifier) que $n = 4$.

Reprenons la courbe $D = C_1 \cup_p C_2$ du paragraphe 1.

On va montrer les faits suivants :

(A) la suite exacte normale de $D \subset X \subset \mathbb{P}^4$ n'est pas scindée.

(B) Soit $S = X \cap X'$, une surface lisse contenant D, avec $\deg X' = k$ suffisamment grand; soit D' une courbe lisse du système linéaire $|\, mH + D|$ sur S, avec m suffisamment grand; alors la suite exacte normale de $D' \subset X \subset \mathbb{P}^4$ n'est pas scindée.

2.2. **Lemme** : (A) $\Longrightarrow$ (B).

Démonstration : On notera $e_D \in H^1(N_D X(-d))$ la classe de l'extension $0 \longrightarrow N_D X \longrightarrow N_D \mathbb{P}^4 \longrightarrow \mathcal{O}_D(d) \longrightarrow 0$, et pour toute hypersurface X' telle que $X' \cap X = S$, on notera $F_D^{X'} \in H^1(N_D S(-d))$ la classe de l'extension : $0 \longrightarrow N_D S \longrightarrow N_D X' \longrightarrow \mathcal{O}_D(d) \longrightarrow 0$

– mêmes notations pour D'.

Considérons la suite exacte : $0 \longrightarrow N_D S \longrightarrow N_D X \longrightarrow \mathcal{O}_D(k) \longrightarrow 0$; elle fournit une flèche $\alpha : H^1(N_D S(-d)) \longrightarrow H^1(N_D X(-d))$, telle que $\alpha(F_D^{X'}) = e_D$.

i) Supposons S fixée, et soit m tel que $H^1(\mathcal{O}_S(-D)(k-d-m)) = 0$: cela entraîne : $H^0(\mathcal{O}_S(k-d)) \longrightarrow\!\!\!\!\!\rightarrow H^0(\mathcal{O}_{D'}(k-d))$, pour toute courbe D' dans le système linéaire $|mH + D|$ sur S. On en déduit immédiatement : si $e_{D'} = 0$, il existe une hypersurface X'' de degré k, telle que $X \cap X'' = S$, et $F_{D'}^{X''} = 0$.

ii) Considérons la suite exacte : $0 \longrightarrow \mathcal{O}_S(-d) \longrightarrow \mathcal{O}_S(D'(-d)) \longrightarrow N_{D'} S(-d) \longrightarrow 0$. Elle fournit une flèche $\delta_{D'} : H^1(N_{D'} S(-d)) \longrightarrow H^2(\mathcal{O}_S(-d))$; on a par ailleurs l'application naturelle donnée par le cup–produit :

$$\beta : H^2(\mathcal{O}_S(-d)) \longrightarrow \mathrm{Hom}(H^0(\mathcal{O}_S(d)), H^2(\mathcal{O}_S));$$

on vérifie facilement que β est injective, dès que $K_S \geq 0$.

Il est alors bien connu que l'image $\beta \circ \delta_{D'}(F_{D'}^{X''}) \in \mathrm{Hom}(H^0(\mathcal{O}_S(d)), H^2(\mathcal{O}_S))$ s'identifie au composé : $H^0(\mathcal{O}_S(d) \longrightarrow H^1(T_S) \xrightarrow{[\lambda_{D'}]} H^2(\mathcal{O}_S)$, où $[\lambda_{D'}]$ est le cup–produit par la classe $\lambda_{D'} \in H^1(\Omega_S)$ de D', et la flèche $H^0(\mathcal{O}_S(d)) \longrightarrow H^1(T_S)$ provient de la suite exacte :

$$0 \longrightarrow T_S \longrightarrow T_{X''|S} \longrightarrow \mathcal{O}_S(d) \longrightarrow 0.$$

Il est alors facile de vérifier que $[\lambda_{D'}]$ ne dépend que de la "classe de cohomologie primitive" de D′, i.e. $[\lambda_{D'}] = [\lambda_{D''}]$ si $\lambda_{D'} = \lambda_{D''} + n\lambda_H$, $n \in \mathbb{Z}$.

On en déduit que $[\lambda_{D'}] = [\lambda_D]$.

iii) Choisissons alors k, (et S), tels que l'on ait : $H^1(\mathcal{O}_S(D(-d))) = 0$ (il est facile de voir que cette condition est satisfaite pour k assez grand).

iv) Supposons par l'absurde que $e_{D'} = 0$, où D' est lisse et choisie comme en i) : il existe alors X'', telle que $F_{D'}^{X''} = 0$. On en déduit que $\beta \circ \delta_{D'}(F_{D'}^{X''}) = 0$, et, d'après ii) que $[\lambda_{D'}] = 0$. Toujours d'après ii), il vient $[\lambda_D] = 0$, d'où $\beta \circ \delta_D(F_D^{X''}) = 0$. Or le choix de k, fait en iii), entraîne que δ_D est injective. Comme β est également injective, on en déduit $F_D^{X''} = 0$, et immédiatement $e_D = 0$, ce qui contredit (A).

2.3.– **Preuve de (A)** : la suite exacte normale de $D \subset X \subset \mathbb{P}^4$ s'écrit : $0 \longrightarrow N_D X \longrightarrow N_D\mathbb{P}^4 \longrightarrow \mathcal{O}_D(d) \longrightarrow 0$; il est clair qu'il suffit de prouver : $h^0(N_D\mathbb{P}^4(-d)) = 0$.

Considérons les suites exactes suivantes : (cf. [6]).

$$(E_1) \qquad 0 \longrightarrow N_D\mathbb{P}^4(-d) \longrightarrow N_D\mathbb{P}^4(-d)_{|C_1} \oplus N_D\mathbb{P}^4(-d)_{|C_2} \longrightarrow N_D\mathbb{P}^4(-d)_{|P} \longrightarrow 0$$

$$(E_2) \qquad 0 \longrightarrow N_{C_1}\mathbb{P}^4(-d) \longrightarrow N_D\mathbb{P}^4(-d)_{|C_1} \longrightarrow \mathbb{C}_P \longrightarrow 0.$$

On a : $h^0(N_{C_1}\mathbb{P}^4(-d)) = 1 = h^0(N_{C_2}\mathbb{P}^4(-d))$.

Il suffit donc de montrer :

i) $H^0(N_D\mathbb{P}^4(-d)_{|C_i}) \simeq H^0(N_{C_i}\mathbb{P}^4(-d))$

et

ii) $H^0(N_{C_1}\mathbb{P}^4(-d)) \oplus H^0(N_{C_2}\mathbb{P}^4(-d)) \hookrightarrow H^0(N_D\mathbb{P}^4(-d)_{|P})$.

i) Par Riemann–Roch et par dualité, $H^0(N_D\mathbb{P}^4(-d)_{|C_i}) = H^0(N_{C_i}\mathbb{P}^4(-d))$ si et seulement si l'inclusion $H^0(N_D\mathbb{P}^{4*}(-d)\otimes K_{C_i}) \hookrightarrow H^0(N_{C_i}\mathbb{P}^{4*}(d)\otimes K_{C_i})$ est stricte; or, pour $d \geq 3$, le faisceau $N_{C_i}\mathbb{P}^{4*}(d)\otimes K_{C_i}$ est engendré par ses sections globales. La conclusion est donc immédiate, au vu de la suite exacte duale de (E_2).

ii La section de $H^0(N_{C_i}\mathbb{P}^4(-d))$ provient de la section canonique de $N_{C_i}P_i(-d)$ pour $i = 1,2$. L'assertion résulte immédiatement du fait que les espaces tangents de P_1 et P_2 sont transversaux au point p, et la description locale de $N_D\mathbb{P}^4$.

2.4.– Remarque : Il est naturel de penser qu'une courbe du type $C_1 \cup_p C_2$ fournisse des contrexemples à v) et v′); en effet considérons la surface réduite $P = P_1 \cup_p P_2$, union des plans P_1 et P_2 se coupant transversalement au point p. Alors son intersection schématique avec X n'est pas la courbe réduite $C_1 \cup_p C_2$, mais possède un point immergé, de sorte que D n'est qu'ensemblistement l'intersection $P \cap X$.

3.– Concernant les autres points de la conjecture de Griffiths et Harris, on peut faire la remarque (peut–être évidente) suivante :

3.1. **Lemme** : ii) $\Longrightarrow$ i).

Démonstration : Supposons qu'une hypersurface X générale contienne une courbe de degré m, et que l'application d'Abel Jacobi φ_X soit nulle.

Il existe une variété irréductible W munie d'une application propre $p : W \longrightarrow \mathscr{X}$, $\mathscr{X} = \mathbb{P}(H^0(\mathbb{P}^4, \mathcal{O}(d)))$, telle que la fibre de p en X paramètre des courbes de degré m contenues dans X; deux telles courbes sont homologues et pour X générale, on a : $\forall\, C, C' \in p^{-1}(X)$, $\varphi_X(C - C') = 0$; en fait, ceci reste vrai pour tout X lisse : En effet, si H dénote une section plane de X, on a, pour X générique, $\varphi_X(dC - mH) = 0$, $\forall\, C \in p^{-1}(X)$. Par irréductibilité de W, ceci reste vrai pour tout $X \in \mathscr{X}$ Donc $\varphi_X(C - C')$ est un point de torsion, constant sur les composantes connexes de $p^{-1}(X) \times p^{-1}(X)$. Mais la normalité de $\mathscr{X}$ et l'irréductibilité de W entraînent que si $W \longrightarrow W_1 \longrightarrow X$ est la factorisation de Stein de p, chaque composante irréductible du produit $W_1 \times_{\mathscr{X}} W_1$ domine $\mathscr{X}$. Ce qui entraîne facilement le résultat

Fixons une droite Δ de $\mathbb{P}^4$, et notons $\mathscr{X}_\Delta$ la famille des hypersurfaces de degré d contenant Δ. Notons $W_\Delta = p^{-1}(\mathscr{X}_\Delta)$; on a alors une fonction normale ν_Δ définie comme suit sur $\mathscr{X}_\Delta$: soit X lisse $\in \mathscr{X}_\Delta$ et soit $C \in p^{-1}(X)$; alors $\deg(m\Delta - C) = 0$ et l'on peut poser $\nu_\Delta(X) = \varphi_X(m\Delta - C)$.

Or il est connu que le groupe des fonctions normales sur $\mathcal{X}_\Delta$ est cyclique engendré par la fonction normale ν_Δ^H définie par : $\nu_\Delta^H(X) = \varphi_X(d\Delta - H)$ (il suffit de généraliser l'argument de [4], § 3). On en déduit qu'il existe un entier k tel que : $\forall\, X$ lisse $\in \mathcal{X}_\Delta$, $\Phi_X(m\Delta - C) = k\,\varphi_X(d\Delta - H)$, pour $C \in p^{-1}(X)$.

Comme Δ se déforme continuement sur Δ' on a en fait $k = k'$; sur $\mathcal{X}_\Delta \cap \mathcal{X}_{\Delta'}$, il vient donc : $(m-kd)\Phi_X(\Delta-\Delta') = 0$. Mais d'après Griffiths [3], si X est générale dans $\mathcal{X}_\Delta \cap \mathcal{X}_{\Delta'}$, $\Phi_X(\Delta-\Delta') \in J(X)$ n'est pas un point de torsion. Donc $m - kd = 0$, ce qui prouve i).

4.– Conclusion : En paragraphe 1 on a dégagé la condition 1.2 nécessaire pour qu'une courbe C soit complète intersection $X \cap \Sigma$ de X et d'une surface Σ de $\mathbb{P}^4$. Si d divise le degré de C, cette condition est automatiquement satisfaite lorsque C est sous–canonique (i.e. $\exists\ m / K_C = \mathcal{O}_C(m)$). De même, il semble difficile de construire par des procédés analogues à celui décrit en paragraphes 1 et 2, des courbes sous–canoniques qui n'ont pas la suite exacte normale scindée. Il n'est donc pas exclu que v), v′) soient vrais pour les courbes sous–canoniques.

BIBLIOGRAPHIE

[1] G. Ellingsrud, L. Gruson, C. Peskine, S.A. Stomme.– On the normal bundle of curves on smooth projective surfaces, Invent. Math. 80, 181–184 (1985).

[2] M. Green.– Griffiths' infinitesimal invariant and the Abel Jacobi map, preprint.

[3] P. Griffiths.– On the periods of certain rational integrals I, II, Ann. Math. 90 (1969) 460–541.

[4] P. Griffiths, J. Harris.– On the Noether–Lefschetz theorem and some remarks on codimension two cycles, Math. Ann. 271, 31–51, (1985).

[5] J. Harris, K. Hulek.– On the normal bundle of curves on complete intersection surfaces, Math. Ann. 264, 129–135, (1983).

[6] R. Hartshorne, A. Hirschowitz.– Smoothing algebraic space curves, dans *Algebraic geometry*, Sitjes, (1983), Lecture Notes in Math. N° 1124.

OPEN PROBLEMS

collected by E. Ballico and C. Ciliberto

(1) (E. Ballico - C. Ciliberto) Let C be a general k-gonal curve of genus g. The philosophical question is if the only restriction on the possible g^r_d on C is related to the g^1_k on C (as for instance in the case k = 2). For "related" we means: " either contained in a multiple of the g^1_k+fixed points, or residual to such a series". Is this a good definition of "related" ? Or are there other universal ways to produce g^r_d's from a g^1_k ?) (a) is $G^r_d(C)$ smooth of dimension ρ away from the set of linear series related to the g^1_k ? If r = 1 and $\rho<0$, this is true ([AC]) (b) Let $G^r_d(C)$ be closure in $G^r_d(C)$ of the set of linear series not related to the g^1_k; is it smooth irreducible of dimension ρ ? (c) Determine the arithmetic genus of $G^r_d(C)$ and (more difficult) compute the class of its image in the Jacobian. In particular if $\rho = 0$ compute its cardinality; if r = 1, Castelnuovo in [C] suggested an iterative formula justified by C. Ciliberto. (d) If $\rho = 0$, what is the action of the monodromy on $G^r_d(C)$? P. Pirola made the following useful remark. If d = g-1, for every C, $W^r_d(C)$ has an involution (send D to K_C-D); in particular if $\rho = 0$ and d = g-1 the monodromy is not doubly transitive (this gives in this case a counterexample to the conjecture raised at the end of [EH2] and contradicts in this case the union of [EH2] and [BP] (e) $G^r_d(C)$ seems interesting also for non general C; (e1) classify all curves with $G^r_d(C) = \emptyset$; if r = 1, this is easy and the curves are on a quadric cone in P^3; (e2) is $G^r_d(C)$ always connected if $\rho>0$? Remarks on this problem are in [B],in which the range of the dimensions of the linear series related to the g^1_k is determined.

(2) (E. Ballico) In [S] (and [BE4], using [HH]) for many r,d,g was costructed an irreducible component V(d,g,r) of the Hilbert scheme of smooth curves of genus g and degree d in P^n with many good properties; in [EH1] it was constructed another component A(d,g,r) of the same Hilbert scheme. If $\rho(d,g,r)\geq 0$ V(d,g,r) = A(d,g,r) since both contains curves with general moduli (if $\rho(d,g,r) = 0$ one have to use also a weak form of [EH2]). The result of [EH3] imples that V(d,g,r) = A(d,g,r) also if $\rho(d,g,r) = -1$. Is this true under more general circumstances ?

(3) (C. Ciliberto) Let C be a general curve of genus g and take d, r with $\rho(d,g,r) = 2$. Determine $NS(G^r_d)$ (see [Pi]).

(4) (C. Ciliberto) Describe the group of rationally determined line bundles on the Severi variety.

(5) (a) (S. Greco) For a curve C put $S_C = \{n \in N \mid$ there is a g^1_n without base points$\}$. Question: Compute S_C for a general smooth plane curve C of degree d≥15. (b) (E. Ballico) Do the same for plane curves with a small number of singularities and say as much as possible when the curves are not general.

(6) (a) (G. Martens) What can be proved about curves C having infinitely many g^1_k not composed with the same involution but no g^1_t with t<k ? If k≥6 and g(C)≥k(k-1)/2 - (k-3), then C has a birational model as a plane curve of degree k+1 ([Co1])(b) (C. Ciliberto) More generally what can be said about curves with a g^1_k such that the Petri map related to the g^1_k is not of maximal rank ? (c) (related to a problem mentioned to C. Ciliberto by E. Sernesi) Determine the locus of curves in M_g with a g^r_d for which the Petri map fails to be injective. References: [Coj],1≤j≤6.

(7) (P. Craighero) Let us call a variety $V \subset A^n$ an m-flat (m<n) if it is isomorphic to A^m .An m-flat F is said to be linearizable if there is $g \in Aut(A^n)$ such that g(F) is a linear subspace of A^n. It has been proved ([Cr2]) that if m≤(n-1)/3, every m-flat is linearizable (even by an automorphism g which is tame (i.e. a product of linear and triangular automorphisms)). This implies that every 1-flat is linearizable in A^n if n≥4. Abhyankar and Moh ([AM]) proved that a 1-flat in A^2 is tamely linearizable. So, what about 1-flats in A^3 ? In the 70's Abhyankar ([A]) conjectured that there are 1-flats in A^3 which are not linearizable and among them the curves $C_n = \{t+t^n, t^{n-1}, t^{n-2}\}$. It has been proved ([Cr1],[Cr3]) that C_5 and C_6 are linearizable. What about C_7 ? Moreover, what about the curves $C'_5 = \{t^5, t+t^4, t^3\}$ and $C''_5 = \{t^5, t^4, t+t^3\}$? If C'_5 and C''_5 are linearizable, what about the problem of linearization of every quintic 1-flat in A^3 ?

(8) (R. Gattazzo) Let D be a smooth quartic rational curve in $\mathbf{P}^3(C)$. E. Stagnaro proved that D is not set-theoretically the complete intersection of two surfaces, one of degree 3 and one of degree 4. Recently ([CG]) it has been proved that D is not the set-theoretical complete intersection of two surfaces, one of them of degree 3. It is known that the curve $C'_4 = \{t^4, t^3a, ta^3, a^4\}$ is not the set-theoretical complete intersection of two surfaces of degree 4. Open question: Show that C'_4 is not the set-theoretical complete intersection of two surfaces, one of them of degree 4.

(9) Is there any "reasonable" (e.g. polynomial) bound for the number of irreducible component of the Hilbert scheme parametrizing irreducible curves of fixed degree and genus in $\mathbf{P}^r$? Related work was done by F. Catanese for the components of the moduli scheme of surfaces of general type.

(10) (P. Pirola) What is the maximal number of nodes for curves C of given degree on a "general" surface of fixed degree in $\mathbf{P}^3$ (hence C complete intersection). (See also [No],3.18, and [T]).

(11) (C. Ciliberto) Is it true the following assertion ? Take an irreducible component X of the locus of surfaces of degree d≥5 in $\mathbf{P}^3$ for which the Noether-Lefschetz theorem fails to be true; assume that codim(X) (in the space of all degree d surfaces) is at most p_g-1 ($p_g = (d-1)(d-2)(d-3)/6$); then the general $S \in X$ contains some non-complete intersection curve which is contained in a surface of degree d-4. (Solved for d = 5 in [G4] and [V1]; related work has been done by Ch. Peskine).

(12) Fix a smooth curve $C \subset \mathbf{P}^n$, deg(C) = d, $p_a(C)$ = g, and let N_C be its normal bundle. Hilb($\mathbf{P}^n$) at C has tangent space of dimension $h^0(C,N_C)$ and has dimension$\geq\chi(N_C)$; in particular if $H^1(N_C) = 0$, Hilb($\mathbf{P}^n$) is smooth at C of the expected dimension. If n = 3, there are asymptotically very strong results about the vanishing of $H^1(N_C)$ ([ElH],[H],[W1]), roughly if $d\geq Kg^{2/3}$ with $K = (9/8)^{1/3}$. Since $\chi(N_C) = (n+1)d+(3-n)(1-g)$, if n>3, $h^1(N_C) \neq 0$ except in a very restrictive (i.e. linear) range for (d,g). Give reasonable upper and lower bounds for the dimensions of the Hilbert scheme of curves of degree d and genus g in $\mathbf{P}^n$ and for their image in M_g; for a discussion of these problems, see [BE7], [BE8] and [Ci3]. A striking related question (Ph. Ellia): is there any component of Hilb($\mathbf{P}^n$) (for g>0) which is the closure of an orbit of Aut($\mathbf{P}^n$) ? It seems very useful to have criteria for the smoothness of Hilb($\mathbf{P}^n$) ; one criterion is given in [K].

(13) (E. Sernesi) Let X be a smooth curve of genus g≥2 in $\mathbf{P}^3$ with general moduli. Does X has finitely many 4-secant lines ? Work on not too far questions is in [H1].

(14) (E. Ballico) Under what assumptions on r,d,g with $\rho(d,g,r)<0$ there is a "good" component X of the Hilbert scheme of smooth curves of degree d and genus g in $\mathbf{P}^r$ such that the general fiber of its map to M_g is an orbit under Aut($\mathbf{P}^r$)? This is not always true (e.g. d = 7,g = 8, r = 2) by Pirola's remark in problem (1)(d). For r = 1,see [AC]. For complete intersections and for g large with respect to d, see [Ci2] and [CL].

(15) (C. Ciliberto) Let V be an integral smooth non denerate variety in $\mathbf{P}^r$,dim(V) = n, with r = 2n+1, V not a ruled by lines. Is it possible that the tangent n-space at the general point of V intersects V at some other points ? (No if n = 1 [Ka] and Sacchiero (private communication)).

(16) (C. Ciliberto) Give informations about the structure of the endomorphism ring of the Jacobian of a general smooth curve of a component of the Hilbert scheme. In particular is there any example in which such a Jacobian is simple but with non trivial endomorphisms ?

(17) (Ch. Peskine) Find an irreducible family of smooth curves in $\mathbf{P}^3$ such that the general one is a complete intersection and the special one is smooth and not complete intersection.

(18) Construct bundles on $\mathbf{P}^n$ with rank≤n-2.

(19) (A. Hirschowitz) Let C be a "generic" curve in $\mathbf{P}^3$ and S a general surface of a given degree≥4 containing C; is Pic(S) freely generated by $\mathcal{O}_S(1)$ and the class of C ? The answer is known to be YES in some cases: see [Lo].

(20) (R. Hernandez - I. Sols) Let C be a non degenerate smooth curve of $\mathbf{P}^3$ (over the complex number field) and let C' be its projection from a generic point of $\mathbf{P}^3$. We conjecture that the nodes of C' (classically called "apparent" nodes of C) are in uniform position. Let e be the maximum integer such that $h^1(\mathcal{O}_C(e)) \neq 0$. It is an obvious fact that no curve of degree n<d-e-3 contains the apparent nodes. The fact they were in uniform position would imply then that in fact no (n+1)(n+2)/2 of these nodes lie in a curve of degree n. According to Harris's method (see [ACGH]) of proving uniform position principia we have to show first that 2 bisecant lines of the curve passing through a generic point can be transformed by monodromy (as the points travels around $\mathbf{P}^3$) in any other 2 bisecant lines, passing through the same point. Second, one transposition at least must be found in this monodromy: it must be proved the following

Conjecture: A non degenerate, connected smooth curves of P^3 has always a stationary bisecant L (i.e. with coplanar tangents to C at the points of $L \cap C$) which is not a trisecant (of couse one excludes the trivial case with deg(C) = 3).

(21) (E. Ballico, J. Rathmann,...) What are the possible monodromies for the family of hyperplane sections of a projective curve in characteristic $p > 0$. References: [Ra].

(22) (E. Ballico) Let $C \subset P^n$ be a reduced curve, C not arithmetically Cohen-Macaulay curve; let $R := k[x_0,...,x_n]$ be the graded ring of P^n and **m** the homogeneus ideal of R generated by $x_0,...,x_n$. The order ord(C) of C is the minimal integer t such that $m^t H^1(P^n, I_C(a)) = 0$ for all integers t; this notion was introduced in [E]. In [E], questions (2), it was asked if the fact that ord(C) is not maximal (i.e. not equal to the diameter of the Hartshorne-Rao module of C) has geometrical consequences. Is it true that all the embeddings of a curve with general moduli have maximal order ? Here "general moduli" may means either " in a Zariski open subeset of M_g " or " outside a countable union of proper subvarieties of M_g ". References: [E],[MM],[B1].

(23) (C. Ciliberto - E. Sernesi) Let d, N be integers with $d \geq 5$ and $0 < N < (d-1)(d-2)(d-3)/6$. Is there any irreducible surface S of degree d in P^3 with arithmetic genus N ?

(24) (E. Ballico) Fix a linear system A on a variety V (smooth, for instance) and a finite number of types of singularities, say $A^1,...A^s$. (a) Find reasonable conditions such that for general points $P^1,...,P^s$ of V there is $D \in A$ with singularity of type A^i at the point P^i for all $i, 1 \leq i \leq s$, and no other singularity. (b) Find reasonable conditions for the existence of points $P^1,...,P^s$, such that the conditions in (a) are satisfied. In (a) and (b) we are interested essentially in the case $V = P^n$ or a surface, and D complete linear system. See [H] and [Al] for (a), without the condition of non singularity outside the points P^i. The condition "no other singularity" (or the condition that the map given by the divisor in D having at P^i the singularity A^i to be an embedding of a suitable blowing-up of V) can be handled sometimes using ideas of [H],[BH] and references quoted there. Of course for (b) see the literature on plane nodal (or cuspidal or with mild singularities) in P^2 ([Ha],[R],[Kn],...) or in rational surfaces or in more general surfaces;compare with problem (10). Of course, (a) and (b) can be asked also for sections of vector bundles of rank bigger than 1.

(25) (E. Ballico) Find the minimal free resolution of important embeddings of important classes of curves (e.g. plane curves with very few singularities with respect to series which differ by a few points by a series $\mathcal{O}(t)$; bielliptic curves, general k-gonal curves). For the canonical embedding, everything should be governed by a conjecture of Lazarsfeld-Green (see [G1] and [Sc]).

(26) (E. Ballico - G. Bolondi) (a) It is known ([GP1]) what are the possible degrees, genera of the arithmetical normal curves in P^3. What are the possible degree, genera and cohomology group for the smooth curves which are specializations of a family of arithmetical normal space curves? (b) For what pair of graded modules of finite length A, B, over $R := k[x_0,...,x_3]$ there is an irreducible flat family T of (smooth) curves in P^3 such that for general $t \in T$, the corresponding curve has Hartshorne-Rao module isomorphic to A, while for one point $0 \in T$ the corresponding curve has Hartshorne-Rao module isomorphic to B? (c) With A and B as in (b), gives some condition (on A, B, d, g) such that there is a (smooth) space curve of degree d and genus with Hartshorne-Rao module isomorphic to B and that any such curve is a specialization of a flat family of space curves with Hartshorne-Rao module isomorphic to A.

(27) (E. Ballico) Let F be a rank-2 reflexive sheaf on $\mathbf{P}^3$. What can be said about the relations between the position and type of the non-locally free points of F and the cohomology modules (Hartshorne-Rao module) of F.

(28) (K. Hulek) Question: Let C be a smooth curve in $\mathbf{P}^3$. Assume that its normal bundle splits as $\mathcal{O}_C(a)\oplus\mathcal{O}_C(b)$. Is C a complete intersection? Remark: The answer to the analogous question for 2-codimensional smooth subvariety of $\mathbf{P}^n$ is true if $\dim(X)\geq 2$ by a result of Faltings. Van de Ven posed the following question some time ago. Question: Let X be a smooth surface in $\mathbf{P}^4$ whose normal bundle splits as a sum of line bundles. Is X a complete intersection? By recent work of Peskine this is true for smooth 2-codimensional subvarieties of $\mathbf{P}^n$ if $n\geq 5$.

(29) (J. Stevens) Let r(g) be the maximal integer such that r(g) general points of $\mathbf{P}^{g-1}$ are contained in a canonical curve of genus g. It is shown in [St] that $r(3) = 14$, $r(4) = 9$, $r(5) = 12$, $r(6) = 11$, $r(7) = 13$, $r(g) = g+5$ if $g\geq 9$ and $13\leq r(8)\leq 14$. Conjecture: $r(8) = 13$.

(30) (C. Ciliberto) Classify all rational surfaces S in $\mathbf{P}^5$ such that at any point $p\in S$ there is a hyperplane of $\mathbf{P}^5$ cutting out on S a curve with a triple point at P. (Recommended reading: [Se]).

(31) (C. Ciliberto) (i) Find the rank of the Wahl map (see [W],[BM],[CHM]) on (the general,any?) curve of genus $g\leq 2$; (ii) compute the rank of the Wahl map on (the general,any?) curve of the following families: coverings of order 2 (n?) of elliptic curves, k-gonal curves (k = 3?); (iii) is the locus in M_g of hyperplane sections of a K3 an irreduciblke component of the locus where the Wahl map fails to be surjective? (iv) same problem for the hyperelliptic locus; (v) what is the rank of the Wahl map for the general hyperplane section of a general K3 ?

(32) (C. Ciliberto) Describe all correspondences of a general curve of genus $g\geq 1$ into itself; in particular find, if (or when) possible, the analougus of the Brill-Noether formula for the correspondences on a curve with general moduli, saying that if a certain number $f(n,m,g,\gamma)$ is ≥ 0, then on a general curve of genus g there is a $f(n,m,g,\gamma)$-dimensional continuos system of correspondences of indices (n,m) and valency γ. This is of course equivalent to studying the Hilbert scheme of all curves on the product (or symmetric product if one deals with symmetric correspondences) of a curve with general moduli by itself. For an approach to this problem, see [F] and [Ci1].

(33) (C. Ciliberto) Count the number of stick (or poligonal, or graph: see [BEi]) canonical curves. It is known an asymptotic formula.

(34) (Ph. Ellia) Let $X\subset\mathbf{P}^3$ be the nodal union of two smooth curves C,C' (i.e. $X = C\cup C'$, the intersection being quasi-transversal). Assume the natural map $H^0(N_X)\to H^0(T^1_X)$ is identically zero. Does this imply that X is not smoothable in $\mathbf{P}^3$?

(35) Let H(d,g) be the subset of Hilb($\mathbf{P}^3$) parametrizing smooth,irreducible curves of degree d and genus g in $\mathbf{P}^3$. For $\rho(d,g,3)\geq 0$ le $P(d,g)$ be the unique irreducible component of H(d,g) which dominates M_g. Show that (except a few exceptions like d = 6, g = 2) the general curve C of $P(d,g)$ satisfies $h^1(C,N_C(-2)) = 0$. (References: [ElH],[Hi],[Pe],[W1]).

(36) Determine $G(d,s):= \max\{g(C): C\subset \mathbf{P}^3$, C smooth connected of degree d and satisfying $h^0(\mathbf{P}^3,\mathcal{I}_C(s-1)) = 0\}$ ([GP1], [BE8]).

(37) Determine $s(d,g):= \min\{k$: every smooth connected curve of degree d and genus g is contained in a surface of degree k$\}$ ([BE8]).

(38) Give more examples of self-linked space curves with indecomposable Hartshorne-Rao module; in characteristic 2 one example is given in [SV],p.190-195 ; for self-linked curves, see also [Rao] and [Scw].

(39) (E. Sernesi) On a general curve C of genus g characterize the multiplication maps in cohomology that are of maximal rank. We know that the Petri map $H^0(D)\otimes H^0(K-D)\rightarrow H^0(K)$ is injective for all divisors D. We also know that for every divisor D of degree $d\geq 2g+1$ the maps $S^nH^0(D)\rightarrow H^0(nD)$ are surjective for all $n\geq 0$ (this is true for all curves) and that they are of maximal for a general D which embeds C in $\mathbf{P}^r$ under suitable hypothsis (see [BEj],$1\leq j\leq 8$).

(40) (E. Sernesi) Since all known examples of obstructed curves in $\mathbf{P}^3$ are not of maximal rank it is natural to ask: does there exist a smooth curve $C\subset \mathbf{P}^3$ of maximal rank which is obstructed? Or at least (asked by Ph. Ellia) is it true that for all smooth curves $C\subset \mathbf{P}^3$ with maximal rank every irreducible component of Hilb($\mathbf{P}^3$) containing [C] has maximal rank?

(41) (C. Ciliberto - E. Sernesi) Let $C\subset \mathbf{P}^3$ be a smooth curve corresponding to a general point of an irreducible components of Hilb($\mathbf{P}^3$); let N_C be the normal bundle of C in $\mathbf{P}^3$. Question: is it true that $h^1(N_C)\leq h^1(\mathcal{O}_C(2))$? (Note that both members vanishes for non special embeddings and that the inequality is true and is an equality for curves on a quadric). Sacchiero has produced an (unpublished) example which shows that the above inequality can be violated by particular curves in a component of Hilb($\mathbf{P}^3$). His example is a curve C of genus 12 and degree 13 with $h^1(N_C) = 1$ and $h^1(\mathcal{O}_C(2)) = 0$.

(42) (E. Sernesi) Do there exist components of $I^r{}_{d,g}$ such that $\rho(d,g,r)<0$ and having a number of moduli strictly smaller than expected ? References: [S] and [CS].

(43) (E. Arrondo - I. Sols) Clifford theorem in higher rank: Let E be a rank two vector bundle of degree d on a smooth curve C of genus g. Set $r+1:= h^0(E)$ and let $-e$ be the minimal self-intersection number of a unisecant curve in the ruled surface $\mathbf{P}(E)$. In an unpublished paper, M. Pedreira proves that if $h^0(E\otimes E^*) = 1$ and E is special, i.e. $h^1(E) \neq 0$, then $r\leq(d/2)+g-1$; for instance $h^0(E\otimes E^*) = 1$ for instance if $e<0$ i.e. if E is stable. We have checked that more can be asserted: if E is semistable, i.e. if $e\leq 0$, and E is special, then $r\leq(d/2)+1$ with equality holding if and only if $E = L\oplus L$, where L is a line bundle for which Clifford inequality is an equality; if $e>0$ and E is special, then $r\leq(d+e)/2$. Thus we state as an open problem the following

Conjecture: If $-e\leq d\leq 4g-4+e$ and $E \neq L\oplus L$, then $r\leq(d+e)/2$.

Note that if d is outside this range, then E i s not special. An evidence for this conjecture to be true is that we can prove it in the case C is hyperelliptic, which is just the extremal case in Clifford's bound for line bundles. For E of higher rank R, if we assume both E and $E^*\otimes K_C$ to be generically generated by global sections (we will refer to this saying that E is strongly special), then we can prove that $r\leq(d/2)+R-1$. Proving this bound for the merely special case, and sharpening this bound by involving degrees of stability are still open problems. Let W^r_d be the closure in the moduli space of semistable rank two bundles of degree d on C of the set of those bundles E which are generated by global sections and have $h^0(E) \geq r+1$. Translating Clifford bounds into Martens theorems we can prove that if $d \leq 2(2g-2)/3$ and $2r \leq d$, every irreducible component A of W^r_d with general element strongly special has $\dim(A) \leq d(R+1)/2 - (R+1)r$. For the other components of W^r_d this bound is an open problem, as well as the corresponding statement involving degrees of stability.

References

[A] S. S. Abhyankar: *On the semigroup of a meromorphic curve,* in : Intl.Symp. on Algebraic Geometry (Kyoto-1977), pp.249-414.

[AM] S. S. Abhyankar - T. T. Moh: *Embeddings of the line in the plane,* J.f.d.r.u.a.M. **276** (1975), 148-166.

[Al] J. Alexander: *Postulation dans P^n avec applications a la classification des surfaces rationnelles dans P^4,* thesis,Nice 1986.

[AC] E. Arbarello - M. Cornalba: *Footnotes to a paper of Beniamino Segre. The number of g^1_p's on a general p-gonal curve, and the unirationality of Hurwitz spaces of 4-gonal and 5-gonal curves,* Math. Ann. **256** (1981), 341-362.

[ACGH] E. Arbarello - M. Cornalba - Ph. Griffiths - J. Harris: *Geometry of algebraic curves,* vol.1 (1985) and vol.2 (to appear), Springer.

[B] E. Ballico: *A remark on linear series in general k-gonal curves,* Boll.U.M.I. (to appear).

[B1] E. Ballico: *On the order of projective curves,* preprint.

[BE1] E. Ballico - Ph. Ellia: *The maximal rank conjecture for non-special curves in P^3,* Invent. math. **79** (1985), 545-555.

[BE2] E. Ballico - Ph. Ellia: *The maximal rank conjecture for non-special curves in P^n,* Math. Z. **196** (1987), 355-367.

[BE3] E. Ballico - Ph. Ellia: *Beyond the maximal rank conjecture for curves in P^3,* in Space curves, pp.1-23, Lect. Notes in Math. **1266**, Springer.

[BE4] E. Ballico - Ph. Ellia: *On the existence of curves with maximal rank in P^n,* Crelle's Journal (to appear).

[BE5] E. Ballico - Ph. Ellia: *On the projection of a general curve in P^3,* Annali Mat. Pura e Applicata (4) **142** (1985), 15-48.

[BE6] E. Ballico - Ph. Ellia: *On the postulation of a general projection of a curve in $P^N, N \geq 4$,* Annali Mat. Pura e Applicata (4) **147** (1987), 267-301.

[BE7] E. Ballico - Ph. Ellia: *Bonnes petites composantes des schémas de Hilbert de courbes l isses de P^n,* Comptes Rend. Acad. Sc. Paris **306** (1988), 187-190.

[BE8] E. Ballico - Ph. Ellia: *A program for space curves,* Rend. Sem. Mat. Torino **44** (1986), 25-42.

[BEi] D. Bayer - D. Eisenbud: *Graph curves, note toward a paper,* preprint.

[BM] A. Beauville - J. Merindol: *Sections hyperplanes des surfaces K3*, Duke Math. J. **55** (1987), 873-878.

[BP] R. Bercov - R. Proctor: *Solution of a combinatorially formulated monodromy problem of Eisenbud and Harris*, Ann. scient. Ec. Norm. Sup. **20** (1987), 241-250.

[BH] J. Brun - A. Hirschowitz: *Le probléme de Brill-Noether pour les idéaux de P^2*, Ann. scient. Ec. Norm. Sup. **20** (1987), 171-200.

[C] G. Castelnuovo: *Opere scelte*, Zanichelli, Bologna (1937).

[Ci1] C. Ciliberto: *Sui sistemi di corrispondenze di una curva algebrica a moduli generali in sé*, Ricerche di Matematiche **28** (1979), 651-682.

[Ci2] C. Ciliberto: *Alcune applicazioni di un classico procedimento di Castelnuovo*, in Seminari di Geometria,Università di Bologna 1982-83.

[Ci3] C. Ciliberto: *Curve algebriche proiettive: risultati e problemi*, Atti Convegno GNSAGA del CNR, Torino 1984.

[CHM] C. Ciliberto - J. Harris - R. Miranda: *On the surjectivity of the Wahl map*,Duke Math. J. (to appear).

[CL] C. Ciliberto - R. Lazarsfeld: *On the uniqueness of certain linear series on some class of curves*, in Lect. Notes in Math. **1092**, pp. 198-213, Springer.

[CS] C. Ciliberto - E. Sernesi: *Families of varieties and the Hilbert scheme*, (to appear).

[Co1] M. Coppens: *Smooth curves having infinitely many linear systems g^1_d,I*, Bull. Soc. Math. Belg. **40** (1988), 153-176.

[Co2] M. Coppens: *Some sufficient conditions for the gonality of a smooth curve*, J. Pure Appl. Algebra **30** (1983), 5-21.

[Co3] M. Coppens: *On G. Martens' characterization of smooth plane curves*, Bull. London Math. Soc. **20** (1988), 217-220.

[Co4] M. Coppens: *Some remarks on the scheme W^r_d*, preprint.

[Co5] M. Coppens: *Smooth curves possessing many linear systems g^1_n*, preprint.

[Co6] M. Coppens: *A study of the scheme W^1_e of smooth plane curves*, preprint.

[Cr1] P.C. Craighero: *About Abhyankar's conjectures on space lines*, Rend. Semin. Mat. Univ. Padova **74** (1985), 115-122.

[Cr2] P. C. Craighero: *A result on m-flats in $A^n{}_k$*, Rend. Semin. Mat. Univ. Padova **75** (1986), 39-46.

[Cr3] P. C. Craighero: *A remark on Abhyankar's space lines*, Rend. Semin. Mat. Univ. Padova **80** (1988) (to appear).

[CG] P. C. Craighero - R. Gattazzo: *No rational non singular quartic curve $C^4 \subset P^3$ can be set-theoretic complete intersection on a cubic surfaces*, to appear on Rend. Semin. Mat. Univ. Padova **81** (1989).

[EH1] D. Eisenbud - J. Harris: *Limit linear series: Basic theory*, Invent. math. **85** (1986), 337-371.

[EH2] D. Eisenbud - J. Harris: *Irreducibility and monodromy of some families of linear series,* Ann. scient. Ec. Norm. Sup. **20** (1987), 65-87.

[EH3] D. Eisenbud - J. Harris: *Irreducibility of some families of linear series with Brill-Noether number - 1,* preprint.

[E] Ph. Ellia: *Ordres et cohomologie des fibres de rang deux sur l'espace projectif,* preprint.

[ElH] G. Ellingsrud - A. Hirschowitz: *Sur le fibré normal des courbes gauches,*C. R. Acad. Sci. Paris **299** (1984), 245-248.

[F] A. Franchetta: *Sulla superficie delle coppie non ordinate di punti di una curva algebrica a moduli generali,* Rendiconti di Matematica e Appl. V, 7 (1948), 327-367.

[G1] M. Green: *Koszul cohomology and the geometry of projective varieties,* J. Differential Geometry **19** (1984), 125-171.

[G2] M. Green: *Koszul cohomology and the geometry of projective varieties,II,* J. Differential Geometry **20** (1984), 279-289.

[G3] M. Green: *A new proof of the explicit Noether-Lefschetz theorem,* J. Differential Geometry **27** (1988), 155-159.

[G4] M. Green: *Components of maximal dimension in the Noether-Lefschetz locus,* J. Differential Geometry (to appear).

[GP1] L. Gruson - Ch. Peskine: *Genre des courbes de l'espace projectif,* in: Proc. Trømso 1977, p.31-59, Lect. Notes in Math. **687**,Springer.

[Ha] J. Harris: *On the Severi problem,* Invent. math. **84** (1986), 445-461.

[HH] R. Hartshorne - A. Hirschowitz: *Smoothing algebraic space curves,* in: Algebraic Geometry,Sitges 1983, Lect. Notes in Math. **1124**, Springer.

[H] A. Hirschowitz: *La méthode d'Horace pour l'interpolation à plusieurs variables,* Manuscripta Math. **50** (1985), 337-388.

[H1] A. Hirschowitz: *Sections planes et multisecantes pour les courbes gauches generiques principales,* in: Space curves,pp.124-155, Lect. Notes in Math. **1266**, Springer.

[Ka] H. Kaji: *On the tangentially degenerate curves,* J. London Math. Soc. (2) **33** (1986), 430-440.

[Kn] P.-L. Kang: *On the variety of plane curves of degree d with δ nodes and x cusps,* Mem. A.M.S. (to appear).

[K] J. Kleppe: *Liaison of families of subschemes in P^n,* this volume.

[Lo] A. Lopez: *Sul gruppo di Picard di superfici liscie in P^3,* tesi di dottorato,Roma 1988.

[MM] J. Migliore - R. M. Miro'-Roig: *On k-Buchsbaum curves in P^3,* preprint.

[No] M.V. Nori: *Zariski's conjecture and related problems,* Ann. scient. Ec. Norm. Sup. **16** (1983), 305-344.

[Pe] D. Perrin: *Courbes passant par m points généraux de P^3,* Bull. Soc. Math. France, Memoire **28/29** (1987).

[Pi] G.P. Pirola: *On the Néron-Severi groups of the surfaces of special divisors,* this volume.

[R] Z. Ran: *On nodal plane curves,* Invent. math. **86** (1986), 529-551.

[Rao] P. Rao: *On self-linked curves,* Duke Math. J. **49** (1982), 251-273.

[Ra] J. Rathmann: *The uniform position principle for curves in characteristic p,* Math. Ann. **276** (1987),565-579.

[Sc] F.-O. Schreyer: *Syzygies of canonical curves and special linear series,* Math. Ann. **275** (1986), 105-137.

[Scw] P. Schwartau: *Liaison addition and monomial ideals,* Ph. D. thesis Brandeis Univ. 1982.

[Se] C. Segre: *Su una classe di superficie degli iperspazi legate colle equazioni lineari alle derivate parziali di 2° ordine,* Atti Accad. Sci. Torino **42** (1906-1907), 559-591, also in: Opere scelte, vol.2, Ed. Cremonese,1958.

[S] E. Sernesi: *On the existence of certain families of curves,* Invent. math. **75** (1984), 25-57.

[St] J. Stevens: *On the number of points determining a canonical curve,* preprint.

[SV] J. Stuckrad - W. Vogel: *Buchsbaum rings and applications,* Berlin,Heidelberg,New York, Springer (1986).

[T] A. Tannenbaum: *Families of curves with nodes on K-3 surfaces,* Math. Ann. **260** (1982), 239-253.

[V] C. Voisin: *Une precision concernant le théorèm de Noether,* preprint.

[V1] C. Voisin: *Composantes du lieû de Noether-Lefschetz en degré cinq,* preprint.

[V2] C. Voisin: *Sur une conjecture de Griffiths et Harris,* this volume.

[W] J. Wahl: *The Jacobian algebra of a graded gorenstein singularity,* Duke Math. J. **55** (1987), 843-872.

[W1] C. Walter: Ph. D. Thesis, Harvard,November 1987.

LIST OF PARTICIPANTS

Alberto ALZATI, Dipartimento di Matematica, Università, Via C. Saldini 50, 20133 Milano, Italy

Marco ANDREATTA, Dipartimento di Matematica, Università di Milano, Via C. Saldini 50, 20133 Milano, Italy. E-mail address: ANDREATTA at ITNCISCA.BITNET

Enrique ARRONDO, c/o Prof. Ignacio SOLS, Departamento de Algebra, Facultad de Matematicas, Universidad Complutense, 28040 Madrid, Spain

Edoardo BALLICO, Dipartimento di Matematica, Università di Trento, 38050 Povo (TN), Italy. E-mail address: BALLICO at ITNCISCA.BITNET

Fabio BARDELLI, Dipartimento di Matematica, Università di Pavia, Strada Nuova 65, 27100 Pavia, Italy

Margarida BARRO MOREIRA, R. Dr. Sousa Rosas 348, 4100 Porto, Portugal

Giorgio BOLONDI, Dipartimento di Matematica e Fisica, Facoltà di Scienze, Università, 62032 Camerino (MC), Italy

Michela BRUNDU, Dipartimento di Scienze Matematiche, Università, P.zzale Europa 1, 34100 Trieste, Italy. E-mail address: TI2TSG22 at ICINECA2.BITNET

Gianfranco CASNATI, Via Golena, 18, 44020 Cocomaro di Ferrara, Italy

Maria Virginia CATALISANO, Via dello Scoglio 28, 16030 Cavi di Lavagna (Genova), Italy

Giuseppe CERESA, Dipartimento di Matematica, Università de L'Aquila, Via Roma, 67100 L'Aquila, Italy

Luca CHIANTINI, Dipartimento di Matematica, Università di Napoli, Via Mezzocannone 8, 80134 Napoli, Italy

Ciro CILIBERTO, Dipartimento di Matematica, II Università Roma, Via Orazio Raimondo, 00173 Roma, Italy. E-mail address: CILIBERTO at MVXTVM.INFNET

Maria Grazia CINQUEGRANI, Dipartimento di Matematica, Città Universitaria, Viale A. Doria 6, 95125 Catania, Italy

Alberto COLLINO, Dipartimento di Matematica, Università di Torino, Via Principe Amedeo 8, 10123 Torino, Italy

Katia CONSANI, Corso De Michiel 14, Chiavari (GE), Italy

Maria CONTESSA, Istituto di Matematica, Università di Palermo, Via Archirafi 34, 90123 Palermo, Italy

Pasqualina CONTI, Istituto di Matematica Applicata, Facoltà di Ingegneria, Università di Pisa, 56100 Pisa, Italy

Alessio CORTI, Scuola Normale Superiore, P.zza dei Cavalieri 7, 56100 Pisa, Italy. E-mail address: ALESSIO at IPISNSVA.BITNET

Pier Carlo CRAIGHERO, Istituto di Matematica Applicata, Università, Via Belzoni 7, 35131 Padova, Italy

Caterina CUMINO, Via Petitti 35, 10126 - Torino, Italy

Alberto DOLCETTI, Via del Naviglio 24, 44100 Ferrara, Italy. E-mail address: T99A at ICINECA.BITNET

Philippe ELLIA, Mathématiques, Université de Nice, Parc Valrose, F-06034 Nice Cedex, France

Carel FABER, Department of Mathematics, University Amsterdam, Roetersstraat 15, 1018 WB Amsterdam, The Netherlands

Barbara FANTECHI, Scuola Normale Superiore, Piazza Cavalieri 7, 56100 Pisa, Italy. E-mail address: BARBARA at IPISNSVA.BITNET

Mario FIORENTINI, Dipartimento di Matematica, Università, Via Machiavelli 35, 44100 Ferrara, Italy

Emma FRIGERIO, Dipartimento di Matematica, Università, Via Saldini 50, 20133 Milano, Italy

Remo GATTAZZO, Istituto di Matematica Applicata, Università, Via Belzoni 7, 35131 Padova, Italy

Alessandro GIMIGLIANO, Dipartimento di Matematica, II Università Roma, Via Orazio Raimondo, 00173 Roma, Italy

Salvatore GIUFFRIDA, Dipartimento di Matematica, Università, Viale A. Doria 6, 95125 Catania, Italy

Silvio GRECO, Dipartimento di Matematica, Politecnico, Corso Duca degli Abruzzi 24, 10129 Torino, Italy

Mark GREEN, Department of Mathematics, University of California, 405 Hilgard Avenue Los Angeles, CA 90024, U. S. A.. E-mail address: MLG at MATH.UCLA.EDU

Rafael HERNÀNDEZ, Department of Mathematics, Brown University, Providence, R.I. 02912, U.S.A.. E-mail address: ST402175 at BROWNVM.BITNET

André HIRSCHOWITZ, Mathématiques, Université de Nice, Parc Valrose, F-06034 Nice Cedex, France. E-mail address: BLH at FRMOP11.BITNET

Jan KLEPPE, Oslo College of Engineering, Cort Adelersgt. 30, N-0254 Oslo 2, Norway

Monica IDA', Dipartimento di Scienze Matematiche, Università, P.zzale Europa 1, 34127 Trieste, Italy. E-mail address: TI2TSG22 at ICINECA2.BITNET

Maurizio IMBESI, Via Tenente A. Genovese 26, 98051 Barcellona P.G. (ME), Italy

Herbert H. LANGE, Mathematisches Institut, Universität Erlangen, Bismarckstraße 1 1/2, D-8520 Erlangen, West Germany. E-mail address: MPMA23 at DERRZE0

Angelo LOPEZ, Department of Mathematics, Brown University, Providence, R.I. 02912, U.S.A.. E-mail address: ST402423 at BROWNVM.BITNET

Renato MAGGIONI, Dipartimento di Matematica, Università, Viale A. Doria 6, 95125 Catania, Italy

Mirella MANARESI, Dipartimento di Matematica, Università, P.zza di Porta S. Donato 5, 40127 Bologna, Italy

Paolo MAROSCIA, Via Corvisieri 17, 00162 Roma, Italy

Gerriet MARTENS, Mathematisches Institut, Universität, Bismarckstraße 1 1/2, D-8520 Erlangen, West Germany. E-mail address: MPMA23 at DERRZE0

Emilia MEZZETTI, Dipartimento di Scienze Matematiche, Università, P.zzale Europa 1, 34127 Trieste, Italy. E-mail address: TI2TSG22 at ICINECA2.BITNET

Juan Carlos MIGLIORE, Department of Mathematics, Drew University, Madison, NJ 07940, U.S.A.. E-mail address: JMIGLIOR at DRUNIVAC.BITNET

Rosa M. MIRO' ROIG, Facultat de Matemàtiques, Universitat de Barcelona, Gran Via 585, 08007 Barcelona, Spain. E-mail address: D3AGRMR0 at EB0UB011.EARNET

Rita PARDINI, Via Verdi 27, 55043 Lido di Camaiore (LU), Italy

Giuseppe PARESCHI, Via degli Armari 13/A, 44100 Ferrara, Italy

Giuliano PARIGI, Viale Toscanini 50, 50019 Sesto Fiorentino (FI), Italy

Christian PESKINE, U.E.R. de Mathématiques, Université Paris VII, Tour 45-55, Pl. Jussieu 2, F-75251 Paris Cedex, France

Monica PETRI, Dipartimento di Matematica, Università di Trento, 38050 Povo (TN), Italy

Luciana PICCO BOTTA, Istituto di Matematica, Università, Facoltà di Scienze, 84100 Salerno, Italy

Gian Pietro PIROLA, Dipartimento di Matematica, Università, Strada Nuova 65, 27100 Pavia, Italy. E-mail address: PIRO23 at IPVIAN.BITNET

Grazia RACITI, Dipartimento di Matematica, Università, Viale A. Doria 6, 95125 Catania, Italy

Alfio RAGUSA, Dipartimento di Matematica, Università, Viale A. Doria 6, 95125 Catania, Italy

Luciana RAMELLA, Dipartimento di Matematica, Università di Genova, Via L.B. Alberti 4, 16132 Genova, Italy

Ziv RAN, Department of Mathematics, University of California, Riverside, CA 92521, U.S.A.. E-mail address: ZIVR at UCRVMS.BITNET

Gianni SACCHIERO, Dipartimento di Matematica, Università di Trieste, P.le Europa 1, 34100 Trieste, Italy. E-mail address: TI2TSG22 at ICINECA2.BITNET

Frank-Olaf SCHREYER, Fachbereich Mathematik, Universität Kaiserslautern, Pfaffenbergstr. 95, D-6750 Kaiserslautern, West Germany

Edoardo SERNESI, Dipartimento di Matematica, Università "La Sapienza", P.zzale A. Moro 5, 00185 Roma, Italy. E-mail address: MARTA at IRMUNISA.BITNET

Rosa STANGARONE, Dipartimento di Matematica, Campus Universitario, Via G. Fortunato, 70125 Bari, Italy

Rosario STRANO, Dipartimento di Matematica, Università, Viale A. Doria 6, 95125 Catania, Italy

Giulio TEDESCHI, Dipartimento di Matematica, Politecnico di Torino, Corso Duca degli Abruzzi 24, 10129 Torino, Italy

Montserrat TEIXIDOR i BIGAS, Balmes 370, 08022 Barcelona, Spain

Gerard van der GEER, Department of Mathematics, University Amsterdam, Roetersstraat 15, 1018 WB Amsterdam, The Netherlands. E-mail address: GEER at UVA.UUCP

Luisella VERDI, Dipartimento di Matematica Applicata, Università di Napoli, Via Claudio 21, Napoli, Italy

Alessandro VERRA, Istituto Matematico, Facoltà di Architettura, Via Monteoliveto 3, 80134 Napoli, Italy

Claire VOISIN, Mathèmatique, Bat. 425, Centre d'Orsay, Université de Paris Sud, F-91405 Orsay Cedex, France

S. XAMBO' DESCAMPS, Dept. Algebra i Geometria, Univ. de Barcelona, Gran Via 585, Barcelona 08007, Spain, E-mail address: D3AGSXD0 at EB0UB011.EARNET